Whitemarsh
Information Systems Corporation

Data Semantics Management
Volume 1
Rationale, Requirements and Architecture

Michael M. Gorman

Whitemarsh Information Systems Corporation
2008 Althea Lane
Bowie, Maryland 20716
Tele: 301-249-1142
Email: Whitemarsh@wiscorp.com
Web: www.wiscorp.com

ISBN 978-0-9789968-4-0

Printed in the United States of America

Table of Contents

Figures

Tables

Forward

This book is divided into two volumes. Both volumes contain this Forward. The Figure and table numbers. The tables of Contents for topics, figures, and tables for both volumes reside in both volumes. Finally, the Index for each volume is contained in both volumes.

The focus of Volume 1 is Rationale, Requirements, and Architecture for Data Semantics Management. In short, this book makes the case why accomplishing data semantics management is so critical to the overall Enterprise-wide success of commonly understood and shared data.

The approach embraced by these two volumes is founded on the idea of top-down and centralized architecture, engineering, policies, and procedures, but bottom-up, distributed accomplishment. If attempted only top-down, the outcomes will mirror familiar centralized czar-like failures of the past. If accomplished only bottom-up, the outcomes will be the vast forests of semantic stove pipes. This book embraces the best of both, and sets down a data semantics management metadata management architecture within the Metabase System that enables some aspects to be accomplished top-down and centralized as is appropriate while allowing individual database[1] efforts to be accomplished in a decentralized manner which is also appropriate.

The Metabase System is a metadata management system from Whitemarsh. It can be requested, downloaded, and installed from the Whitemarsh website, www.wiscorp.com. All the screen shots in these two

---

[1]. The term, database, used in this book, can have two meanings. There is a third use, but that is a wrong use. Under the first "good" use, database is something that is constructed, that is, a Customer Database, or a Marketing Database. In that state, it is something that is real in that it contains a dictionary of all its parts (called the database's schema), records of actual data for Customers or Sales, relationships among the data records, and indexes that speed access to individual records. In this context, a database consists of: Dictionary, Indexes, Relationships, and Data.

The second "good" meaning for database is that it is something that is accomplished such as Enterprise Database. That use implies a state of organization across large collections of data that is characterized by consistent semantics, common meanings, easy to understand data structures, integration, and non-redundancy. Thus, it is a state of success and achievement with respect to an enterprise's data.

The third meaning for database is DBMS. That is, database management system. This is bad use of the term database. A DBMS is a very large software system for defining, updating, reporting from, securing, and managing databases. Oracle is a DBMS company. It's DBMS is called Oracle. Similarly, IBM has the DB2 DBMS; Sybase as the Sybase DBMS; and Microsoft has the DBMS SQL Server 2005.

In this book, the first two definitions are employed and are distinguishable by their context. The third definition of database is never used because that causes confusion.

---

volumes are examples from the Metabase System. The Metabase System user guides can be downloaded from the Metabase System's web-pages. The Metabase System brings Data Semantics Management alive.

Volume 2 is the logical follow-on volume, Deployment. It sets out a complete set of data semantics management applied examples using the single-user, single-metabase database version of the Metabase System.

Throughout both volumes of this book, the following very simple data modeling convention was employed. 1. A line from a meta-entity to itself with a one arrowhead is a recursive relationship. Ex. Entity contains entity. 2) A line between two meta-entities with a single arrow head is a one-to-many relationship. Ex. Entity has zero, one, or more attributes. 3) A first component that has two one-to-many relationships to a second meta-entity, and a third meta-entity that has a single one-to-many line between it and the second meta-entity is a bill-of-materials data structure. Ex. concept (first meta-entity), concept-structure (second meta-entity), and concept structure type (third meta-entity).

This two-volume book was published subsequent to the three Whitemarsh books:

- Data Interoperability Community of Interest Handbook.
- Enterprise Architectures.
- Strategy for Successful Development of Business Information Systems[2].

The Data Interoperability Community of Interest Handbook book sets out the engineering and architecture of the Communities of Interest that engineer the databases essential for enterprise-wide shared data. This book provides the organizational, and functional engineering. This book also provides the policies, procedures, guidelines, job descriptions, duties, and responsibilities.

Collectively this book identifies the "who" and "how-organized" aspects of successful data sharing.

It is important to note that this book is based on three complementary Communities of Interest (COI). One supports 26 NATO nations in the creation

[2]. A business information system is general term for software systems that employ data to achieve some business purpose. Included are specialized packages such as Customer Management, Accounts Payable, Human Resources, and even Internet browsers. Typically business information systems access data through SQL DBMSs via ODBC or JDBC. Systems such as WinSQL could also be considered business information systems as they allow users to access and report business data. Excluded from this definition are software systems such as operating systems and database management systems.

of a shared-data specification for Command and Control databases; another sets out the program, organization, and management of standards for all American National Standards in the United States; and the last, arguably the most successful Community of Interest, ANSI INCITS H2 Technical Committee on Database, defines and standardizes the SQL language.

The second book, Enterprise Architectures, sets out and describes the critical components that comprise Enterprise Architectures. Each of these components is thoroughly defined and interrelated one to the other. Collectively these Enterprise-Architecture components identify "what" has to be engineered before any real database and/or business information system efforts can be started.

The third book, Strategy for Successful Development of business information systems, provides a very well-proven and successful approach to business information system development. This book contains real examples, work products, and the like. Collectively, this book is the "how-process" book.

This third book is based on highly successful business information system development efforts. In one such effort, a $2.4 Million business information system was created for just more than $300 thousand. How? It was accomplished via prototyping and a business information system generator. This business information system was 6600 function points and through the use of prototyping and the business information system generator it was implemented for $50 per function point instead of the industry average, $400.

While all these three books are important, none of them has focused in any intensive way on data semantics management. Data semantics management is important because without it, there is a very high probability of failure for shared-data efforts. There is also a very high cost to be paid if enterprise-wide business information systems are not data-based. For example, a study has shown that there is a 4.6 times quantity increase in what has to be implemented if a process-first approach is followed rather than a data-first approach.

The lessons of not having proper data semantics management are both painful and costly. The United States Department of Defense (DoD) has been trying to accomplish DoD-wide data standardization for more than 30 years. Failure has generally been the result. These failure costs exceed $500 million.

In contrast, the business information systems of the U.S. Navy are highly successful because a unified data-centric approach that has been championed and made mandatory by the Navy Office of the CIO.

Another data-centric success has been the Multilateral Interoperability Program (MIP) that has built and mandated a shared-data model for all Command and Control data across all 26 NATO nations. This MIP effort provides the organization-engineering structure for the Data Interoperability Community of Interest book described above.

There is a multi-hundred-**billion** dollar effort within the United States Department of Defense focused on the development of Future Combat Systems (FCS). While this effort will eventually be declared successful through a sheer force of will, its cost and quality will have been dramatically affected–negatively–because the effort is hardware and process focused almost to the exclusion of any data-focus or data-engineering. It seems that there are no "data oriented" deliverable requirements in the contracts that govern the FCS effort.

This book is therefore 100% focused on data semantics management. As the picture of Captain Ahab shows in his battle against Moby Dick, successfully achieving data semantics management is a "whale" of a battle. Failing to achieving it in your enterprise is sort of like getting eaten by Moby Dick.

Essentially, we're all Captain Ahab. We're determined and we know the journey. In the book by Herman Melvile, Ahab's battle against Moby Dick ultimately fails. For us, that is failing to achieve data semantics management.

Unlike Captain Ahab, however, we now have a much more sophisticated understanding of objectives. Our methodologies and approach are tested. We are supported by sophisticated tools. This book is about winning this "whale" of a battle.

Preface

If you have ever asked the question, "What do you mean by that," you get the whole point of this book. There are also two follow on issues: The frequency of that question during an exchange, and the frequency of any follow-up correction and re-questioning. The goal of Data Semantics Management is to both minimize the first question and to eliminate any of the follow-up questions.

When verbal exchanges occur, there is commonly a most sophisticated computer on either end of the exchange: the brain. As the message receiver's brain takes in the words, the brain's semantic processor is ingesting and sending out messages such as smile, nod, frown, or "huh?" The message sender's eyes, ears, and brain have an opportunity to stop, reverse, explain, and so forth until there is either a terminal error (e.g., "I give up, go away!") or some other gesture of acceptance.

For example, this author and a newly met professional colleague were assigned to a project. A key element of the project was "work breakdown structure." The colleague thought of a work breakdown structure as the physical parts' explosion of a "work." That it is, a noun which means for example, the rear axle of a truck. To the colleague, the work breakdown structure looked like Figure 1. In his mind, each work breakdown structure element was a noun phrase as in the Figure 1 engineering diagram of a 1994 Ford Ranger Rear Axle.

In contrast, this author thought of a work breakdown structure is a hierarchically organized set of work tasks where in each work break down structure element began with a verb. An example would be to "Write the preface of the Data Semantics Management book." It was several days of working together before one of us reached that terminal error of "Stop, what do you mean by work breakdown structure?" Once the meaning of the commonly employed terms was synchronized, the exchanges were recalibrated and productive exchanges restarted.

Unsaid are the constantly occurring interpretations and transformations. Now, add to these exchanges, language (e.g., French to English, or Chinese to German), and cultures (E.g., European to American, or Asian to African). In either of these cases the quantity of initial questions and gestures, and of the followup questions and different gestures dramatically increases because the semantic disconnect in both areas must first be identified, analyzed, and resolved. Only then can meaningful exchanges occur.

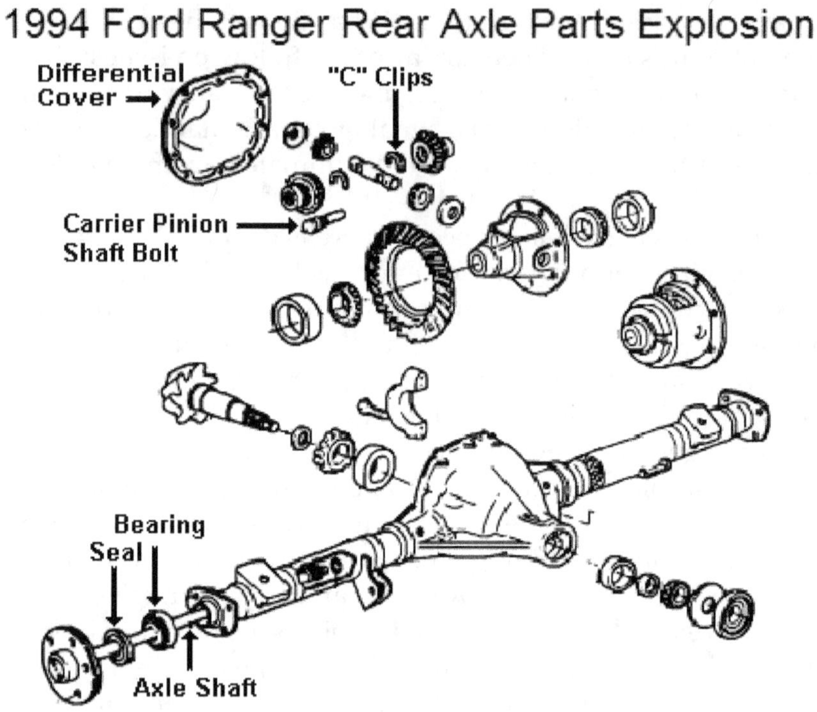

Figure 1. "Work" breakdown structure.

For example, again, this author traveled to and from elementary school sixth grade (note, this was in the early years of the first decade of the last half of the last century of the previous millennium) with the grandson of the ambassador from Brazil. The grandson was learning English. One day he exclaimed, "English is so hard." I inquired what he meant. He said, "I understand what is meant by 'cutting a tree down,' but what is meant by 'and now you cut it up?" Here's another: What's the difference between "slow up" and "show down?"

In contrast, it is often said that couples who have been married many decades can finish each other's sentences. Why? Could it be because they have completely absorbed each other's cultures and mannerisms? Likely. When couples meet for the first time there is commonly many long phone calls or conversations. Why? It's probably because there's an ingesting and synchronization of their individual cultures and mannerisms. There's a thirst for a relationship bridge constructed of common semantics.

In all these cases, it is also common that as time goes by from days to months to years and even decades, the exchanges become fewer. It may not be because they have grown distant, Rather, it may be because there is now an acceptable environment for exchange without all the copious culture and mannerism synchronization activities that occurred when the couple first met and were dating.

With all that said, and hopefully agreed to, changing the environment of exchanges from "human" to information technology greatly complicates the whole process of exchange. From the verbal exchange scenario there are the following components:

- Commonly recognized words or phrases.
- Mutual semantic trust of inferred meanings.
- Interactive posting and receiving of "error messages" and responses.
- Previous uses of previously engineered and agreed-to semantic maps.
- Language matches.
- Culture matches.
- Context matches.
- Content matches.

It is the premise of this book that all information technology data exchanges are just surrogates for verbal exchanges. The information technology data exchanges exist merely to increase velocity, or to reduce costs. Consequently, all IT information exchanges naturally follow the same rules and requirements as do verbal exchanges. A critical and key difference between human versus computer-based exchanges is that for human exchanges, the volume and velocity are far smaller and slower than computer-based exchanges, and further, on a per-exchange basis, the human-exchanges are employed through a much larger and profoundly more capable processor (the brain) than is the computer for the computer-based exchanges.

This book is all about engineering high velocity, semantically harmonious information exchange environments which are critical because "human brains" do not exist on either end of the IT exchange. This book is not about the engineering high velocity, semantically harmonious information exchange environments that are perfect, automatically malleable, and endlessly mutable. That's because this book is not fiction. At best, the engineered environments, just like the human environments, work well only after significant periods/efforts in which the word/phrases, meanings, language, culture, context, and content that frame the information exchanges have all been satisfactorily addressed. Perfection and total "automation" cannot be achieved, so you must stop dreaming and get down to the real work of establishing a productive information exchange environment.

The chapters in this two-volume book are:

Volume one

1. Rationale for Data Semantics Management
2. Why Semantics
3. Data Semantics Failures
4. Engineering Data Semantics
5. Semantics of Names
6. Semantic Hierarchies
8. Fact Specification Cases

Volume two

9. Data Element Model
10. Specified Data Models
11. Implemented Data Models
12. Database Object Classes
13. Operational Data Models
14. Interface Data Models
15. Work Plans
16. Summary

Appendix 1, Data Architecture Classes appears at the end of Volume 1.

Acknowledgments

My journey to an understanding of data semantics management started almost 40 years ago. Over this 40-year span six individuals stand out. I would like to acknowledge them here. I would also like to acknowledge that this book represents materials presented to many companies through the Whitemarsh Data Interoperability Seminars. Questionnaires were issued after every day long (225 slide) presentation. The results' tabulation is provided at the end of this section.

At the very start of this trip, I was a database analyst for the System Development Corporation (SDC). SDC was one of the original inventors of database, data management, and database management systems. SDC, along with Lincoln Labs and the MITRE Corporation invented this entire discipline in the late 1950s and early 1960s. SDC created and deployed a database management system (DBMS), TDMS (Time-shared data management system), for the U.S. Air Force in the early 1960s.

As an employee of SDC in the Washington, D.C. area in 1969, I was introduced to the "programmer replacement system." That is, a DBMS called Commercial Data Management System (CDMS). It was a commercial version of TDMS. With CDMS you could design hierarchical data structures, write data conversion scripts, load extremely large databases, and create and execute large collections of ad hoc reports to prove or disprove assertions about bodies of knowledge. In 1969, a 3,000,000 character database was extremely large.

My boss was Alec Bumstead. He was the first to question whether I had any cogitative abilities at all. I read the CDMS manuals, and designed the database. He read the design and started to ask questions. For every question I had only "ah, er, I hadn't thought of that" for answers. "Think, think, be disciplined, think, think," Alec would say. So I did. After much interaction, we managed to develop a database design, that when it was loaded with data and exercised, disproved several of the major conclusions of the "Coleman Report." What the DBMS, via a good database design and sophisticated queries, was able to prove and disprove in a mater of minutes took researchers weeks. CDMS's capabilities were mind boggling. Database, DBMS, and data management has been my life's work ever since. So, here's to the memory of Alec Bumstead.

In 1971, I started to work with Ted Ziehe. Ted was the application developer manager for MRI Systems Corporation of Austin, Texas. MRI had developed System 2000, which was a "grandson" of TDMS. At the time I was the PMIS architect. PMIS, a Planning and Management Information System, a demonstration prototype funded by the U.S. Office of Education, took weekly data extracts from operational files and configured this data into a database of statistics about Students, Teachers, Schools, and Educational Programs. This data was used by Dallas Texas educators to measure the progress of students, teachers and programs against yearly educational objectives. Clearly this was a data warehouse. During this contract, we also built a database about PMIS. This "about-PMIS" database stored all the design artifacts, query specifications, the database designs, and the like. Clearly this was a metadata database. PMIS was very successful as a prototype and demonstration project. All during the PMIS database design phase, Ted Ziehe worked with us for days to get the database design just right. His strategy was to take 3x5 cards and put a data-subject title on each, and list the data facts about each subject. We would write the "short story" about the 3x5 card on the back. Once we had all the cards done, we put them onto a wall, and with string, make relationships between them in a one-to-many fashion. Ted would pepper us with all sorts of questions that addressed granularity, precision, and time synchronization. Eliminating redundancy was Ted's big objective. To Ted, Database meant never having the fact in more than one place. To Ted, everything had to have a clear, unambiguous definition. It had to be clear, crisp, and once only. So, thanks Ted for being my first real data modeler and data semantics management teacher.

From 1979 through 1981 I had the privilege of working with Matt Flavin. Matt, the inventor of database object classes, worked for Ed Yourdon of New York City. Matt's course and workshop, Fundamentals of Information Modeling, started with the premise that data is executed policy. Matt didn't say it like that, but that's what he meant. Matt was very critical of the relational model as he felt that in the name of table-based simplicity and rigor, the relational model had done away with the overall greater objective of database. That is, coherent, state-based expressions of enterprise policy. My contribution to Matt's effort was end-to-end database project methodology, and database administration. Matt's workshop showed that a rigorously engineered database design methodology would almost always result in the same database design, regardless of the team applying the methodology. Matt provided science and engineering to database designs, while at the same time restoring a critical and lost component: database object classes. So, here's to the memory and contribution of a database pioneer and legend, Matt Flavin.

In 2001, I met Hank Lavender. Hank (actually Col Henry C. Lavender, USAF Retired) was the functional data administrator of the United States Defense Logistics Agency. Hank got interested in "correct and quality" based logistics because he could never seem to get enough aluminum patches for all the bullet holes that "mysteriously appeared" while flying around the Vietnam countryside in his Douglas A-1 Skyraider. Hank was a leader of the Hobo Squadron. After the war, Hank retired and focused all his energies on logistics management. Hank and I worked together on the development of a strategy and methodology to achieve interoperability through shared-data. Hank and I presented this strategy over a two-year period to the DAMA International Conferences. We employed the Metabase System as the metadata management system. During this time there were many changes to the Metabase System so that it could accomplish shared-data specifications with maximum effect and minimum effort. So, thanks to Hank for all his contributions and constant reminders that "if it doesn't make common sense, it probably just makes non-sense."

During the last eight years, and specifically from 2001 through 2005, Bruce Haberkamp and Jim Blalock of the Office of the CIO of the United States Army and I have worked very hard in the development of the Army's Data Interoperability Program. This program has resulted in Army policy, and a large series of documents that included seminars, workshops, methodology, and the employment of the Metabase System. The Data Interoperability Community of Interest Handbook, this Data Semantics Management book, modifications to the Metabase System, and a series of Data Interoperability workshops and seminars are the result. So, thanks to both Bruce and Jim.

From the above, you would be correct to understand that much of the data semantics management work has been focused on U.S. Federal Government efforts, especially the Department of Defense. That is certainly true. But during 2006 and 2007, I took the Data Interoperability Seminar on the road and have delivered it to a large quantity of "commercial" organizations. At the end of each very long day, the attendees were invited to fill out a questionnaire. These questions represent their report card of me on this whole approach. What follows are the questions and a tally of the questions. Following that is the list of companies that attended the seminar and filled out the questionnaire.

Results of the Questionnaire Given to Attendees of the Data Interoperability Strategy Seminar				
Question	**Strongly Agree**	**Agree**	**Disagree**	**Strongly Disagree**
Organizations would benefit from a comprehensive data management program including courses, methodologies, workshops, etc.	66	31	3	0
Organizations would benefit if databases (not DBMS) had consistent semantics for names for table columns across schemas.	80	17	0	3
Organizations would benefit from automatic data element, attribute and column naming and abbreviations built with consensus-based word lists and phrases (taxonomies).	65	35	0	0
Organizations would benefit from an automatic data element, attribute and column definitions derived from contexts, valued domains, etc.	69	39	2	0
Organizations would benefit if databases (not DBMS) had consistent semantics for value domains.	75	22	3	0
Semantics should be inheritable from iso 11179 data elements to attributes, and from attributes to columns.	42	53	5	0
There is value in having data model templates that can be used in defining column collections within tables.	61	39	0	0
XML schemas should be generated from a foundation of well designed and integrated data models.	63	33	4	0

Results of the Questionnaire Given to Attendees of the Data Interoperability Strategy Seminar				
Question	**Strongly Agree**	**Agree**	**Disagree**	**Strongly Disagree**
Organizations would benefit from metadata repositories that support metadata integration and non-redundancy across projects, departments, etc. that is, define-once and use many-times.	87	13	0	0
Organizations would benefit from metadata repositories that are part of the everyday work process rather than a post implementation documentation effort.	90	10	3	0
End-users would benefit if they had access to metadata repositories for names, definitions, and the like.	83	14	0	0
Organizations would benefit from metadata repositories if the repositories could capture, report, and support updates of analysis and design work products in an integrated non-redundant manner.	70	30	0	0
Project quality and work-speed would likely benefit if supported by comprehensive metadata repositories.	75	25	0	0
Organizations would benefit from metadata management strategies supported by both bottom-up reverse engineering, and top-down new database (not DBMS) manufacturing.	53	47	0	0

Data Interoperability Seminar Attendees who filled out questionnaires.

American Airlines
American Family Insurance
Bank of Montreal
Blue Cross
BTE Corporation
Burlington Northern Santa Fe
Canadian Tire
Cap Gemni Energy
ChartWell
Chevron Texaco
CIBC
Cingular Wireless
Circuit City
CitiFinancial
CitiGroup
CUNA Mutual Group
DOFASCO (Canadian Steel Corporation)
Economical Insurance
Essential Futures
Federal Express
Great Lakes Higher Education Loan Services
Hudson Bay Company
IESO (Independent Electric System Operator, Ontario Canada)
Kohls Department Stores
Landsend
RBC Capital Markets
Royal Bank of Canada
Science Applications International Corporation
Sigma Systems
Sprint
U.S. Department of Homeland Security Customs and Border Protection
Wall-Street Consulting
Washington [State] Courts
Wisconsin Physicians Service Insurance Corporation
Wisconsin Education Association Trust
Wisconsin DOT

It is on the shoulders of all the individuals and organizations I have acknowledged that I commend this book to your reading. What is more important, I commend this book to your adoption and implementation. Go forth and do data semantics management. You will have only increased productivity, increased quality, decreased costs and decreased risk to show for your efforts. Not too bad an accomplishment, I would suggest.

Michael M. Gorman
September 2008

1

Rationale for Data Semantics Management

The objective of this chapter is to set out the rationale for data semantics management. The rationale is presented by first citing the clear state of the world, that is, that we live in a world economy. With the Internet, there are no borders.

Commerce and communications of virtually all abstract goods are exchanged at the speed of the Internet. Good examples are all the Internet-based jokes that travel around from person to person and organization to organization almost instantaneously. Commerce too is similarly engineered. The computer on which this manuscript was created was built in Asia. The operating system is from Microsoft in Washington, the word processor from Canada, and the Metabase System that is the demonstration vehicle for all this book from a company in Florida with supplementary products from Arizona, Australia, Germany, South Africa, and England.

Simply stated, the most common languages throughout the world are not English, French, or Spanish. Rather, they are SQL, C, C++, because all communication throughout the world are first founded upon these universally accepted and unambiguously understood computer-based languages.

This chapter presents a clear statement of the problem that must be solved and several examples to illustrate it. Thereafter are examples of the effects of not having sufficiently engineered data interoperability.

This chapter sets out the remaining chapters in the book and conveys how each chapter contributes to addressing the two classes of problems that characterize the achievement of data interoperability through data semantics management.

This chapter, as does all chapters, concludes with both a summary and a set of questions and exercises that should be able to be accomplished by practitioners of the book. This book is not just to be read and thought about. Rather, it is to be acted upon. The former will provide intriguing thoughts and concepts to consider. The later will provide real action plans coupled with a down-loadable production class metadata management system, the Metabase System, to manage data semantics in support of creating the interoperable data environments critical to the world economy.

1.1 Rationale

The rationale for data semantics management is simple. We live in a world economy. Operational aspects of commerce are so large and the requirements for commerce processing are so high that the world's economy cannot operate without a very high degree of the semantic trust that is essential for large volume and high velocity information exchanges.

Humans are natural semantic translators. Whenever an exchange fails, error messages, e.g., "huh?" are posted. The speaker almost always adjusts in real time and corrects the understanding errors in the exchange.

Computers, however, almost certainly do not have the ability to dynamically switch the appropriate words/phrase, meanings, language, culture, context, and content, that is, recalibrate the exchanges so as to make it semantically harmonious. Attempts to create computer-based semantic processors have existed for 50 years or more without much success. An early experiment was "Eliza" in which a human questioner post a question, for example, "Why do you hate me?" The computer responded with, "Why do you feel that you are hated?" It was "off to the races," with questions that were merely turned into syntactically transformed responses. In this environment, the transactions moved only as fast as the human respondent allowed. Information-technology-based information-exchanges however move at the "speed of light" without any reasonable manner of resolving semantic error messages. At best, the information exchange transaction is set aside, a message is posted, and it is dealt with by a human at some later time.

If there is too large a volume of human-intervention required messages, the whole process of information exchange grinds to a halt. Examples of high velocity information exchanges include credit-card approval processing, insurance claims processing, payroll creation and processing, material logistics management, tax return processing stock exchanges, check, and financial instrument clearing houses, and the cellular phone system. The very existence of each of these examples depends on highly engineered semantics that are agreed to by all parties to such an extent that the volume of error messages falls within the ability of the human systems to process them.

1.2 The Problem

The problem that exists is to create information exchange environments such that the semantics exchange error rates and the volumes of such errors are within the acceptable limits of human correction processing. The level of effort that needs to be expended to achieve this is inversely proportional to the quantity and velocity of semantic errors that can be tolerated. For electronic commerce, the tolerable error rates must be low, hence the investment must be high. In contrast, entry of an application for employment may both generate errors and also permit human-corrective interaction because both the volume of applications is low, and the acceptable duration for processing can be relatively long.

1.3 Effects of The Problem

The effects of human information-exchange errors can be severe. The gun-battle at the "OK Corral" might have resulted from such an error. In contrast, semantic-errors as a consequence of information technology can range from having a slight effect to a severe one. A rejection of a misread credit card at a store likely has small effect, while mixing up units (Metrics versus English) caused NASA to lose a $125 million Mars orbiter. This was because one engineering team used Metric units while another used English units for a key spacecraft operation.

Information exchange errors are just reactions to a lack of semantic mapping efforts that occurred much earlier in the analysis, design, and development of IT systems. That a credit card reader failed to correctly read a card as a consequence of a worn magnetic strip is not a semantic exchange error and thus not attributable to a poor design of some aspect of an IT system. Semantic exchange errors originate somewhere within the IT system's design. For example, the "real" data types (versus Integer or Decimal) of the fields that were employed were not checked. Such checking should not be a human controlled process. Rather, the fundamental data types (e.g, metric vs English) of the fields should have been an absolutely required entry in the IT system's metadata[3] support environment. Similarly, strong data typing would

---

[3]. Metadata is a generic term that identifies all classes of information technology specifications across the enterprise. Hence all data and process specifications are metadata. All (continued...)

prevent an object's weight to be multiplied by an hour value. While such a multiplication would be mathematically possible, it would make no semantic sense. Consequently, such mistakes must be automatically prevented.

As stated above, the ability to tolerate and/or recover from semantic errors must be assessed and employed in the determination of the scope and volume of work that must be employed to prevent information exchange errors. In the case of the credit-card machine, manually cleaning the reading head, or manually entering the credit card number that is embossed is certainly within acceptable limits. The Mars orbiter, however, is very much the opposite. There was no opportunity for error processing and recovery for two reasons: By the time the error was noticed, the orbiter would have already crashed. Second, the error wasn't understood until some weeks after the orbiter was destroyed.

1.4 Contents of this Book

This section of the book is much more than just an enumeration of the chapters and a one-or-two liner introduction to each chapter. Rather, it contains the overall architecture, strategy and even "battle plans" to achieve data semantics management on an enterprise-wide basis. Procrastination in accepting the dictums and strategies of this book will only result in greatly increased costs and significantly more difficult efforts in the future.

With that in mind, every chapter of this book is targeted to the reduction of the two semantic exchange errors:

- Initial semantic disconnects.
- Follow-up semantic synchronization activities.

The first error class, initial semantic disconnects, is also complicated by the two classical statistical error types:

---

[3](...continued)
requirements could also be considered metadata. A metadata database is a database within which all metadata is stored. A metadata database is a metabase. Whitemarsh has employed the term, Metabase, in this context since 1982. A metadata management system is a software system that captures, stores, reports, and manages all metadata. The Whitemarsh system that manages the Metabase is the Metabase System. Sophisticated metadata management systems, like the Metabase System, are multi-user and support the capture and reporting of metadata in a non-redundant, integrated manner across the enterprise.

- Type I Error: An error which occurs when a true hypothesis is rejected.
- Type II Error: An error which occurs when a false hypothesis is accepted.

A Type I error would occur in an information exchange in which a Social Security Number was represented with dashes and the receiving computer system was expecting a pure integer. A Type II error might be the acceptance of a Gender = 1 as acceptable when the sending system had a reference table of 0 = Male, and 1 = Female, while the receiving system has 1 = Male, and 2 = Female. These Type I and II errors can and must be prevented through real specification and implementation engineering.

The second error class, follow-up semantic synchronization activities, are those activities that an organization must perform to reestablish productive data interoperability. In the example above, these activities would first require that there are effectively, three gender codes. That is, 0 for Male, 1 for Male and Female, and 2 for Female. That would mean that while the total could of persons is the count of all instances from the three gender codes, some of the employees associated with code 1 are male and others are female.

One set of followup synchronization activities might be to fix source the systems so that all employees are either 0 and 1, or are 1 and 2.

Another fix might be to transform the understanding of the genders by means of a views that would query one database and underlying system using values 0 and 1 and query the other database and underlying system using values 1 and 2.

Which followup synchronization activities should be followed depend entirely on what is possible and practical. If the source systems are home-grown, the first approach might be both practical and possible. Alternatively, if the source systems are all ERP or commercial vendor type packages, the second approach might be the only one possible and practical.

Where this book can help with respect to the two error classes is that this books enables the recognition of both, the ability to effect a strategy to determine the one, right enterprise approach, and then the ability to accomplish this enterprise strategy in an incremental approach rather than an all or nothing approach.

This book not only addresses these two error classes including architecture and strategies to deal with these two error classes, it is also highly engineered. Chapter 1 introduces it all; Chapter 16 summarizes it all. Every other chapter has its own introduction, unique material, summary and next chapter introduction.

Chapters 2 through 8 provide the overall rationale, requirements and architecture for a comprehensive data semantics management solution. Chapters, 6, 7, and 8 within this set enable the construction of very key data semantics management components. Chapters 9 through 15 provide detailed strategies, examples, and screen shots of how to deploy data semantics management on a project, organization, function, and enterprise basis.

The book contains a number of "can do" chapters. These chapters, 7 through 14 contain an introduction, significant architecture and engineering material, and a detailed set of Metabase System based examples that illustrate the implementation of that data semantics management component.

Every chapter contains questions and exercises that directly bear on the accomplishment of data semantics management.

Because of the chapters' engineering, some material that is essentially the same. That is not an accident. Rather, it is purposeful reinforcement. Additionally, this repetition enables many of the chapters to stand alone.

1.4.1 Chapter 2, Why Semantics

Chapter 2, Why Semantics, addresses the fundamental need for highly engineered semantics as a way to control the error classes cited above. This need is first explained within the context for data management. Key to understanding data management is understanding the role of enterprise-level policies and procedures. Because these need to be enterprise-wide, there must be consistency in their definitions at all levels and across all enterprise organizations.

Another section of Chapter 2 identifies the need for database object classes. The specifications of database object classes are set squarely within corporate policy specifications. Data, the by-product of policy executions, are the "what" that should be shared within and across corporate organizations. Database object classes are the coherent specifications of shared data. Unique to database object classes are the definitions of the states and state transformations through which a given object proceeds from birth through all life cycle stages until the database object becomes fully mature.

A key construct in Chapter 2 is the graphical representation of the various layers that start with policy specification and proceeds down through data-based evidences of policy executions.

Chapter 2 continues with a simple definition of data semantics and illustrates this definition through several examples. Key to the explanation is the simple fact that there is no silver bullet that addresses all the needs of data interoperability across the enterprise. Because there is no silver bullet, the management of data semantics must proceed through several different, but thoroughly interconnected models of metadata. These models address: data elements, data structure templates, the employment of these data structure templates within databases, and finally as specifications of semantically complete data exchanges. Each of these metadata models is generally defined in terms of their critical components.

Chapter 2 concludes with a section on the return on investment (ROI) from effective data semantics management. The basis for the ROI is in two parts: Data and business information systems. While the ROI for data is significant, the largest ROI is on the development, evolution, and maintenance of business information systems. Simply put, the business case for data semantics management is in four parts: increased productivity, increased quality, reduced costs, and reduced risk. Not accomplishing data semantics management is not only a bad business decision in the short run, it likely is a terminal business decision in the long run.

With respect to the two error classes, initial semantic disconnects, and follow-up semantic synchronization activities, Chapter 2, Why Semantics, the first overall framework for addressing these two error classes. The ultimate strategy is to design both these error classes out of existence from the very start. If there are no initial errors, there is no need for follow-up activities. To that end, first there are the Semantic Hierarchies that support the creation, meaning, and allocation of all the semantic-based words that make up critical elements of fact names.

Thereafter there is the Data Element Model, the Specified Data Model, and finally, the two database data models, that is, Implemented Data Model and Operational Data Model. Each of these provides the ability to define the data semantics that control their immediate lower contained levels. Each lower contained level provides the ability to provide a more refined set of semantic definitions and/or restrictions. The parent level of any given semantic definition, for example, concepts within which data element concepts are defined, provide the ability to posit semantic affinities due to common parentage.

If all this is accomplished top-down, there will be a great reduction in semantic errors. The problem will still exist with bottom-up development. That, however, is where most of reality exists. There is no choice then but to have top-down connections to bottom-up developed or imported data models. As this is done, compromises is inevitable. In such cases there at least exists evidence of the mapping decisions. Over time as more and more one-off databases and business information systems are replaced with enterprise-based semantic-driven databases and business information systems, the quantity of semantic errors will be driven out of existence.

1.4.2 Chapter 3, Data Semantics Failures

Chapter 3. Data Semantics Failures, is a very critical chapter. That is because there has been spent 100s of millions of dollars in the attempt to achieve enterprise-wide data semantics, and in virtually all the cases, the result has been failure. A common reaction to these failures, however, has not been to understand the causes and try again. Rather it has been to abandon the efforts because it seems to be too hard or takes too long. The problem with that solution is that the next time data semantics is tried, the problem area has grown and the existing ad hoc, stove pipe solutions have become more entrenched. The right time is now. Procrastination is never the correct solution because waiting only makes it harder and more costly.

Another reaction to failure is to jump on the next "silver bullet" that whistles through. Silver bullets are nonexistent in this area, so again, waiting only makes the problem space bigger and more costly. Therefore, the time is now, not the tomorrow that never seems to come.

Chapter 3 goes to significant length to thoroughly diagnose the common data semantics failures, including some that will surprise you. The lessons learned are clearly understood and have been completely incorporated in the strategies and examples of this book. Everything that needs to be done to engineer and deploy interoperable data environments is set out in this book and in other Whitemarsh books. Additionally, every example from this book is able to be operationally demonstrated via the Metabase System.

A key ingredient to the accomplishment of enterprise-wide data semantics management is the establishment of Communities of Interest. The Whitemarsh book, *Data Interoperability Community of Interest Handbook*, sets out the architecture of these groups and the work plans, deliverables, and all necessary policies and procedures for the establishment, operation, and evolution of these groups.

Excuses of not knowing what causes failures, and once known, not knowing what to do have been removed. The time is now, not the tomorrow that never seems to come.

With respect to the two error classes, initial semantic disconnects, and follow-up semantic synchronization activities, Chapter 3, Data Semantics Failures, clearly outlines the most common sources of data semantics management failures that ultimately cause instances of the two error classes. These failures must be carefully studied and solutions to these failures must be set into place. This book's dictums, strategies, and action plans represent a solution set.

1.4.3 Chapter 4, Engineering Data Semantics

Chapter 4, Engineering Data Semantics, is another key preparatory chapter. Data semantics management efforts cannot be slapped together and set into motion. The right environment must exist. Ironically, even if all the factors were completely known 20 to 30 years ago, enterprise-wide interoperable data environments were not possible due to immature technologies. If, however, the dictums and strategies of this book and those from the *Data Interoperability Community of Interest Handbook* were executed, there would have been integrated data across all the stove pipes of database and business information systems. Had this been done in advance of technology being available, it would have all fallen into place.

It was no accident that one division of a world-wide corporation had a Y2K bill of virtually zero while another division of the same corporation spent many millions repairing and redeploying their ad hoc databases and business information system implementations.

A key set of material in Chapter 4 is the exposition of the evolution of bindings that existed between data, programs, database management systems, and users. There has been a long and gradual trek in the evolution of bindings from the 1950s through to the present day that has enabled, for example, service oriented architectures. It is critical to know, however, that success in service oriented architectures require careful attention and deployment of the fundamentals of this book. Disregarding the dictums, strategies, and products set out in this book is to court failure.

With respect to the two error classes, initial semantic disconnects, and follow-up semantic synchronization activities, Chapter 4, Engineering Data

Semantics, sets out the overall domain within which the solution must reside. Again, the ultimate goal is to define both these error classes out of existence.

The very first line of defense is of course the operational databases and information systems that are represented by the Operational Data Models of Chapter 13. Those if defined in isolation form islands or stove pipes of semantic consistency. While these are valuable, what is even more valuable are broader areas across whole business functions and organizations. These would be guided by the higher levels of semantic specifications and would exist within the Implemented Data Models of Chapter 11. Backing those up would be the Specified Data Models of Chapter 10 and finally the Data Element Models set out in Chapter 9. If these models are implemented, not only will individual databases and business information systems be more semantically acceptable, but so will all other related databases and business information systems because they are drawn from common higher semantic levels.

1.4.4 Chapter 5, Semantics of Names

Chapter 5, Semantics of Names, squarely addresses one of the key reasons for data semantics management failure: Names. An example, the name string, Social Security Number, is all that's needed to understand the problem. According to the traditional naming strategy, Social is a modifier, Number is a class word, and therefore Security is the "real part." If the complete semantics are not instantly obvious, the point's been made. Further Number, in this case is a lousy choice for a class word. What's the meaning of an average Social Security Number. The minimum or maximum. What are those dashes, subtraction signs? Finally, Social Security Numbers are often used as identifiers. Thus, the real class word should be Identifier. That would be OK except a Social Security Number is really a set of three codes. Now, onto Social. If there are so many law firms advertising their services to "fight" the Social Security Administration, they must not be very "Social." And as far as security is concerned, the Social Security Trust funds (which really do not even exist) rushing headlong into insolvency. So much for Security.

Similar examples can be made for Reservation Numbers that seldom contain only numbers or even numbers at all. Telephone numbers are another example. Are they really the serial numbers of the telephone instruments? Of course not. The point to all this is that there is a general tendency to overload the semantics of names.

There also seems to be a unyielding demand for single definitions for given name strings. At one conference during the early 1990s, a speaker was questioned about the meaning for a given name string. The questioner indicated that a different definition had been used for the previous 2000 years. The speaker stated that they would just have to change. Oh yeah, right. The point to Chapter 5 is to set out a formal strategy and infrastructure for the construction of names that does not fall apart under even the slightest of scrutiny.

With respect to the two error classes, initial semantic disconnects, and follow-up semantic synchronization activities, Chapter 5, Semantics of Names, is very important. That's because there's a commonly accepted but mistaken notion that names are the alpha and the omega of data semantics management. Far from it. Names, while clearly the most visible semantic component, can be very misleading and can give rise to both Type I and Type II errors within the first error class, and can also lead to delays and confusion as the second error class attempts to be sorted out. One U. S. DoD agency spends many millions every year just sorting out errors in logistics labels due to reference data errors. These are clearly a second error class activity caused by Type II errors such as sending M1 Abrams battle tanks to Bimini versus Bosnia.

This is not to say that names are unimportant. What is critical is that there is almost automatic strategy for name construction that ensures that when names are the same they have the same meaning, and vice versa. If and when this is done, there will be a great reduction in both error classes.

1.4.5 Chapter 6, Semantic Hierarchies

Chapter 6, Semantic Hierarchies, is a can-do implementation chapter. That is, it defines and illustrates the actual establishment of data semantics management. This chapter identifies and provides real screen-shot examples of how to define data semantic hierarchies and how to deploy them in real-world examples.

The real world examples come from the Metabase System and examples from the Metabase System metadata database, Movies.

The only restriction on the use of this downloadable "Free" Metabase System is that it only allows one metadata database beyond the example metadata database, Movies, and only allows one concurrent user. The Metabase System, unencumbered from the "Free version" use restrictions, is an enterprise class system capable of supporting a very large enterprise system of integrated database information systems for interoperable data sharing, and is available for purchase at a very modest price compared to other popular enterprise class metadata management system systems

Semantic hierarchies are a critical element in the development of enterprise-wide semantics because they form the building blocks for name construction. What is very important about these semantic hierarchies is that they can be allocated to a number of different components in the data semantics management space including: data element concepts, data elements, attributes, and columns. Not only can these be allocated, each semantic name string can be defined singly or in context with its hierarchy.

Further, the Metabase System software ensures that semantic nesting is enforced. Users are not able to allocate to a lower level data component, for example, a column of a table, a semantic that is the same as the one already assigned to the attribute of an entity. Nor are users able to assign to the column a semantic that is at a higher level of generalization as was assigned to the attribute of an entity. Attributes are semantically superior to columns. Thus, assigning the same or a lower level semantic to an attribute than is assigned to a column would simply be a mistake. The Metabase System prevents such mistakes.

Finally, not only are the assigned semantics employed to "automagically" (not really by magic of course, the strategy is set out in Chapter 7) construct names, but they are also employed to "automagically"define data element concepts, data elements, attributes, and columns. Both these capabilities are critical to the efficient and effective management of enterprise-wide data semantics.

With respect to the two error classes, initial semantic disconnects, and follow-up semantic synchronization activities, Chapter 6, Semantic Hierarchies, is a very critical component. That is because the words that comprise fact names within an interoperable data environment can be constructed through the allocations of specific strings from these hierarchies.

Simply put, words, within their context have single meanings. When definitions are automagically constructed they will always be generated the same way. Hence, they will be reliable and repeatable regardless of the person or organization, or function within which they are generated. What is the same is the meant to be the same, and vice versa. This strategy, if followed will greatly reduce the first error quantity and it naturally follows that the second error class is reduced as well. All these same words and same meanings are carried down through the entire set of Metabase System data models, that is, Data Element, Specified, Implemented, Operational, and View.

A very important additional benefit from these semantic hierarchies is the ability of the Metabase System to do automatic name construction and automatic definitions of all business facts from data element concepts within the Data Element Model down through to DBMS columns within the Operational Data Model. This enables all uses of databases constructed through these strategies to know the enterprise-wide and consistent definitions of all facts. Clearly this addresses both classes of errors.

1.4.6 Chapter 7, Value Domain Management

Chapter 7, Value Domain Management, is a "can do" chapter. Increasingly, value domains are important to data semantics management. The Type II error cited above, that is, the ability to specify Gender = 1 as meaning Female in one environment while it means Male in another environment can be a critical source of errors that must be prevented. The first step in prevention is having the ability to know about these possibilities, and being able to control and manage value domain value collections. The strategies set out in this Chapter enable this to take place.

Demanding that there are completely congruent value domain collections is just not possible, however. Nice, yes. Recommended, absolutely. But absolutely possible? No. This chapter sets out strategies for setting down value domains, associated value collections, and the ability to map one value collection with another. There is, however, no silver bullet here. If one collection has three values for a concept and another five values for different instances of that concept, there can only be three mappings, and none of these might be exact. This is illustrated in the case of mapping "one-third's based values" to a set of "one-fifths-based values," or academic "letter" grades to precise "numeric" grades. Miracles are not possible. Some value domain mappings are going to be unsatisfactory whenever square pegs are shoved into round holes. At least this chapter provides a strategy to define the square to round mappings, and a Metabase System solution to know when and where the mappings are occurring.

With respect to the two error classes, initial semantic disconnects, and follow-up semantic synchronization activities, Chapter 7, value domain Management, is also a very important chapter because it provides a very explicit first line of defense in stopping invalid values from entering any database.

This kind of error is the root cause of the Type I and Type II example in Section 1.4. While solving the "word" and name problems are important, so too are the problems associated with restricted value domains. When this is done, a great deal of the hidden value-based error class errors will go away. The key problem with value domain errors is that they can go undetected for a long time. They must be both prevented at the outset and ruthlessly sought out on a continual basis. All is lost if the data cannot be trusted.

Value domains and their accompanying values can take on both type and value forms. For example, in a data warehouse database, the column Region might take on the value, New England. While in another database, the column name might be New England Region. It all depends on the data modeling strategy. In either case, the value-domains component of the Data Element Model has the values and their definitions from which either the names are constructed or the values that would become part of reference data are produced. Thus, there would be a good effort to prevent the first class of errors and to therefore prevent the second error class all together.

1.4.7 Chapter 8, Fact Specification Cases

Chapter 8, Fact Specification Cases, is a chapter focused on understanding all the different cases associated with fact specification. This is also a "can do" chapter from the point of view of knowing the meta model specifications of data integrity rules, how these rules are bound into the Metabase System, and finally, how and where business facts are implemented across the Metabase System.

Some of the fact specification cases are simple, for example, single values for fields like Birth Date. Other cases require groups of fact collections such as a Full Name (First Name, Middle Name, and Last Name). Still others required facts like Life Span which would be from Birth to Death. Many of these fact cases result in virtual values such as a person's age ((Today - Birthdate)/365.25). Still others require sophisticated processes. Examples of all these data integrity rules that govern the fact cases are set out.

With respect to the two error classes, initial semantic disconnects, and follow-up semantic synchronization activities, Chapter 8, Fact Specification Cases, enables the specification of the rules under which facts are know singly or in combinations. If facts exist in combinations, then, given that DBMSs are employed as the "protectors" of data, processes can be created and embedded in the DBMS schema language that enforce many different classes of fact based rules.

SQL:1986 through SQL:1992 were simple. Each column represented a single atomic value. SQL:1999 and more recent SQL versions started to allow more sophisticated data structures. Returned were the "good old days." Some, however say those days were not all good. For example, in SQL:1999 an array has persistent cell value ordering. Thus, there might be semantics inferred by being the "first" or the "last" value in an array. If so, that semantic has to be defined, stored, and retrievable somewhere. Each of the fact cases, if properly specified and enforced can reduce error quantities. For example, by never allowing a U.S. Address to be entered without a valid combination of City-State-Zip.

Fact specification classes beyond the simple, single value facts, are best specified in a Business Rules component of the Metabase System. Currently (as of this book's publication), these rules are set out within the database object classes model that is the subject of Chapter 12. Properly engineered and embedded in sophisticated DBMSs, both classes of errors can be comprehensively addressed.

1.4.8 Chapter 9, Data Element Model

Chapter 9, Data Element Model, is a "can do" chapter. Data elements are not facts that are "valued." Rather, data elements are the specifications of fact-based semantics that are, in turn, employed to infuse these semantics into the attributes of entities, and columns of database tables. Data elements are also not simple. Data elements can be set within derived and compound data elements. They are set within data element concepts value domains. Finally, data element concepts are set within concepts and conceptual value domains. This is quite an extensive context for each data element.

Semantics, as represented by the Semantic Hierarchies set out in Chapter 7 can be allocated to data element concepts and data elements. When they are allocated to one level of abstraction, the semantics are presumed inherited at all lower levels. If the allocation to a data element concept is sufficient for a data element, the data element's semantics are "automagically" bound by those allocated to the data element concept. If however a data element's semantics are to be further restricted, allocation can additionally be to the data element. In this case the only allocation allowed must be a semantic subset. Prevented as well are same-semantics allocation or recursive semantics allocation. Finally, if semantics are already allocated to attributes from within entities of a Specified Data Model, or to columns of tables of an Implemented Data Model, the same rules apply. In this case, only super-sets would be allowed to be allocated.

Finally, value domains can also be allocated to data elements. These too must be super sets if there are value domains already allocated to attributes and/or columns.

The Data Element Model set out by this book is a valid and correct implementation of the ISO Standard, 11179 for data element metadata and registries. The data element module of the Metabase System is a data element registry. While having an ISO 11179 standard's conforming registry is important, it is only really valuable when it is interconnected and integrated with the ability to employ those data element semantics within data structure templates (Chapter 9), or database data models (Chapter 10 and 11). The ISO standard, 11179 is a conceptual specification. That's unfortunate because this is not a "rubber meets the road" standard. Only real and valid implementations of ISO 11179 are "rubber meets the road" uses. Finally, Chapter 9 sets out examples of valid and invalid implementations of ISO 11179.

With respect to the two error classes, initial semantic disconnects, and follow-up semantic synchronization activities, Chapter 9, Data Element Model, is a very important component. Data elements provide the fundamental and overarching semantics for all the business facts that are ultimately captured within the Operational Data Models.

While this overarching data element architecture can be specialized via attributes and columns, all these lower level specifications are still just variants of the fundamentally defined data element.

If data elements are seen as a define-once use many-times component, the work will be able to be completed and will be of real and lasting value.

This one-to-many approach enables a greater ability to deal with both classes of errors starting with first class because a DBMS column from an Operational Data Model can be researched to find its column, attribute, or data element parent. Similarly, a different DBMS column from a different Operational Data Model can also be researched. The sooner these two DBMS columns have a common parent the greater the ability to say that the two DBMS columns fundamentally represent the same data. Affinity lessens the need for the second class of error research.

1.4.9 Chapter 10, Specified Data Models

Chapter 10, Specified Data Models, is a "can do" chapter. Just having standardized data elements is not sufficient to have data semantics management. Included in the management is also granularity, precision, and temporal aspects. By having standardized data structure templates, enterprises can specify the granularity, precision and temporal aspects of the data that will ultimately be specified within the various database data models and that finally represents standard data that is exchanged either through SQL view-based exchanges or, for example, XML-based exchanges.

The data models set out in Chapter 10 are data models of concepts. A data model of a concept is not a conceptual data model. The former is a fully defined data model, and the later represents a "fuzzy" form of a data model that, through data model "baking" becomes more well-formed, that is, a logical data model, and thereafter, through more "baking," becomes a physical data model. This is a critical difference. In Whitemarsh, Specified Data Models are data models of concepts. This is a difference with a real distinction.

There can be relationships among the various data structures participating in these subject-based data models. Each subject-based data model consists of entities, attributes, and relationships. The relationships are expressed as primary and foreign keys. Candidate primary keys can also be defined. Because every attribute is related to a "parent" data element there is a walk-back to enterprise semantics across the entire set of concept data models.

Relationships can cross subjects just as it does in real situations. By having data structure templates, there can be standardized uses of these structures. For example, if a structure exists for Address, wherever an address is employed throughout all the different database data models, they can be mapped back to the template. This provides the ability to know that data really is the same even when named differently, and really is different when named the same. This is very critical when assessing whether different database models have the same granularity and temporal aspects.

For example, one database data structure might have a real-time characteristic for Sales data while another has a daily close-of-business characteristic. In both cases the majority of the database table columns would be named the same and might exist within similarly named tables. These data cannot really be shared because in one case, the database records are real-time while in the other case, the collection of records is daily close-of-business based. If sharing must occur notwithstanding the significant differences in granularity and temporal aspects, there would have to be some level of summarization and transformation of sets of the real-time data records into a "week-COB" record.

Semantics and value domains can be allocated to attributes but only if they are subsets of those allocated to data element concepts or data elements. The Metabase System, as it should, ensures correct allocation.

With respect to the two error classes, initial semantic disconnects, and follow-up semantic synchronization activities, Chapter 10, Specified Data Models, provides two major assists in reducing the quantity of errors. The first is the standardization of data structure templates that will dramatically accelerate the creation of database data models. The second is the bringing forward a great deal of standard semantics from the data element to the attribute, and then from the attribute through to the DBMS column. All this work, done in advance, will dramatically reduce semantic error quantities.

This will greatly improve the ability to better know whether a given DBMS column, as a semantic implementation of an attribute has the same precision and granularity as does another DBMS column as it is related possibly to the same attribute.

As with this same analogy above, the sooner two DBMS columns share a common parent, the greater the ability to say that the two DBMS columns fundamentally represent the same data. Affinity lessens the need for the second class of error research.

1.4.10 Chapter 11, Implemented Data Models

Chapter 11, Implemented Data Models, is a database data model "can do" chapter. Implemented Data Models are very different from Specified Data Models because Implemented Data Models are schema-based. Thus, relationships are not allowed from one table to the next between different schemas. Implemented Data Models through down-stream Operational Data Models are generally designed to serve a restricted set of business information systems. There are generally five distinct data architecture classes:

- Original Data Capture.
- Transaction Data Staging Area.
- Operational Data Stores.

- Data Warehouses (wholesale and retail (sometimes called data marts)).
- Reference Data.

Each of these data architecture classes have certain characteristics, styles of design, and specific audiences and users. A recent popular class of database, master data, is a class of database that is to represent enterprise-wide definitive value sets for certain data structures such as customers, products, and the like. Thus, master data can be considered to be reference data given an expanded definition of the word, reference. The data architecture classes are described in Appendix 1, Data Architecture Classes.

Each schema-based data model consists of tables, columns, and relationships. The relationships are expressed as primary and foreign keys. Candidate primary keys can also be defined. Because every column is related to a "parent" attribute, there is a walk-back to enterprise-based data models of concepts. If these models do not exist, columns can be mapped to data elements. Again, this ensures there is a walk-back to enterprise semantics across the entire set of schema-based data models.

There can be multiple data structure templates represented in a single table as would be the case of person-data and address-data. There could also be multiple sets of person-data as would be needed if there were multiple person-roles contained in one table. Of course, there can be multiple uses of data structure templates in different tables of the same schema and within different schemas. Collectively, all this define-once use many-times enables a comprehensive mapping from enterprise data model templates of concepts to various aspects of database schemas across all the different data architecture classes.

With respect to the two error classes, initial semantic disconnects, and follow-up semantic synchronization activities, Chapter 11, Implemented Data Models, is a great leap forward. Every column in every table can be related to enterprise-wide data elements, and the granularity and precision of the table can be related to the enterprise's standard for granularity and precision that is set out in the Specified Data Models of concepts. These standardized walk-backs to enterprise semantics greatly reduce the first class of error because of standardized semantic mapping. With that class of error reduced, the second class of follow-up semantic synchronization activities is also greatly reduced.

1.4.11 Chapter 12, Database Object Classes

Chapter 12, Database Object Classes, is a "can do" chapter, and is a key chapter with respect to data semantics management. When database first began in the late 1950s and when DBMSs started to appear in the early 1960s, the concept of a "database record" was more comprehensive than what would be considered appropriate for a database table. The database record of "yesteryear" had complex structures, and some DBMSs had embedded processes. This was because a database record was to represent a complete and coherent definition of a corporate policy. For example, all the data structures associated with a person, or a contract, customer, company, invoice, shipment, or order.

With the advent of the relational model, the concept of a table as a collection of single-valued columns took on prominence. Tables, in the relational model are very simple. Properly constructed, they consist of single valued columns with only one column type in each table. For example, if there are four telephone numbers, there should not be TelephoneNumber-1, TelephoneNumber-2, TelephoneNumber-3, and TelephoneNumber-4. There should, instead be a separate Telephone Number table. This type of data modeling is classified as Third Normal Form (3NF) and is written about extensively. The problem with 3NF tables is they break up single policy-based database records into multiple tables. Thus, there is data model fracturing. Generally, this is not an ideal situation.

The SQL data model, contained in the ISO/ANSI standard SQL for 1986, 1989, and 1992 all conformed to 3NF tables. Starting in 1990 and continuing until 1997 market pressures were brought to bear to restore database records back into DBMS standards.

During this time frame, this author created a concept called database objects. Every database object is represented by its definition, i.e., its class, and by its instances, that is, the database objects themselves. Created is therefor the complete definitions for constructs like person, contract, customer, company, invoice, shipment, or order.

Every database object class consists of four parts: database object data structure, database object table processes, database object states, and database object information systems. The database object data structure can consist of nested structures, which in relational data model terms would be nested tables. The database object table process is that set of constraints and processing rules to ensure absolute integrity of a database object table row's insert, modify, or deletion.

Database object states represent the value-based states across a complete database object data structure that conforms to a corporate policy. For example, there might be Employee Requisition, Employee Candidate, Employee New Hire, Employee Assigned, and the like. Each name represents a state that is unambiguously achieved in terms of values across the database object class's data structure.

A database object state is achieved through the successful execution of database object information system. Each of these database object information systems, in turn, execute insert, delete, or modify database object table processes.

The ISO/ANSI standard for SQL incorporated much of the concepts set out in database object classes in the SQL:1999 standard. It was however accomplished in a unique way. Instead of creating a whole new schema object called Object, the committee decided to expand the data structures contained in tables. Column data types were freed from just being single value types such as integers. Added were arrays, groups (e.g., Address that contains Street, City, State, etc.), and nested groups such as Company contain Employees which contain Benefits, Assignments, and the like. Processes were able to be established within the confines of these newly created SQL:1999 constructs. Essentially, after an absence of about 25 years, the database record rose from the dead like the proverbial phoenix from Phoenicians Legends and became part of the SQL:1999 standard. It has since been expanded in SQL:2003. No significant extensions are part of SQL:2008.

With respect to the two error classes, initial semantic disconnects, and follow-up semantic synchronization activities, Chapter 12, Database Object Classes, serve two key roles. First, they become coherent objects of shared data. That is, a whole customer, order, shipment, invoice, and the like. Contained in each is a precise specification of integrity, state, precision, and granularity. Second, every database object table contains columns from traditional 3NF tables, which in turn are mapped onto attributes and data elements. There is, therefore, enterprise-wide integrity across the database object data structure. Database object classes represent wholly contained structures within Implemented Data Model databases. The effect on the two classes of errors is that if data interoperability is based on these database object structures there is an automatic acceptance of precision, granularity and temporal aspects.

If the SQL DBMS does not support SQL:1999 or later data structures, the database object classes and all their components provide a template for constructing the Implemented Data Model design within the Implemented

Database. Faithfully implemented and followed either through embedded SQL DBMS processes or through outer rings of data integrity subsystems that all business information systems must proceed through, both classes of errors will be greatly minimized.

1.4.12 Chapter 13, Operational Data Models

Chapter 13, Operational Data Models, a "can do" chapter, serves to represent the database models that are closest to the reality of actual data. Each Operational Data Model schema, implemented via the DBMS' SQL data modeling language represents the actual structures, integrity, data types, rules and processes through which data is allowed into databases, is queried from databases, and is maintained within databases. Every Operational Data Model schema cannot be semantically more rich than what is allowed by the SQL DBMS through which it is defined. It is, therefore, not unheard of nor unacceptable to have different SQL DBMSs of different capabilities implementing different databases. Neither shoes nor DBMSs come in one size to fit all.

The data structures contained in Operational Data Models serve two masters. The first is the Implemented Data Model from which they are derived. There may be multiple Implemented Data Model "parents" as would be the case for a Data Warehouse database, or an even more refined data mart database. The other master would be the one or more business information systems that are the main users of the database.

If legacy business information systems and databases already exist, it may not be practical or politically possible to make massive changes to an existing database. In this situation, the only practical solution is to provide mappings to the one or more Implemented Data Model databases in terms of columns, data types, and value domains to the maximum extent possible. Then, over time, slowly transform the Operational Data Model database from stove-pipe semantics to one of enterprise-semantics.

A very practical approach in accomplishing Operational Data Model transformations is to transform the business information systems from stove-pipe engineering to enterprise-process engineering. Transforming these legacy systems through business information system generators is often much cheaper and faster than trying to retrofit an existing legacy, stove-pipe engineered business information system.

Enterprises have been procuring Enterprise Resource Planning (ERP) software packages in the hope that they will gain enterprise interoperability, industry best-practices, and enterprise-wide standardized semantics. Seldom, however has this been true. Different packages from even the same vendor can have conflicting semantics.

Often the vendors of these packages just view DBMS and database as operating system access methods, which has the unfortunate result of embedding all data and process semantics into the application layer. Even though that kind of embedding was discredited more than 20 years ago, it seems to live on in ERP packages.

The effect of this is that there are many tens of thousands of database tables, no viable database schema, very expensive maintenance, an almost impossible task of efficient evolution, and finally, an almost impossible ability to employ ERP data in everyday business class report writers such as Crystal Reports.

One possible solution to all this self-inflicted chaos is to create a set of shadow databases that read and write the ERP databases. These shadow databases conform to enterprise semantics and common-sense-based data structures. Interfaces are built between the shadow databases and the ERP databases such that any data going to or coming from the ERP package goes through the shadow database. The practical effect of this strategy is that the ERP package is fire-walled off from the rest of enterprise applications such that the shadow databases are able to be included in enterprise-wide databases just as if the ERP package's databases did not even exist. The alternative to this is to attempt to integrate the often semantically discordant and conflicting semantics from many different ERP packages into a set of enterprise-wide semantics. This effort can range from very difficult to totally impossible.

With respect to the two error classes, initial semantic disconnects, and follow-up semantic synchronization activities, Chapter 13, Operational Data Models, the effort to prevent these errors with respect to enterprise-wide semantics can be very difficult if the role of these databases is to serve the needs of ERP packages. In this case, all the rules that would prevent these errors are embedded inside the ERP software package itself. So either the prevention rules are there, or the package has to be modified to execute them.

If the Data Element, Specified, and Implemented Data Models are constructed from an enterprise-wide mission perspective, at least when the Operational Data Models are imported, the semantic mismatches will be obvious. Compromises will have to be made to accommodate these

Operational Data Models. Likely while databases and business information systems will have to be built to transform nonstandard data to what is required by the enterprise. While the way ahead will not be easy, at least it can be known.

Otherwise, if the Operational Databases are built in a top-down fashion from data elements through Specified Data Models and Implemented Data Models including the assignment of layers of value domains, the probability of preventing these error classes is quite high.

1.4.13 Chapter 14, Interface Data Models

Chapter 14, Interface Data Models, a "can do" chapter, is focused on the interfacing of database-based data to business information systems. There are many levels of sophistication that can exist in interfaces starting with just simple one-to-one relationships between database tables and the business information systems, to whole interface specification processes that cause DBMS selection, combination, and calculations prior to the business information system receiving the database data. In either of these two cases, there is a tight binding between the DBMS and the business information system. Each must know of the other, and protocols must exist to ensure that they work well together.

Service Oriented Architectures, are entirely different. The business information system does not know when, where, or how the DBMS will get the data-based transaction. Similarly, the DBMS does not know when, where, or how the business information system will get the database data. How are the interconnections made? Magic? For sure not. There must exist what would effectively be a Cellular Phone Network.

The key to making such an environment work is highly engineered transactions wherein every transaction at least knows the ultimate comes-from and goes-to address for the message that is being transmitted. Today's Internet is essentially an SOA environment because every transaction is sent with sufficient "goes-to" information to get to the necessary location, be processed, construct a response, and sent that response back with sufficient return "goes-to" information for the waiting business information system to process the response.

Since virtually all SOA transactions are sent in just plain ASCII, using XML-based transactions is a clear and obvious choice. These transactions can be greatly compressed with existing technologies such as those found in WinZip or other software applications.

Key to the SOA environment are well-known semantics wrapped up in highly engineered transactions. In short, nothing's really changed except that the transaction, instead of being hardwired from sending receiving location must be enhanced to have that information inside it. If the transaction arrives and is misunderstood or not understood at all, the only possible response back is "Huh?"

With respect to the two error classes, initial semantic disconnects, and follow-up semantic synchronization activities, Chapter 14, Interface Data Models, the ability to catch and/or prevent errors depends greatly on whether the Operational Data Model is interfacing with business information systems perceive that database and the DBMS to be merely as a set of files and a file access method, or as a part of the semantic fabric of the enterprise. If the former, the options range from limited to none.

The Interface Data Models enable the forward deployment of enterprise-wide semantics despite Operational Data Models that might not conform. It's not that the Operational Data Models are avoided, it's that views can be constructed to some extent to transform the Operational Data Model's granularity, precision, names, data types, and value domains to be that of the enterprise.

Most ERP packages, however, have installation steps that require the specification of all the legal values, so catching errors is entirely under the control of the ERP package. The advice, therefore, is to evaluate these packages very carefully. What can be of great value is to have accomplished all the enterprise semantic steps of creating the Data Element Models, Specified Data Models, and Implemented Data Models and use these as requirements specifications for evaluation the ability of proposed software packages to trap and/or prevent errors.

If, however, the business information systems enable the creation and effective use of SQL Views, many of the first class of error can be programmed directly into views. Incorporated can be valid value selector lists, column renames, value transformations, and the ability to create new values from values that have been provided in the columns. When these are comprehensively addressed, the quantity of the second class of errors drops dramatically.

1.4.14 Chapter 15, Work Plans

Chapter 15, Work Plans, consists of work breakdown structures specially designed to fit the data models that are set out in this book. That is, those for Data Element Models, Specified Data Models, Implemented Data Models, Operational Data Models, and View Data Models. Supplementing these data model structuring work plans are specialized work plans for discovering value domains, reference data, and performing overall enterprise-wide reference and master data development.

With respect to the two error classes, initial semantic disconnects, and follow-up semantic synchronization activities, Chapter 15, Work Plans, form a key component that ensures that database and business information systems projects are on the right path and accomplish the right steps in the right sequence. The work plans focus on discovering the right metadata and placing that metadata into the Metabase System so that it can be brought to bear and at the right time.

1.4.15 Chapter 16, Summary

Chapter 16, Summary provides a summary section for every chapter. It also provides a set of sections that addresses how every chapter addresses the two error classes that are described in every chapter. Finally, this chapter contains a set t of concluding statements and a challenge to the reader.

1.5 Rationale Summary

This chapter set out the rationale for data semantics management. The rationale is presented by first citing the clear state of the world, that is, that we live in a world economy. With the Internet, there are no borders.

Commerce and communications of virtually all abstract goods are exchanged at the speed of the Internet. A good example is all the Internet-based jokes that travel around from person to person and organization to organization almost instantaneously. Commerce too is similarly engineered. The computer on which this manuscript was created was built in Asia. The operating system is from Microsoft in Washington, the word processor from Canada, and the Metabase System that is the demonstration vehicle for all this book evidenced is from a company in Florida with supplementary products from Arizona, Australia, Germany, South Africa, and England.

Simply stated, the most common languages throughout the world are not English, French, or Spanish. Rather, they are SQL, C, C++, because all communication throughout the world is first founded from these universally accepted and unambiguously understood computer-based languages.

This chapter presented a clear statement of the problem that must be solved and several examples to illustrate it. Thereafter were examples of the effects of not having sufficiently engineered data interoperability.

This chapter set out the remaining chapters in the book and conveyed how each chapter contributes to addressing the two classes of problems that characterize the achievement of data interoperability through data semantics management.

This chapter, as does all chapters, concludes with this summary and a set of questions and exercises that should be able to be accomplished by practitioners of the book.

This book is not just to be read and thought about. Rather, it is to be acted upon. The former will provide intriguing thoughts and concepts to consider. The later will provide real action plans coupled with a down-loadable production class metadata management system, the Metabase System, to manage data semantics in support of creating the interoperable data environments critical to the world economy.

1.6 Questions and Exercises

1. In your review of the Chapter Sections (1.4.1 through 1.4.15) are there any missing chapters that are important to data semantics management? Which ones? How will these missing chapters affect your data semantics?

2. What are examples in your organization of dysfunctional semantics? How would you quantify the cost and lost opportunity? What factors constitute cost? How much does each factor cost? What would be these costs be determined at a project level? At an organizational level? At the enterprise level? How would you present this to management so that your efforts are funded and supported?

3. How different would these costs be determined at each different level? Are there additional costs as you proceed through the different levels? Why do these additional costs exist? How can you place a value on them? What are the benefits from incurring these extra costs? How can you present these additional costs to management so they can see the benefit from proceeding from a stove pipe semantics management environment to an enterprise semantics management environment?

4. What are examples in your organization of good semantics management? How would you quantify the benefits and gained opportunities? How much did each factor save? Extended to a complete functional area and the enterprise, how much would good semantics management have saved? How would you present this to management so that your efforts are funded and supported?

5. Why should there be semantics management? What's the business case? How can you make the business case in your organization? What are the political and cultural inhibitors to data semantics management? What strategies can be put into place to overcome these inhibitors? How would you present this to management?

6. After a review of the semantics effort failures, have any occurred at your organization? What was the consequence? Were the failures swept under the rug? How were the failures explained? What remediation was put into place? How can you make the case that all lessons are not free?

7. In regard to the semantics of names, how should names be engineered? Is it possible to have multiple levels of names from the most generalized to the most specialized? What is the consequence of forcing a single set of names onto all database columns in the enterprise? Discover examples from within your project, organization, function, and enterprise of badly constructed names and show what the costs have been from these bad constructions. How would you present this to management so that your efforts are funded and supported?

8. What is the value of semantic hierarchies? Have these been tried in our organization? How have they been constructed and used? Have these

semantic hierarchies been integrated into the processes that construct names? If so, how? What was the value? If not, why not? Discover from within your project, organization, function, and even the enterprise how there have been the same words with different meanings, and different words for the same meaning. Quantify the costs and the confusion. Determine how these might have happened and can now be prevented.

9. What is the role of value domains? Are they not just semantics with a different name? How can values from restricted value domains be used in name construction? In data warehouse dimension construction? Have value domains been standardized in your project, organization, function, and enterprise? What has been the rewards when this has happened? The costs when this has failed. What has been the effects on database design, program and business information system construction. Compute the costs at the project, organization, function, and enterprise level. How would you present this to management so that your efforts are funded and supported?

10. In an examination of the various fact specification cases, are there any that are missing? Which ones? How would you define them? How often do you find these fact specification cases in your enterprise? Have you considered identifying and managing collections of these cases? What would be the benefit from such management? What has been the costs from not managing collections of facts? How would you present this to management?

11. What is the role of the Data Element Model? Its value? Do you have such a model? If so, why? What has been the value? Has your Data Element model suffered any of the data semantics failure above? If so, what has been the consequence? How does your organization distinguish a data element? How many data elements versus columns does your enterprise have? Why? Compute the cost of column-based versus data element based semantics models. How would you present this to management so that your efforts are funded and supported?

12. What is the main role of the Specified Data Model? Has your Enterprise attempted to standardize data structures of concepts? Does it make sense and have value? What's the difference, and why is this

important? What value could be gained by having standardized data models of concepts during the construction of databases?

13. What is the main role of Implemented Data Models? Is there value from having a DBMS independent version of a database model? What role does it play? How can and should it be related to Specified Data Models? What are the possible relationships between a Specified Data Model component and an Implemented Data Model component?

14. What is the main role of Operational Data Models? How can and should these be related to Implemented Data Models? What are the possible relationships between an Implemented Data Model component and an Operational Data Model component?

15. Compare and contrast the engineering, architecture and interrelationships among Specified, Implemented, and Operational data models versus the same characteristics of conceptual, logical and physical data models. Which alternative leads to integrated and nonredundant enterprise data semantics management and which leads to a large volume of semantic stove pipes? Compare and contrast the cost of each alternative. How would you present this to management so that your efforts are funded and supported?

15. The Interface Data Models are defined as containing three components. SQL Views, XML schemas and data, and SOA. How are these the same? Different? Are they alternatives, one to the other or "three peas"in a pod? When do you need each? More than one? Is there value in having these defined in terms of database models that are supported by all the other upper levels of metadata? If so, why? If now, why not? How would you quantify the costs and the benefits?

16. In a brief review of the work plan descriptions, are there any really important ones missing? What value is derived from standardizing work plans? How does the Metabase System of standard artifacts fit within these work plans?

2

Why Semantics

The objective of Chapter 1 was to set out the rationale for data semantics management. The rationale was presented by first citing the clear state of the world, that is, that we live in a world economy. With the Internet, there are no boarders.

Chapter 1 presented a clear statement of the problem that must be solved and several examples to illustrate it. Thereafter are examples of the effects of not having sufficiently engineered data interoperability.

Chapter 1 continued by setting out the remaining chapters in the book and conveyed how each chapter is to contribute to addressing the two classes of problems that characterize the achievement of data interoperability through data semantics management. The key message of Chapter 1 was that readers must not only read the chapters, but also act upon them in their enterprise by downloading the production class Metabase System and seeing how enterprise data semantics can be accomplished. Wether you continue with the Metabase System or merely employ it to confirm that data semantics management is practical is your choice.

This chapter, as does all the other chapters, concludes with both a summary and a set of questions and exercises that should be able to be accomplished by practitioners of the book.

This chapter, addresses the fundamental need for highly engineered semantics as a way to control the error classes. This need is first explained within the context for data management. Key to understanding data management is understanding the role of enterprise-level policies and procedures. Because these need to be enterprise-wide, there must be consistency in their definitions at all levels and across all enterprise organizations.

Another section identifies and describes the different models that contribute enterprise semantics by means of graphical representations and textual explanations of their contents. Included in these models are database object classes. The depicted models are all high-level modules. Thus, they are not third-normal form (3NF), table-depicted graphics. The 3NF diagrams for each of these models is provided in Chapters 9 through 14. It also needs to be noted that these models comprise the data modeler component of the Metabase System. There are other Metabase System models that enable complete definition of enterprise architectures.

The specifications of database object classes are set squarely within corporate policy specifications. Data, the by-product of policy executions, are the "what" that should be shared within and across corporate organizations. Database object classes are the coherent specifications of shared data. Unique to database object classes are the definitions of the states and state transformations through which a given object proceeds from birth through all life cycle stages until the database object becomes fully mature.

This chapter continues with a simple definition of data semantics and illustrates this definition through several examples. Key to the explanation is the simple fact that there is no silver bullet that addresses all the needs of data interoperability across the enterprise. Because there is no silver bullet, the management of data semantics must proceed through several different, but thoroughly interconnected models of metadata. These models address: data elements, data structure templates, the employment of these data structure templates within databases, and finally as specifications of semantically complete data exchanges. Each of these metadata models is generally defined in terms of their critical components.

At the end of this chapter is a section on the return on investment (ROI) from effective data semantics management. The basis for the ROI is in two parts: Data and business information systems. While the ROI for data is significant, the largest ROI is on the development, evolution, and maintenance of business information systems. Simply put, the business case for data semantics management is in four parts: increased productivity, increased quality, reduced costs, and reduced risk. Not accomplishing data semantics management is not only a bad business decision in the short run, it likely is a terminal business decision.

To achieve [4]enterprise-database, data semantics management is essential. Data semantics management embraces:

---

[4]. As defined within this book, an enterprise is merely a term to relate to a collection of organization units that have common collections of data, processes, activities within a business or a company and sometimes beyond corporate affiliations as in the case of data interchanges. An enterprise is therefore not just a synonym for business or a company. Rather, it is intended to convey a common data, process, and activity view across the organizational units sharing that view.

- Mechanisms for defining reusable and inheritable semantics.

- Approaches for defining enterprise data elements.

- The ability to construct standardized a standard data structure templates.

- An approach to accommodate diverse data semantics.

- The ability to enforce standard cultured and language independent value sets.

- An overall semantics and value set enforcement environment that is unavoidable and work enhancing.

Without effectively defined and systemically implemented enterprise data semantics management, the enterprise is severely crippled in today's world-wide commerce.

Data is executed policy. Data semantics are the technical representations of policy specifications. Unless data is represented through homogeneous semantics policy adherence cannot be measured or evaluated. Thus, a key component of quality data management is data semantics.

2.1 Rationale for Data Management

A central tenant of this book is that data is executed policy. Corporations and policy makers create the policy through which organizations operate. Rules, and regulations are established to enforce the execution of policy. Processes exist as mechanisms for policy execution. Data is the most often persistent product of policy execution. Hence, data is executed policy.

Clearly, policy and procedures run the business. Corporations, to have "memory" about their policy executions, build databases within which the policy execution data is stored. Databases are thus, the persistent memory of the organization.

"Data" specifications are thus policy definitions, and process specifications are procedure definitions.

In keeping with the general concept of metadata, that is, that which exists one or more abstraction/generalization levels above "reality," all data (i.e., policy) specifications are metadata. Similarly, all process (i.e., procedure) specifications are metadata.

Since all data is best kept in highly-engineered, expertly constructed databases, metadata should be stored in a database, and the metadata database, i.e., a "metabase," is a database for all policy and procedure specifications.

Finally, data management manages both metadata and data, and a data management program is what creates, evolves, and maintains data management.

Quality-databases represent a technological expression of organization, clarity, and precision. It may or may not be computerized. If it is, it may exist on a desktop, a server, or mainframe. Finally, a database may or may not be centralized. Despite the mode, the mechanism, or the form of database implementation and operation, the codification of and adherence to data semantics, which are the rules for meaning, validity, and usage, are prerequisite to successful database.

When a database is on a computer, it represents the automation of the knowledge component of a business, which is manifest through the business's quality operation, planning and management. With quality-database, management of a business can research the past, organize the present, and plan for the future.

Most databases are defined through collections of simple data structures called tables. Policy specifications are seldom, if ever, neatly wrapped up into simple database tables. Rather, policy specifications may have to be defined across a clearly identified data structure collection of database tables. Each collection is defined through four parts:

- A well-defined data structure.
- Internal processes to ensure integrity.
- Clearly identified value-based states through which the data structure must be transformed.
- Well-defined process collections that create, transform, and delete data structure instances from one business-policy-based state to another.

These policy-based information technology constructs are called database object classes. They are called such because they are defined, created, transformed, and finalized within the domain of a database that is operated through a DBMS (Database Management System). Therefore, to have quality-database is to have defined policy because,

- Each database object class's data structure within the database is the data representation of a policy's definition.
- The database object class's processes through which rows are added, deleted, and modified are the mechanisms necessary for policy execution, that is, the policy's procedures through which database objects are transformed from one valid state to another.
- A fully defined policy includes both its complete definition and its necessary steps for coherent execution.
- Interrelated collections of rows across multiple tables of one object class and across multiple database object classes form more comprehensive policies.

When data is seen as executed policy, and is realized through database object classes, quality databases support the following within the enterprise:

- Business information systems that are a coherent union of policy that in turn support the execution of procedures that represent the accomplishment of the policy.
- Consistent collection and/or modification of policy instances through the life cycle of the policy.
- Consistent execution of policies whenever, wherever, and however deployed as the essence of the policy and the totality of its critical procedures are encapsulated within the database object class itself.
- Minimized redundancy and consistent policy implementations across distributed environments as the database object class can be distributed through encapsulated strategies.

- Comparable instances of deployed policies as they are independent of hardware architectures and operating systems.

What forms the basis of a database object? Simply, it is a business' policies and procedures. While policies can exist without procedures, the converse is not true. This ontological priority dictates that procedure is dependent on policy. Not only do they go together like hand and glove, the glove (procedure) serves no useful purpose without the hand (policy).

A database object is a person, place, or thing that has internal consistency, and is transformed from one valid, predefined state to another through well-defined rules. The minimum value states are null and valued. The internal behavior of a database object class as it transforms database objects from one state to another is immaterial to its user. Database object classes conform to the requirements of business rather than the converse. Database objects are the corporate memory of the enterprise. All the rest is anecdotal.

Policies and procedures, that is, database object classes, bring order, consistency, and predictability. The larger the enterprise, the greater the dependence on policies and procedures. Data is the evidence of policy execution. An employee's record is proof that the Human Resources policies have been carried out. Procedures are the techniques, methods, or processes by which policies are carried out. If an enterprise's policy is to be profitable, its balance statement, produced by processing all the general and subsidiary journals are the measure of adherence to the policy. If policy is met, the enterprise must be profitable.

The procedures are named, and their data actions are associated with specific subsets of the named data structure. The names of the procedure sets represent data structure transformations from one recognizable state to another. Each state represents a determined value set within the business. Procedures include, for example: establishing an employee requisition, accomplishing employee hiring, and performing employee assignment.

Enterprise-database is an organizational operating condition in which there are both defined policy coherence and integrity as well as consistency in policy transformations throughout the enterprise. Enterprise-database is achieved irrespective of functional and organizational style and irrespective of policy transformation technology (that is, computers, operating systems, programming languages, and database management systems).

Organizations not pursuing database object class specification, implementation, and management information system (MIS) evolution through database objects will never achieve enterprise-database. Rather, they will be left with complicated, redundant MIS specifications, expensive MIS implementation and inconsistent difficult MIS operation, evolution and maintenance.

Enterprise-database is the expression, population, use, and manipulation of all database objects. Enterprise-database begins first with quality-database object classes founded upon the policies and procedures surrounding their specification, implementation and evolution. Value proceeds not first from the database object, but from the database object class. The consequences of quality database object classes are "real" database objects. The information technology assets of the enterprise are both its database objects and its enterprise-database object classes.

For example, if only database objects are valued, only musical notes would be needed for a great symphonic score. Performances are differentiated however, from the grade-school band to a first-rate orchestra because of the musicians' talent that is coupled with the quality of the orchestra director's interpretation of the score's rhythms, tempo, articulation, and dynamics.

Policy executions, that is data, are the medium of business communication. Figure 2 illustrates this point. This figure shows subject areas as the basis for one or more business policies. If the subject area is Human Resources there would be policies regarding the application for employment, the information that must be provided on the first day of hiring, and the like. In turn, these policies result in the technical specifications of data that must be provided. These specifications ultimately map onto actual data that exists. The subject area also maps onto database schemas, for example, the database schema for employees.

In this schema, there would be a number of tables, including for example, those that would contain critical information from the application for employment, the first day of hire benefit selection documents, tax withholding documents, dependent information, department assigned, and the like. Again, these tables map onto the actual rows of data that must reside in the database. In short, data is the proof that the policy was executed.

Figure 2 shows not only hierarchies of instances, but also contains multiple levels of abstraction. For example:

- Policy is represented through policy technical specifications.
- Policy technical specifications are represented through rows of data.

- Schema tables are represented through rows of data.

Traditionally, databases contain two levels of abstraction: the schema, and the rows of data. The schema represents *types* and the rows of data represent *instances*. Whenever there are multiple levels of abstraction there is, by definition, types and instances. Metadata management systems built on a DBMS foundation not only contain both a schema and rows of data, they may contain data across multiple levels of abstraction. Thus, metadata management systems may contain multiple sets of interleaved metadata (type) and data (instances).

To effectively represent the data required to store and control data architectures across the enterprise, the metadata management system must capture data from across multiple levels of abstraction, as shown in the database represented by the DBMS tables in Figure 2. In database, there is no restriction on the quantity of levels of types and instances. If one pair is

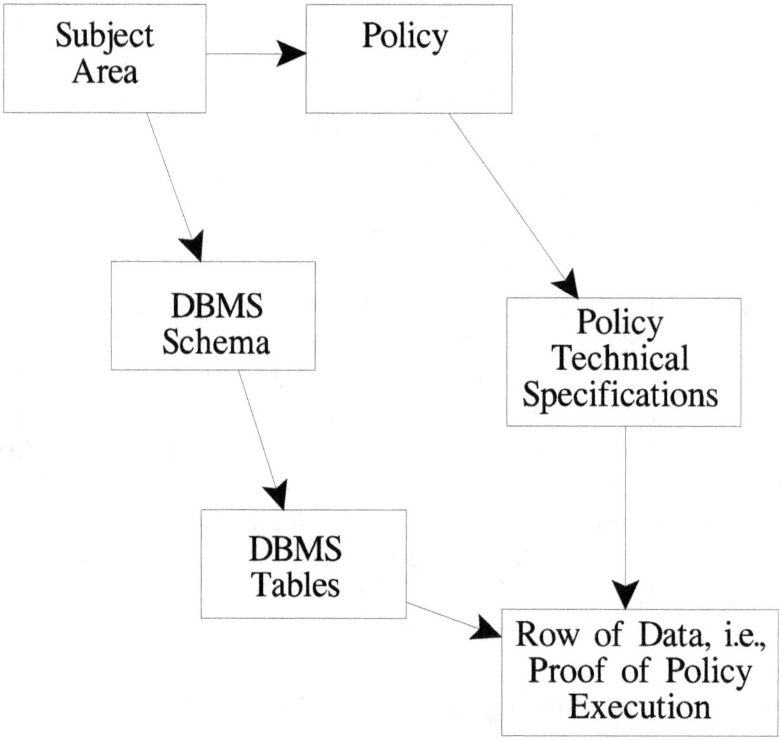

Figure 2. Levels of abstraction required for data standardization.

allowed, all pairs, including interleaving pairs are allowed (type then instance, which in turn is type then instance, which in turn, . . .).

Because data is so critical, it is essential that it be commonly represented and understood. Database is a mechanism of common data definition, common data loading and update, and common reporting. What about databases that cross different cultures, languages, and finance systems? If all data is defined to specifically reflect cultures, languages, and finance systems it becomes virtually impossible to accomplish effective business communications.

Data, the medium of effective business communication, must both reflect localized culture, language, and finance systems yet able to be amalgamated into broader contexts for country, region, and world-wide decision making. The alternatives are simple: one single data standard for all reasons, or a set of mapping rules and mechanisms that allows data to be recast from local culture, language, and finance systems to the broader country, region, and world contexts. While the former is simpler and less costly, it has never been successful. Thus, we are only left with the later.

Database does not solve this problem, rather, it exposes it. Prior to database objects, data definition and data value incompatibilities were hidden by technology mismatches. Because of the database language SQL, these technology inconsistencies have been removed. Exposed for all to see are the data semantics mismatches. These must be solved prior to meaningful data sharing.

2.2 Data Semantics

Data semantic problems exist when the same concept is known under two different names, is described differently, or is different in some way that causes the semantic sameness to be hidden. For example, the EMPLOYEE_ID of 527-49-2396 as an ASCII string of 11 characters is really a Social Security Number of nine integers. While the concept is the same, their names, data types and its lengths are different. The common concept is hidden by the differences.

Another example of semantic irregularities is found whenever concepts hide intrinsic rules and/or common sense. For example, the default number size for the Oracle DBMS is Decimal 12.2. For the column ANNUAL_OVERTIME_HOURS, there could be 999,999,999.99 hours.

Considering there are traditionally only 2080 work hours per year, the amount of accruable annual overtime seems excessive.

A third example of problems associated with data semantics is the different data structure representations for the same concept. For example, in one inventory data file there were many sets of fields all dealing with inventory valuation. Each set of fields consists of 12 related fields, one for each calendar month. The field names were only five characters long. Each had to detail the type of data, the calendar month, and whether the data was for the current, prior, or next year or period.

In another inventory file there were only four fields: year, calendar month, item number, and inventory quantity. The first inventory data structure contained 39 different fields to represent a very specialized version of inventory valuation data while the alternative data structure only had four fields representing a very generalized version.

Data semantics irregularities are most commonly evidenced through differences in:

- Data names.
- Data types.
- Data lengths.
- Data structures.

Data Semantics is not something accomplished with a silver bullet. If it were, only one place needs to be shot, with only one gun, and with only one silver bullet. Such an approach, the magic silver bullet, has been tried over the years and has consistently failed. Not for lack of trying, but for a lack of understanding the breadth and depth of the required effort.

Accomplishing the correct silver bullet approach requires many guns, bullets, and many targets, and not just in one fusillade of shots, but a continuous attack over a sustained interval of time. Is it worth the effort? Ask the folks at NASA to relate the cost of having the Mars lander crash because of a data type error. Ask the folks in Hartford, CT in the 1980s to relate the cost of having an arena roof collapse because of FORTRAN coding error. Ask the folks at the U.S. Department of Defense who spend hundreds-of-millions of dollars every year correcting materiel order and transformation errors.

While every one of these data-centric errors is different, they relate to one reason: a lack of quality data management. The totality of effecting quality data management, which has at its fundamental core, data semantics management, requires careful attention to the classes of data models set out in Table 1.

Data Semantics Component	Description
Semantics Management Model	A Semantics Management Model provides the ability to define and allocate semantic modifier and class words to the fact-based components of the data models: data element, Specified, and Implemented Data Models.
Data Element Models	Data Element Models provide the basis and engineering for all business facts and semantic elements that are to be deployed throughout the various models that form data structure templates (Specified Data Model), or database models (Implemented and Operational Data Models).
Specified Data Models	Specified Data Models provide data models of concepts that can be employed to build enterprise level standardized data structures in database data models.
Implemented Data Models	Implemented Data Models are database data models that are independent from any given database management system (DBMS). Source data models for data warehouses, subset databases, and consensus data models for existing legacy DBMS databases, and Enterprise Resource Planning (ERP) databases. May draw data structures from multiple Specified Data Models.
Database Object Class Models	Database Object Class Models are policy based models that enable Implemented Data Models to coherently represent collections of state-based transformations of critical enterprise data. Database objects represent the data specifications of shared and interoperable data across database data models.
Operational Data Models	Operational Data Models are DBMS, operating system, hardware dependent, and software package dependent data models. May include data structures from multiple Implemented Data Models. May act as repositories and/or exchanges across multiple XML interfaces. May act as DBMS centric view-based interface mechanisms.
View Data Models	View models, regardless of form, that is, DBMS based or technology independent view models such as those based on XML.

Table 1. Data semantics components

2.2.1 Taxonomies and Ontologies

Much has been made in recent years over the definition and use of taxonomies and ontologies as critical and essential elements in semantics management. It's not that either of these are inappropriate or without use, it's that each is often touted as "the" magic silver bullet. From the Wikipedia,

> A taxonomy is the practice and science of classification. Taxonomies, or taxonomic schemes, are composed of taxonomic units known as taxa (singular taxon), or kinds of things that are arranged frequently in a hierarchical structure, typically related by subtype-supertype relationships, also called parent-child relationships. In such a subtype-supertype relationship the subtype kind of thing has by definition the same constraints as the supertype kind of thing plus one or more additional constraints.

The Data Element Model in the Metabase System has a taxonomy for data elements. That's all right for classifying data elements. It must also have taxonomies in support of concepts, conceptual value domains, data element concepts, and value domains. These are shown in Figure 3. These are, however, insufficient because they only control aspects of data elements and none of the other critical data semantics component models such as the Specified, Implemented, Operational, and View Models.

Needed as well are the semantics sets for managing the construction and meaning of names within each data models. This names-component semantic management mechanism is also constructed as a taxonomy, and is depicted in Figure 3. It consists of three nested components: meta category value type classifications, meta category value types, and meta category values. As Figure 3 shows, meta category value types and meta category values are also able to be hierarchically organized. The relationship lines from meta category value to data element concept, data element, attribute, and column enable the proper allocation of semantics. The Metabase System properly manages the allocation to prevent recursion, duplication, and generalization level conflicts.

The very existence of the hierarchies of concept to data element concept, to data element to attribute to column to DBMS column is itself a taxonomy. This taxonomy more comprehensively fits the Wikipedia definition of a taxonomy. The conclusion to all this is that taxonomies have value but they must be appropriately engineered and allocated across the collection of all various data models that are necessary to have effective data semantics

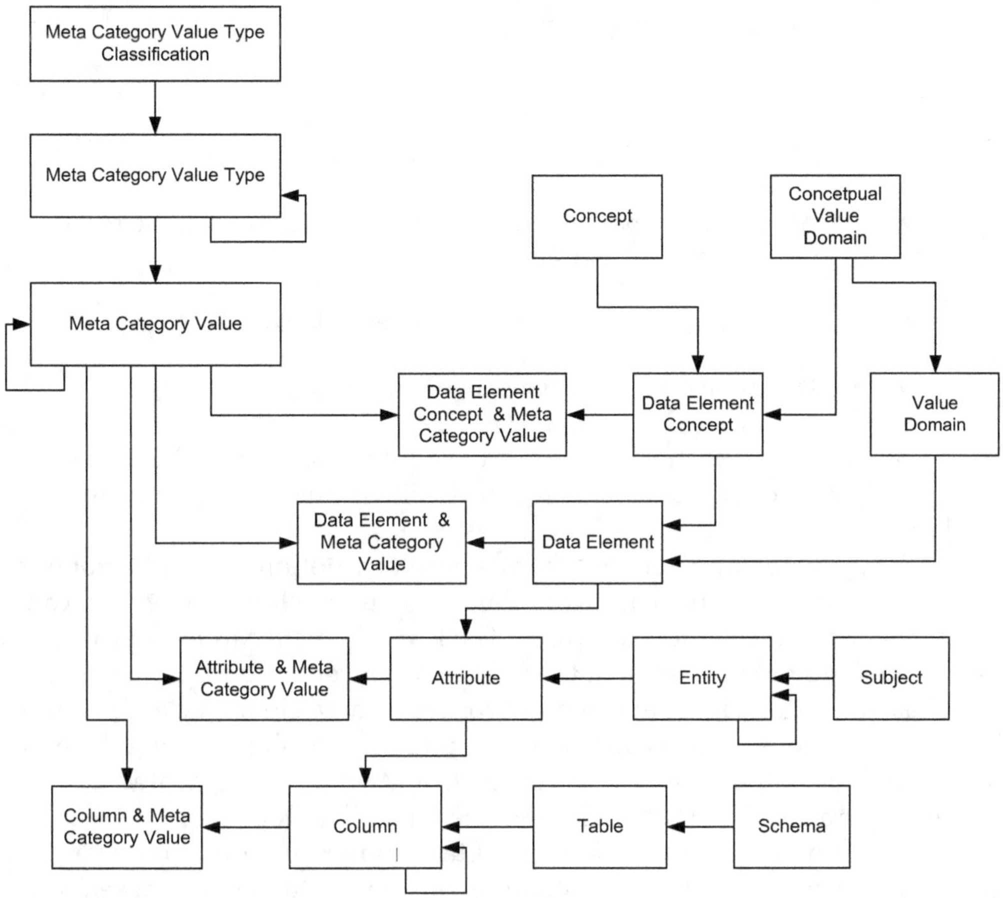

Figure 3. Semantics management model components.

management. The Metabase System has a whole network of taxonomies that are critical to the effective management of data semantics.

The second silver bullet, Ontologies, has been trumpeted as the solution to all the data semantics management problems that exist. Ontologies are, however, more problematic than are taxonomies. From the Wikipedia,

> In both computer science and information science, an ontology is a data model that represents a set of concepts within a domain and the relationships between those concepts. It is used to reason about the objects within that domain. Ontologies are used in . . . Information

architecture as a form of knowledge representation about the world or some part of it. Ontologies generally describe:

- Individuals: the basic or "ground level" objects.
- Classes: sets, collections, or types of objects.
- Attributes: properties, features, characteristics, or parameters that objects can have and share.
- Relations: ways that objects can be related to one another.
- Events: the changing of attributes or relations.

In the broadest sense of the definition cited above, the entire collection of Metabase System data models, not just those illustrated in this book, are an ontology.

A key distinction between formal ontology definitions and taxonomies is that an ontology can have network data structures while a taxonomy can only have hierarchical data structures. The Data Element Models of this book feature the following network data structures: concepts, conceptual value domains, value domains, data element concepts, data element classifications. Additionally, there are network data structure relationships among the entire data models of Data Elements, Specified Data Models, Implemented Data Models, Operational Data Models, and View Data Models.

In summary, when proponents of taxonomies and ontologies suggest that these are both needed in any comprehensive treatment of data semantics management, they are correct. Where the proponents are not correct is that taxonomies and ontologies should be accomplished in a stand-alone fashion, like monoliths. Instead, the best features and objectives of the taxonomies and ontologies need to be incorporated into the overall topologies of data models that enable comprehensive data semantics management. In that regard, these objectives have been accomplished in the Metabase System.

2.2.2 Semantic Management Models

The Metabase System provides independently definable semantic hierarchies via meta category value type classifications, meta category value types, and meta category values. Data modelers can define and allocate these semantic modifier and class words to the fact-based components of the data models: Data Element, Specified, and Implemented Data Models. Within these models, the data components that can reflect semantic attachments are data element concepts, data elements, attributes, and columns.

Automatically flowing from this capability is the ability to have standardized definitions and semantics across many different data domains. This is the explicit objective of both the taxonomy and ontology definitions set out at the start of this section. The component parts of Figure 3 are listed and defined in Table 2.

Semantic Management Model Components	
Component	**Description**
Meta Category Value Type Classifications	Meta category value type classifications are the broadest discrimination of the meta category value types and in turn, meta category values. There are two: prefix and suffix. The prefix meta category value type classification causes all assigned meta category values to appear before the common business name of the data component. Essentially, prefix semantics serve as a collection of "modifiers" to the base semantics of the data component. The suffix meta category value type classification causes all assigned meta category values to appear after the common business name of the data component. Essentially, suffix semantics serve as a collection "class words" that characterize the intended data type, and use of a data element.
Meta Category Value Types	Meta category value type are collections of categories within which meta category values reside
Meta Category Values	Meta category values are the individual semantics assigned to data components in support of the complete specification of the data component. Meta category values are in turn collections within meta category value types. Meta category value collections can be hierarchically organized. The specifically assigned meta category value becomes an explicit part of a data component's name.
Meta Category Value and Data Element Concept	Data element concept & meta category value represents the assignment of one or more data element concepts to one or more meta category values.
Meta Category Value and Data Element	Data element & meta category value represents the assignment of one or more data elements to one or more meta category values.
Meta Category	Attribute & meta category value represents the assignment of one or

Semantic Management Model Components	
Component	**Description**
Value and Attribute	more attributes to one or more meta category values.
Meta Category Value and Column	Columns & meta category value represents the assignment of one or more columns to one or more meta category values.

Table 2. Semantic management model components.

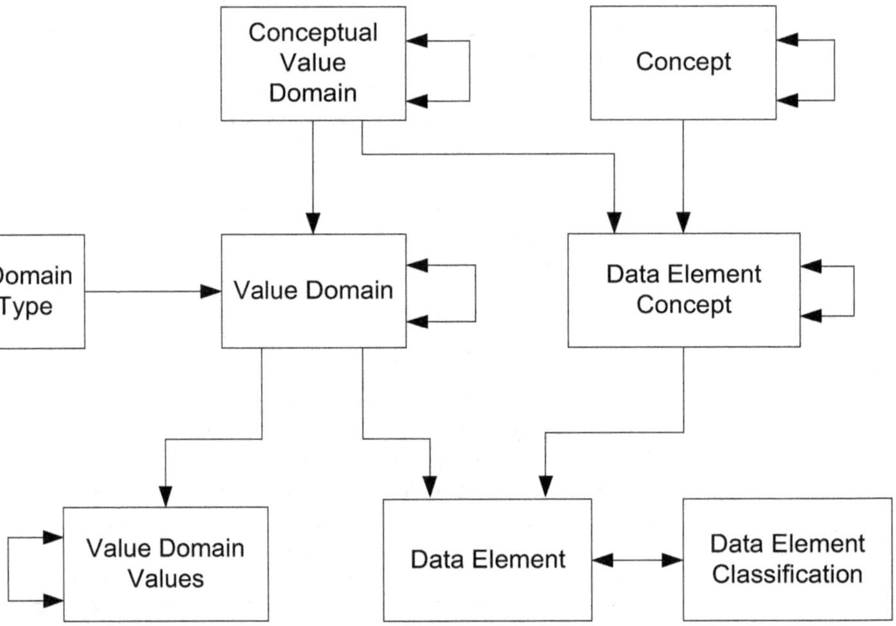

Figure 4. Key components of the Data Element Model.

2.2.3 Data Element Models

Data elements are key to understanding and administering data across the enterprise. It is common for a business fact, no matter what it is called, to be reused between 30 and 100 times across a collection of database data models. It would clearly be a waste to require that every reuse be redefined. Data elements are thus both integral and critical to the proper management of enterprise data. Data element metadata is defined according to the ISO 11179 standard. Figure 4 shows a high level diagram of the key components of the Data Element Model. Included in a comprehensive ISO 11179 Data Element Modeling effort are the creation of classes of data element metadata found in Table 3.

Data types for data elements are defined from the ISO standard for language independent data types. A data element can be assigned both semantic and data use modifiers as can attributes of Specified Data Models and columns from tables. Data elements can also be assigned a value domain.

As with the Implemented and Specified Data Models, if an attempt is made to assign a semantic that is either the same as, or that is not a proper subset of the already assigned semantic, the assignment is disallowed. There is thus a proper hierarchy of semantic modifiers, data use modifiers, and value domains enforced across the Data Element Model and the Specified, Implemented and Operational data models.

Data Element Model Components	
Component	**Description**
Data Element Concepts	Data element concepts are identified, described, and interrelated. Data element concepts are the conceptual forms of data elements. A data element concept is the joining of a concept and a conceptual value domain that, in turn, is a generalized representation of a collection of data elements. An example of a data element concepts is real property characteristics. The data element concept contained data elements that would result would be a real property name, description, dimensions, valuations, and the like. Data element concepts can be either simple, or complex. If complex, they can be networks or hierarchies. Data element concepts can be assigned semantic and data use modifiers to specialize their general nature. Together, these enable automatic data

Data Element Model Components	
Component	**Description**
	naming, definitions, and abbreviations.
Concepts	Concepts are identified, described, and interrelated. Concepts are the root-semantic source for all data elements. These concepts may be similar to the concepts named in the subjects of the Specified Data Models, but that is merely a coincidence. Concepts may be complex in their relationships with other concepts forming either, hierarchies, networks, or both. An example might be resources, of which an example might be a materiel resource such as equipment, or supplies. There also might be a non-material resource such as intellectual property or intelligence.
Conceptual Value Domains	Conceptual value domains are identified, described, and interrelated. Conceptual value domains are the concepts behind value domains from which both data element concepts and value domains are derived. An example of a concept behind a value domain might be numbers, and within that, integers, floating point. Another might be descriptions, text and codes. Conceptual value domains can be either simple, or complex. If complex, they can be networks or hierarchies.
Data Elements	Data elements are defined within the context of their data element concepts and their assigned value domain. Data elements are the semantic-laden business facts used to create either attributes of entities from subjects within Specified Data Models or columns of tables from schemas within Implemented Data Models. Data elements can be assigned semantic and data use modifiers to specialize their general nature. Together, these enable automatic data naming, definitions, and abbreviations. All semantic, data use, and value domains assigned to a data element must be a subset of any previously assigned to a data element concept.
Data Element Classifications	Data element classifications are a way to assign data elements to certain classification schemes. A data element may be assigned to more than one classification scheme.
Meta Category Values	Meta category values are words or phrases that have a specific and controlled meaning in the enterprise. These words/phrases are either prefixed to the common business name of a data element concept, data element, attribute, or column, or are suffixed to the common business name. For the former, these are semantic modifiers, and for the later they are data use modifiers. An example of semantic modifiers is geography such as United States, New England, Rhode Island, and Providence. There can be other classes of semantic modifiers including for example, temporal, or precision. Only one semantic modifier of each class can be

Data Element Model Components	
Component	**Description**
	assigned. Data use modifiers are for example, data type or role. Only one data use modifier of each class can be assigned. Assignments are always checked to ensure that proper semantic nesting is enforced.
Value Domains	Value domains, along with their value domain values and relationships among value domain values, are identified, defined, described, and if appropriate, enumerated. Value domains are also associated with their parent Conceptual value domains. Value domains can be assigned to data elements, attributes of entities of Specified Data Model subjects, columns of tables of Implemented Data Models schemas, and DBMS columns of DBMS tables of Operational Data Model schemas.

Table 3. Data Element Model components.

This kind of semantic management is generally impossible with traditional ER modelers because their foundation is the diagram, not a highly engineered metadata database.

2.2.4 Specified Data Models

The Specified Data Model is a collection of data models of concepts. The triple for the Specified Data Model is: Subject, entity and attribute. Each concept, represented as a hierarchically organized subject, can have multiple entities and attributes. Examples of subject might include location, address, telephone numbers, education information, part's information, and the like.

Figure 5 shows a high level diagram of the key components of the Specified Data Model. Included in a Specified Data Modeling effort are the metadata components found in Table 4.

Subjects can be hierarchically constructed. For example, Human Resources might contain persons, evaluations, hiring, and the like. Each contained subject, for example, Person might have a collection of entities for the person such as Biographic Information, Skill Sets, Education, and Benefits.

Entities can have subtypes to any depth. An example might be for education that might have a common set of attributes for all classes of education, and subtypes for each different education class such as high school, college, or trade school. Attributes are the use of the semantics of data

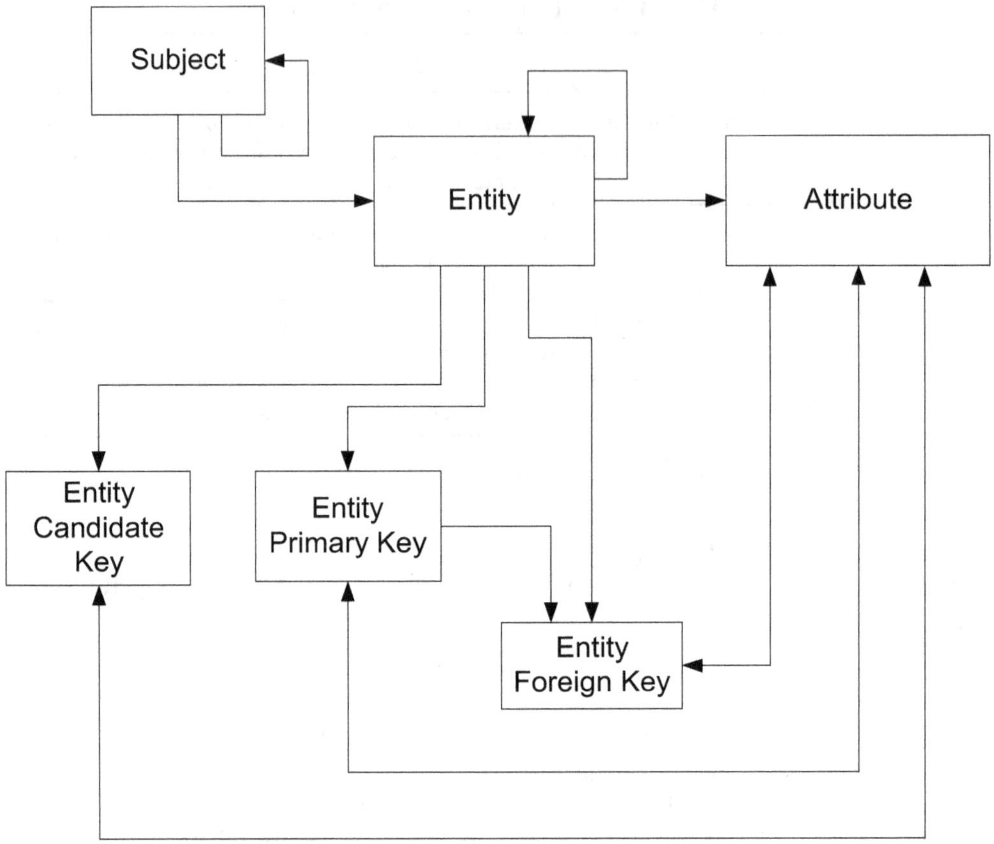

Figure 5. Key components of the Specified Data Model.

elements. Because of this one-to-many relationship between data element and attribute, users can see which attributes employ a data element's semantics no matter the entity or subject.

Attributes can also be assigned various modifier semantics and data use semantics. This enables automatic name construction for attributes. Automatic definitions and abbreviations are also enabled because there is a persistent relationship between the assigned semantic and the attribute. This too enables finding and reporting attributes regardless of entity and subject on the basis of either a semantic or data use modifier.

Relationships can exist among the entities within one subject, and relationships can exist between entities from multiple subjects. If this were just a conceptual data model, relationships could not be across different subject-based data models. Each would be a stove-pipe that can only be transformed into a logical and physical form.

Specified Data Models are not conceptual forms of a data model. Instead, they are well-developed data models in their own right. Each Specified Data Model should be in third normal form. Specified Data Models are persistent. Their form remains as defined. Its structures, that is, the data models of the concepts, are available as templates for the construction of one or more Implemented Data Models.

Most of the above is not possible with traditional ER modeling tools because they are fundamentally diagram-based with a supporting metadata database rather than metadata management system with the ability to generate ER diagrams.

Specified Data Model Components	
Component	**Description**
Attribute Assigned Value Domain	An assigned value domain is a value-based constraint on the attribute. The assigned value domain cannot be the same as that assigned the data Element. Nor can a value domain be more expansive than that assigned the data element.
Attribute Assigned Meta Category Values	An assigned meta category value, whether a semantic or data use modifier is a "meaning" constraint on the attribute. The assigned meta category value cannot be the same as that assigned the data element. Nor can a meta category value be more expansive than that assigned the data element.
Attribute	An attribute is intended to be a data-value based partial descriptor of an entity. There are three classes of attributes. Those that represent the entity's content, for example, if the entity is Address, the content attributes would be Street-1, Street-2, City, State, and Zip. Attributes are related to the instance of the entity. For example, the set of attributes than has a combined value such that only one row from the entity is found. This is commonly called the entity's primary key. The third set of attributes represents infrastructure needs such as who created or updated an entity's instance, or what project caused the entity's instance to be added or changed, and finally, what was the date of the creation or change of the entity's instance. Note: Specified Data Models are never database models and thus, can never be "instances" of data. This third set of attributes is just for completeness of the entity's definition.

Specified Data Model Components	
Component	**Description**
Entity	An entity is a collection of business facts that are commonly called attributes. Every entity should relate back to a specific policy within the domain of the entity's subject. Entities can be subtyped to represent collections of attributes that have a common set and several non-intersecting sets.
Specified Data Model Relationship	A relationship is a defined mechanism to relate one instance of one entity to a set of instances of another entity. While there are eight classes of relationships, the most common are the one-to-many, and one-to-one. In the case of one-to-many, the relationship mechanism is the primary key attributes of one entity related to the foreign key attributes of the related-to entity. The term, foreign key, comes from the fact that it's really a "transplant" of the "related-from" entity's primary key. Relationships should always be named and defined to the extent that the real purpose of the relationship is clear.
Subject	A subject is an area of interest from within the enterprise that is to be represented through structures of data values. Examples can include address structures, person name structures, contracts, purchase orders, and the like. Subjects can be hierarchical. Commonly, the subjects are related to policies accomplished by knowledge workers that require "proof of execution." Subjects are not databases, however. Nor are subjects the concepts from within the Data Element Model.

Table 4. Specified Data Model components.

2.2.5 Implemented Data Models

The Implemented Data Model is a collection of data models of databases. Each Implemented Data Model is composed of the triple: schema, table, and column. Figure 6 shows a high level diagram of the key components of the Implemented Data Model. Included in an Implemented Data Modeling effort are the metadata components found in Table 5.

There can be relationships among the tables within one schema plus data types, assigned value domains and other appropriate metadata for complete data models. There can also be assigned semantic modifiers, data use modifiers, and value domains assigned just as in the Specified Data Model. Thus, as with the Specified Data Model there can be automatic name,

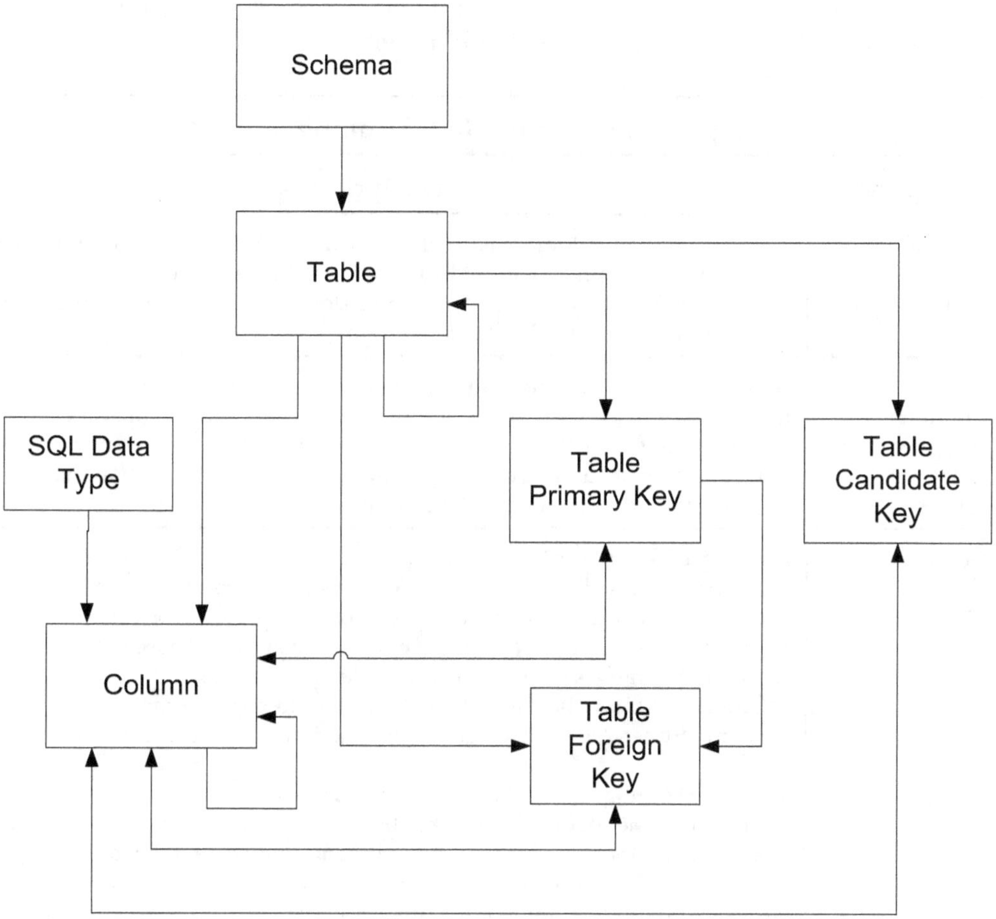

Figure 6. Key components of the Implemented Data Model.

definition, and abbreviation construction. However, because the Implemented Data Model comes from a metadata management system, any semantic assignment conflicts between the specified and Implemented Data Models are immediately discovered and prevented.

Each table may be constructed through the use of a collection of attributes from a single Specified Data Model entity, or collections of attributes from multiple Specified Data Model entities within one or more subjects. An entity or sub-collection of attributes from one entity can be used to form the data structure in multiple tables. The relationship between the

Specified Data Model and the Implemented Data Model is thus, many-to-many, as it exists in real-world data modeling environments.

Implemented Data Model Components	
Component	**Description**
Column Assigned Value Domain	An assigned value domain is a value-based constraint on the column. The assigned value domain cannot be the same as that assigned the data element or the attribute. Nor can a value domain be more expansive than that assigned the data element or column.
Column Assigned Meta Category Values	An assigned meta category value, whether a semantic or data use modifier is a "meaning" constraint on the column. The assigned meta category value cannot be the same as that assigned the data element or an attribute. Nor can a meta category value be more expansive than that assigned the data element or the column.
Column	A column, like an attribute of an entity, is intended to be a data-value based partial descriptor of a table. Columns can be complex, and thus represent arrays, groups, repeating groups, and nested repeating groups. Like attributes, there are also three classes of columns. Those that represent the table's content. For example, if the table is CustomerAddress, the content attributes would be CustomerStreet-1, CustomerStreet-2, CustomerCity, CustomerState, and CustomerZip. There are columns also related to the instance of the table. For example the set of columns that have a combined value such that only one row from the table is found. This is commonly called the table's primary key. The third set of columns represents infrastructure needs such as who did the table's creation or update; what project caused the table to be added or changed; and finally, what was the date of the add or change. Note: Implemented Data Models are never database models and thus, can never be "instances" of data. This third set of columns is therefore just for completeness of the table's definition.
Implemented Data Model Relationship	A relationship is a defined mechanism to relate one instance of a table to another. As with the Specified Data Model, the most common are the one-to-many, and one-to-one. The keys are primary and foreign keys. Relationships cannot exist between tables in different schemas.
Schema	A schema is a database structure that encapsulates all its contained tables plus containing other classes of schema objects such as data types, procedures, constraints, including the interrelationships of the various schema objects.

Implemented Data Model Components	
Component	**Description**
	Implemented Data Models are cast within the domain of a schema. The set of all tables within a schema is not required to be taken from a single set of entities within a subject area.
Table	A table is intended to be a well-defined expression of one policy within a schema. Ideally, the collection of all the tables within a schema area should define a coherent collection policy. Although unlikely, some tables and even some schemas may never be represented within Operational Data Models. Additionally, some columns within a table may never be employed. A table may contain columns that map to attributes from multiple entities. Tables can be sub-typed. A table represents a collection of columns, which when bound to a particular DBMS and valued, represents a set of rows of data for those columns. Tables have precise specifications including constraints, primary and foreign keys, and other table centered features. Within the scope of the Implemented Data Model generalization level for data architectures, table specifications do not represent actual rows of data because this level is not bound to a particular DBMS and is not related to any specific business information systems. These tables, and the entire Implemented Data Model is intended to be a data architecture design layer. Tables can be subtyped to represent collections of columns that have a common set and several non-intersecting sets.

Table 5. Implemented Data Model components.

Tables in the Implemented Data Model can be subtyped. Columns can also be complex to support nested structures of arbitrary depth. The data types assigned to an Implemented Data Model are those allowed by the ANSI SQL standard.

Finally, because the Implemented Data Model is built from a foundation of the data models of concepts from the Specified Data Model, a complete cross reference between the two classes of models is easily reported.

Again, most of the above is not possible with traditional ER modeling tools because they are fundamentally diagram-based with a supporting metadata database rather than metadata management system with the ability to generate ER diagrams. Traditional data modeling tools are for modeling databases as separate independent metadata stove-pipes. For enterprise-wide data management this is entirely unacceptable.

2.2.6 Database Object Models

Database objects "live" entirely within the domain of the DBMS. Database objects can be both persistent and non-persistent, and can be either single or multiple-table objects.

A persistent database object is stored and is retrievable over long periods. An example is an insurance policy along with its full compliment of payments, renewals, and claims. Another example of a persistent object is rotating-three dimensional views of a mechanical part.

A non-persistent database object is materialized and displayed but cannot be rematerialized because some of its components are not retained after the database object's display is ended.

Non-persistent objects are dynamically produced from database data and exist only for the life of their "display." An example might be evening news weather displays. The weather map, that is, the states, cities, streams, rivers, etc. are all persistent database data. The actual streams of clouds, high and low pressure fronts, cloud formations, and the like are time-sequenced BLOBS that are dynamically displayed across the screen. While the displayed database objects may be recorded via videotape and redisplayed at a later time, the detailed components, which upon retrieval make up the non-persistent database objects, are not stored.

By the time the news cast is over, the BLOB parts are discarded. Other than for a videotape replay, the complete set of the non-persistent database objects are gone. The persistent part is traditional data structures with the appropriate quantity of indexes. The BLOBs are just non-indexed streams of binary data stored in a very primitive format.

Persistent database objects are those stored in a database on a permanent basis. Included are traditional "relational" data, abstract data types of complex structures (like an entire auto accident claim that might include BLOBs, free text streams, etc.).

Single table database object classes are those that are fully defined within a single row of an SQL table structure. The database object may further be stored within a single cell within a column. With SQL:1999, very complex structures can be defined within a column. This capability is quite common in hierarchical DBMSs like System 2000 and in independent logical file data model DBMSs such as Adabas, Model 204, Inquire, and Datacom/DB. Not only can a single column support single valued items, it can also support lists, sets, multi-sets, and abstract data types of arbitrary complexity.

For example, in a product sales database, the single table called sales has product number as the primary key with other columns for product name, product description, and the like. The Sales column in contrast, contains Product Sales by Year by Month by Region, District and Territory by Salesman. That is a single column with six dimensions of values. Prior to SQL:1999 such product sales information would require multiple tables. Since the salesman's object identifier is contained as an integral component of the sales data, the salesman's full set of data is accessible through normal SQL language processing. A referenced database object, that is, the referenced salesman's data is not considered a formal part of a single table database object.

Multi-table objects are implemented across multiple tables. For example, an insurance policy may have several dozen tables that make up its full definition. One and only one table is considered as the database object's root table. The database object root table contains among many things, the object's identity column. A row from the root table is the head-row of the database object. All other related tables within the multi-table database object contain other information related to the object. In the insurance policy example, a claim might be contained in one or more database object tables. Other database object tables would contain the underwriting information regarding the person about whom the policy was issued. The person is a different single or multi-table database object. Each table within a multi-table object may consist of single valued columns. It can also support lists, sets, multi-sets, and abstract data types of arbitrary complexity.

Not only do most businesses contain multi-table database objects, most business applications are examples of uses of multi-table database objects. A quick look at business applications reveals that inescapable conclusion.

Database objects, as illustrated in Figure 7, no matter persistence and regardless of whether single or multi-table contain the same four-part composition:

- Database object structure: the set of data structures that map onto the different value sets for real world database objects. A database object structure normally maps on to multiple tables. An example would be the set of database tables for a complete auto accident, vehicles and emergency medicine incident.

- Database object table process: the set processes that enforce the integrity of the data for each database object table, that is, the rows of data that comprise the database object, including references between database rows that interrelate one database object table to another, the proper computer-based rules governing database object table insertion, modification, and deletion. For example, a database object table process would accomplish the proper and complete storage of an auto accident's basic data.

- Database object information system: the set of specifications that control, sequence, and iterate the execution of various database object processes that cause changes in database object states to achieve specific value-based states in conformance to the requirements of business policies. For example, this might include the reception and database posting of data from business information system activities (screens, data edits, storage, interim reports, etc.) that accomplish entry of the auto accident information.

- Database object state: The value state of a database object that represents the after-state of the successful accomplishment of one or more recognizable business events. Examples of business events Are Auto Accident Initiation, Involved Vehicle Entry, Involved Person Entry, and Auto Accident DUI (driving under the influence of alcohol/drugs) involvement. Database object state changes are initiated through named business events contained in business functions. The business function, auto accident investigation includes the business event, auto-accident-incident initiation, which in turn causes the incident initiation database object information system to execute. It, in turn, causes several database object processes to cause the auto accident incident to be materialized in the database.

The components of database object classes are identified and described in Table 6. A database object is specified to the SQL DBMS through the SQL definition language (DDL). All four components of a database object operate within the "firewall" of the DBMS. This ensures that database objects are protected from improper access or manipulation by 3GLs, or 4GLs. A DBMS that only defines, instantiates, and manipulates two-dimensional data structures is merely a simplified functional subset of the DBMS that defines, instantiates, and manipulates database objects.

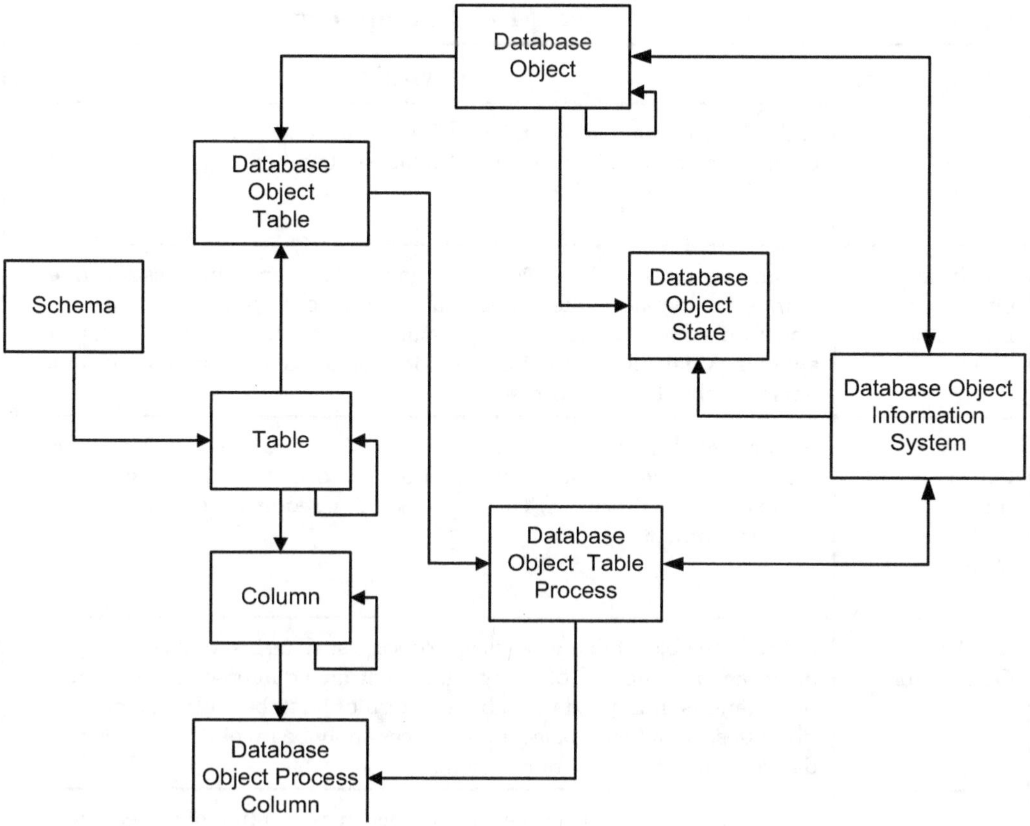

Figure 7. Database Object Model.

Database Object Model Components	
Component	**Description**
Database Object Class	A database object class is a collection of one or more tables. Database objects, the rows from a collection of database object class tables are transformed from one valid state to another via database object table processes and database object information systems. Database objects are related to one or more database domains.

Database Object Model Components	
Component	**Description**
Database Domain and Database Object	A database domain and database object is the association of one or more database objects with one or more database domains.
Database Object Information System	A database object Information System is a collection of processes defined within the domain of the DBMS usually as a stored procedure that transforms one or more rows of a database object from one valid state to another. A database object information system accomplishes one or more database object table processes.
Database Object Information System and Database Object Process	A database object information system and database object Process is the association of one or more database object information system and database object Processes. These are set into a sequence for proper accomplishment.
Database Object Table Process	A database object table process is a process such as insert, change, or delete against one row of a single table within a database object. A table owns (and is thus acted upon by) one or more database object table processes. A database object table process may be invoked by one or more database object information systems.
Database Object Table Process Column	A database object table process column is an association of a specific database object table process and a specific column of a table.
Database Object State	A database object state is a well-defined value state of a database object. States occur in a particular sequence, typically from the null state through a set of value states and returning to a null state. A database object state is accomplished through one or more database object information systems.
Database Object State and Database Object Information System	A database object state and database object Information System is the association of one or more database object states and database object information system. The association exists within a sequence so that the state is properly achieved.

Database Object Model Components	
Component	**Description**
Database Object Table	A database object table is an association of a table with a database object. Membership rationale classifies the reason why a table belongs to the database object.
Membership Rationale	Membership Rationale is a classification of the reasons why a table belongs to a database object as evidenced through the database object table.
Schema and Database Object	Schema and database object is the association between a schema and a database object. This means that a database object may reside in more than one schema.

Table 6. Database Object Model components

2.2.7 Operational Data Models

The Operational Data Model is similar to the Implemented Data Model in its capabilities. The triple for the Operational Data Model is DBMS schema, DBMS table, and DBMS column. Figure 8 shows a high level diagram of the key components of the Implemented Data Model. Included in a Specified Data Modeling effort are the metadata components found in Table 7.

A key difference in the Operational Data Model from the Implemented Data Model is that it represents the data structures that result in SQL DDL that feeds the DBMS to actually make the databases.

Multiple Operational Data Models can exist for a given Implemented Data Model. Each may be a different subset of the Implemented Data Model's tables and columns, or in some cases, it is all the tables and columns, but some tables may be brought up under a parent table. For example, Implemented Data Model table might exist for Telephone Numbers that, in the Operational Data Model, are transformed into TelephoneNumber-1 through TelephoneNumber-5. This is a kind of denormalization created to improve performance.

Multiple Implemented Data Models can exist for a given Operational Data Model. This commonly occurs in data warehouse data model designs

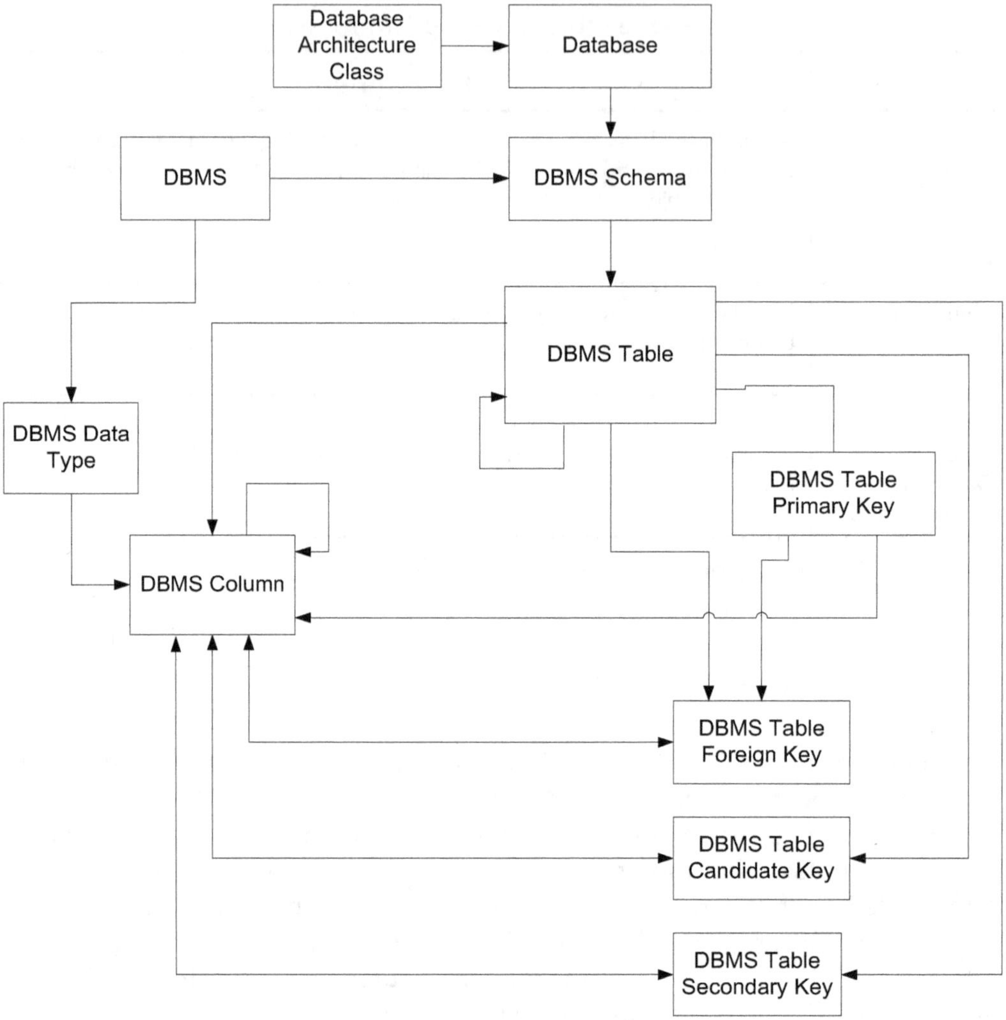

Figure 8. Key components of the Operational Data Model.

where the data model for the data warehouse is sourced from multiple Implemented Data Models.

The string, "DBMS" prefixes these three terms because this data model is directly tied to a specific DBMS. The Operational Data Model can contain subtyped DBMS tables, and substructures in DBMS columns. It can also have an assigned value domain and the data types reflect that of the containing DBMS.

Each DBMS table may be constructed through the use of a collection of columns from an Implemented Data Model table, or collections of columns from multiple Implemented Data Model tables within one or more Implemented Data Model schemas. A table or sub-collection of columns from one Implemented Data Model table can be used to form the data structure in multiple DBMS tables. The relationship between the Implemented Data Model and the Operational Data Model is thus, many-to-many as it is in real data modeling environments.

Operational Data Model Components	
Component	**Description**
DBMS Column Assigned Meta Category Values	There are no assigned meta category values because at this data model generalization level there should be no additional semantic and data use modifiers.
DBMS Column Assigned Value Domain	An assigned value domain is a value-based constraint on the DBMS column. The assigned value domain cannot be the same as that assigned the data element, attribute, or column. Nor can a value domain be more expansive than that assigned the data element attribute, or column.
DBMS Column	A DBMS column, like an attribute of an entity is intended to be a data-value based partial descriptor of a DBMS table. DBMS columns can be complex, and thus represent arrays, groups, repeating groups, and nested repeating groups. Like columns, there are also three classes of DBMS columns, and those definitions are not repeated again.
DBMS Table	A DBMS table is a database construct that exists within a DBMS schema to represent a collection of DBMS columns that are bound to a particular DBMS and represent a set of rows of data across those DBMS columns. DBMS tables have precise specifications including constraints, primary and foreign keys, and other table centered features. Within the scope of the Operational Data Model generalization level for data architectures, these DBMS table specifications represent actual rows of data because this level is bound to a particular DBMS and is related to any specific business information systems. Every DBMS table should relate back to a specific policy within the domain of the DBMS table's schema. DBMS tables can be subtyped to represent collections of DBMS columns that have a common set and several non-intersecting sets.
DBMS Schema	A DBMS schema is a DBMS structure that encapsulates all its contained

Operational Data Model Components	
Component	**Description**
	DBMS tables as well as containing other classes of DBMS schema objects such as data types, procedures, constraints, including the interrelationships of the various DBMS schema objects. The string, "DBMS" is attached to these components because at this data architecture level, these components are directly representative of the real and operating databases under the control of a specific DBMS.
Operational Data Model Relationship	A relationship is a defined mechanism to relate one instance of one DBMS table to another. As with the Implemented Data Model, the most common are the one-to-many, and one-to-one. The keys are almost always primary and foreign keys. Because the Operational Data Model supports the SQL DDL for an actually operating database, a fourth key type, Secondary, is supported. This key type consists of one or more DBMSs columns such that when the values are supplied, a subset of rows is materialized from the table.

Table 7. Operational Data Model components.

2.2.8 View Data Models

View Data Models and its high level components are shown in Figure 9. Views, which represent the interface between Operational Data Models and business information systems, consist of views and view columns with the associated hierarchies of joins and selects to give applications a flat "record set" for processing.

Included in the views can be rename clauses and on-the-fly calculations. View elements map to DBMS columns and as appropriate compound data elements and derived data elements. Included in a View Data Model effort are the metadata components found in Table 8.

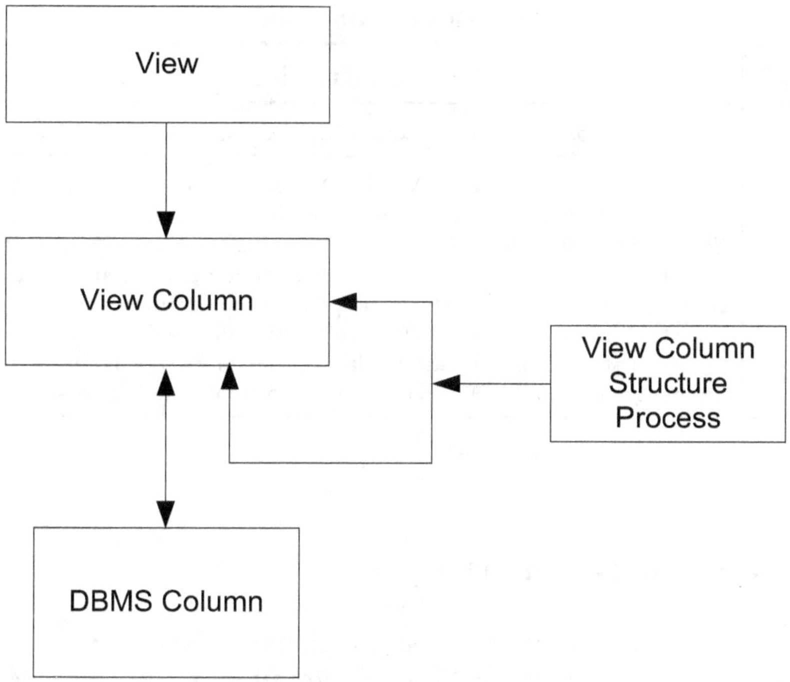

Figure 9. Key components of the View Data Model.

View Data Model Components	
Component	**Description**
View Column	A view column is a "data" component of a view. Some view columns are the result of a computation that "goes on" inside the view involving multiple DBMS columns.
View Column and DBMS Column	A view column and DBMS column is a relationship between a view column and the Operational Data Model. This enables a given view to be specified in the Metabase System such that it is applicable to multiple DBMSs and multiple databases. An executing view, of course, is applicable to only one DBMS and one database.
View Column Process Mapping	A view column mapping process is a process-specific mechanism that maps between view columns of different views. This enables the specification of an input view from one database to an output view to

View Data Model Components	
Component	**Description**
	another database, or even to several databases.
View	A view is the interface between the Operational Data Model and the business information system. Views are defined within the scope of the Operational Data Model for a specific database within a specific DBMS. A retrieval view can contain complex selection logic, and multi-table normalization logic. This shields application programs from having to know about the database's structure to navigate the database. An update view can also contain constraint clauses that check the quality of the data before it is sent to the DBMS that, in turn, updates the database.

Table 8. View Data Model components

2.2.9 Data Semantics Summary

A database and information system's specification may be transformed many different times for different combinations of computer hardware, DBMS and operating systems. Thus, there is a many-to-many relationship between the specified and implemented forms. In addition, there may be implemented databases that are compositions of different parts of specified databases. This makes the relationship many to many. This commonly occurs in data warehouse environments.

It is critical to preserve both the data and process semantics, and to preserve the mappings between these two forms. When this is done, database and information system specifications merely have to be re-transformed whenever there is a need to significantly change the computer hardware, DBMS or operating system. The analogy is essentially the same between the implemented and operational forms of a database and information system.

In a recent resystemization effort a senior executive complained that since the fundamental business model of the corporation had not changed in 40 years that it was a complete waste of time and resources for the information technology department to rediscover and respecify the fundamental business model every 5-10 years just because the technology of business systems' implementation was changed. The executive made his point even more forcefully by not only *rejecting* his cost allocation from the effort but also he charged data processing for *his* time.

2.3 Life Cycle Costs

This section, Life Cycle Costs, is appropriate because there is probably no greater cause for the expenditure of information technologies monies than on the capture, storage, transformation, retrieval, and processing of data. Why else have information technology if it is not for the proper management of information through technology? It is the fundamental strategy of this book to maximize the reuse of defined information technology components. That means the reuse of defined data and process specifications, as well as the reuse of the technology infrastructure that can be brought to bear to manufacture databases and business information systems.

In a well-controlled study in the middle 1980s, it was shown that employing data-driven techniques to create business information systems results in 4.6 times fewer components to actually implement than process driven techniques. It has also been shown that using business information system generators coupled with data-driven techniques enable the business information systems to be built using five times fewer resources than with process driven and custom-coding techniques.

Notice that in both descriptions the common component was "data-driven" techniques. Data driven techniques are most optimally accomplished if they are engineered through the techniques set out in this book. For example, a good metric for the complete development of a database table column is two staff hours. That is about $200 per column. Tables average about 15 columns per table. The cost to define the table is thus, $3,000.

Databases average about 150 tables. Thus, the analysis and design cost of a database is about $450,000, or about a little more than two staff-years in 2008 dollars.

If data driven techniques are employed as set out in this book, there are about 1/30th of the quantity of columns to originally define because research has shown that a well-defined data element is deployed as a column about 30 times. While the quantity of tables is the same, the key cost factor is the cost of defining the columns has been significantly reduced to about five minutes each. The table definition cost is reduced to about $150. The overall database design cost has been reduced to about $22,500. That is just 5% of the original cost. Even more important, is the fact that the database design now consists of tables that have standardized semantics because of the reuse of data elements.

There is now an even greater impact on business information system development. Through common development strategies, every table requires about five business information system processes in support. It is common to take about two weeks for the analysis, design, and coding of each process. That is 80 staff hours, or about $8,000 per process table. For a database of 150 tables, the overall cost is thus, $6,000,000. In contrast, if a business information system generator is employed, the cost per table for the same quantity of functionality is about $4,000 per table, or about $600,000.

Not only is there a dramatic impact on the development costs of a single business information system, the impact on families of business information systems is even greater. That's because the developed enterprise-wide standardized data elements, and reusable data model templates from the Specified Data Model enable the business information system generator to have reusable collections of screen-specifications, data validation rules, and processes. Thus, when a new business information system is to be created, the designers and developers can pick these reusable specifications from the libraries and bring them to bear in the generation of the business information systems.

Note well, this is different from reusing "code." Reusing code efforts have never been really successful. Rather, this process employs the reuse of specifications that, with the business information system generator, causes the creation of the code. This is not a theoretic strategy. It is the common everyday business practice of about 15,000 developers around the world.

These overall cost numbers are for the first cycle of development. The remainder of this section describes the cost effects on the remaining life cycle costs. In general, if the costs associated with requirements and design are $1, the activities associated with detailed design through initial business information system implementation costs another $4. That is $5 in total for a first implementation cycle.

Figure 10 illustrates the main phases associated with the traditional first implementation life cycle of an enterprise-wide business information system such as Human Resources, or Accounting and Finance. The scale on the bottom represents the quantity of months spent in each phase. The vertical scale that is not shown would be staff hours. The purpose of the chart is to show the relative quantities of staff hours expended across time. The curve clearly shows that the bulk of the effort (commonly about 70%) is expended before any demonstration is possible. Figure 10 shows that:

- About 70% of all effort is expended before the first real demonstration of a business information system. That is, before the Operation and Maintenance parts of the last major phase shown in Figure 10. This percent may be much higher if requirements changes are discovered during System Test. Some multi-hundred million dollar efforts are scrapped because of requirements changes even before System Test.
- The 2nd and any subsequent versions require recycling of the first two phases (20-60%). That is, a repeat of part or all of Requirements Analysis and Design, and Detailed Design, Coding, and Unit Testing.

The "problem" may have also significantly changed before the solution arrives. This is the "hazzard" that the U.S. General Accountability Office (GAO) concludes is intrinsic to the very process. This can result in perpetual recycling of requirements without ever getting to System Testing. In an analysis of 13 multi-hundred million dollar U.S. Federal Government business information system failures, it was discovered that 41% of all errors occur in the requirements analysis and architecture stages. 50% of all errors occur after

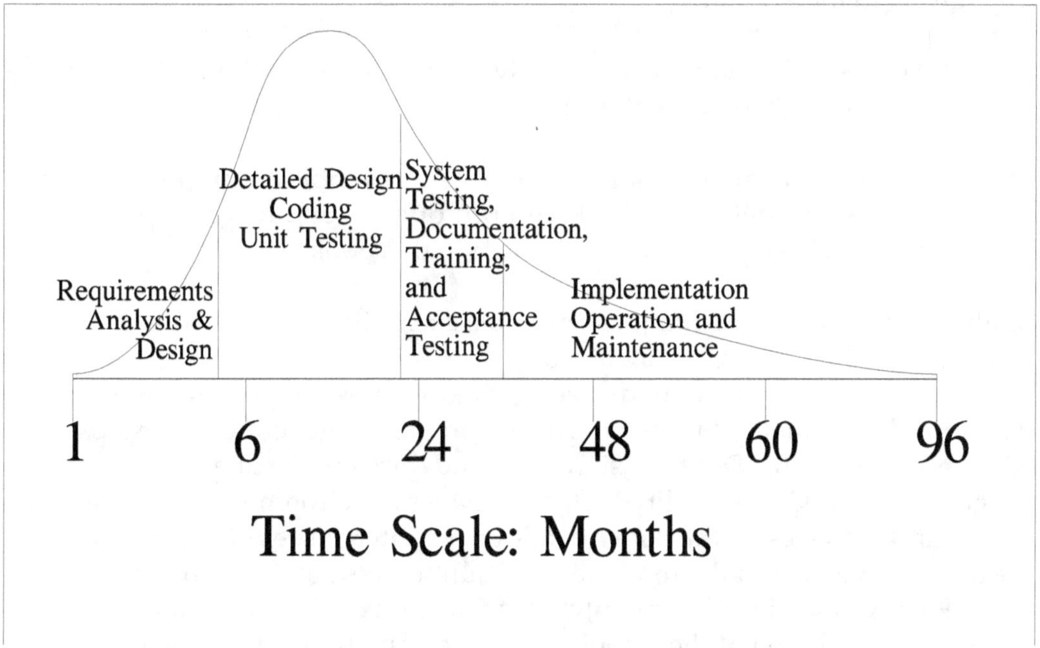

Figure 10. Phases for the traditional information systems development life cycle.

a business information system is deployed because the business processes, functions, and organizations have not be re-engineered, and only 8% of errors are allocatable to IT.

Beyond the first-cycle implementation cost, the total life cycle expenditure for business information system revision cycles (not shown in Figure 10) commonly costs five times more. The total life cycle cost is thus, 30 times the design cost. The problem, however is not that requirements change. Rather, the real problem is that the effects of requirements changes that occur after System Testing et al are complete are too costly to reflect in the implemented business information system.

It must be assumed at the very start that requirements will change, even though this assumption is seldom folded into the methodologies and estimates of today. Requirements changing can be especially fatal to procured software packages if these packages have not been designed for change from the very beginning.

Because $1 in requirements' changes causes $29 in additional life cycle costs, the exhortation is simple: Get the requirements right the first time because the cost of change is prohibitive. The reason the costs are so high and the time is so long is that:

- The tools that either dramatically shorten, or even eliminate steps have only recently become demanded.

- Traditional data modeling approaches are employed in preference to the data modeling coupled with prototyping and recycle approach that has been proven and is shown in this book.

Only by adopting the techniques and tools exemplified in this book can a reduction of 60% or more be realized.

For example, if a data-driven methodology is adopted and is supported by a quality business information system generator, the steps, shown in Figure 10, "Detail Design . . . ," and "System Testing . . . " can be largely eliminated. If a quality Metabase System environment is installed that stores and manages all the data models on an enterprise-wide basis, there can be an improvement in the first phase, Requirements Analysis and Design as well. Finally, if we implement only after four to six design iteration cycles of a prototype, a good bit of the last phase, Implementation, Operations, and Maintenance can be eliminated.

Quality business information system generators can import a database's design and generate a working business information system in a few hours. From that first generation, the generated business information system can be "electronically pruned" to a form that is suitable for demonstration. After several iterations of demonstration, modification, and regeneration, a quality specification can be created.

The change to the process is illustrated in Figure 11. Shown are four prototype development cycles, and one full-implementation cycle. The first "D" (i.e., development) cycle contains the activities noted (that is, Design, . . . , Feedback). Each subsequent "D" cycle is dramatically reduced because it is only focused on design changes, regeneration, and the next demonstration. The first cycle contains a requirements analysis and design step. However, that step is reduced to mainly producing the data model.

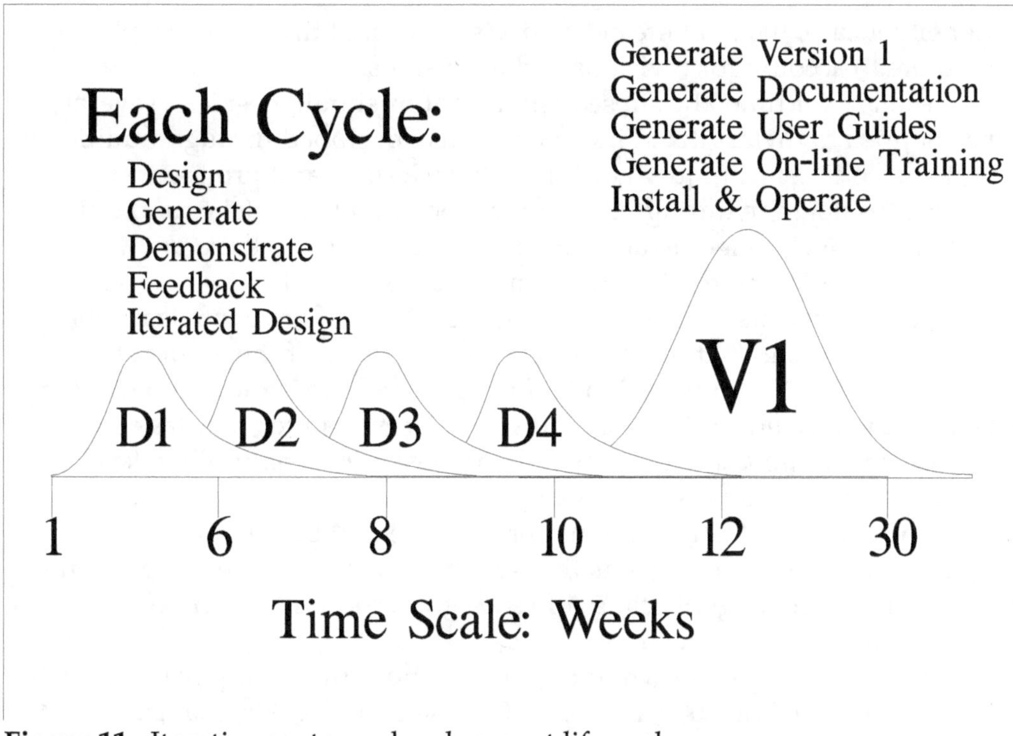

Figure 11. Iterative systems development life cycle.

To arrive at the prototype, the business information system generator takes in the data model's design and generates the first-cut business information system. A day or so is spent to make it "pretty." Then, the prototype is demonstrated. In all, only about six weeks will have passed. Each "D" thereafter can occur in one or two weeks. Under this changed approach:

- "Real" business information systems are available to run at the end of every "D" cycle.
- Requirements are iterated until fully known, but are determined in a very condensed time-period.
- Full development occurs only after requirements are completely validated.

Under either case, the total life cycle cost is the cost of the first cycle plus five times more to account for revisions and extensions.

While over time, the subsequent revisions should cost less, several things happen to thwart that. First, as systems developed through traditional design and coding techniques get "old," their designs and program constructions degrade through the application of ad hoc patches and quick fixes. It takes much longer to accomplish maintenance work than new construction work. The result is the same volume of work.

Second, because of staff turnover, the new staff have to relearn the system from scratch. That is almost a complete recycle of the Figure 10 processes. It is both painful and long. In addition, there is the "I'd never have done it that way syndrom" to contend with. Every maintenance task seems to get transformed into a set of fix and enhance tasks. This naturally takes longer and enhancements may install new bugs.

Third, as new features are required, some of these features are real design breakers. Because a production system has already been implemented, changes take much longer because of redesign, recoding, and the very tedious and error-prone data conversion.

All the costing is shown in Figure 12. Under the traditional approach, if the first box costs $400,000, the total life cycle costs are 30x, that is, 5 + (5 * 5) or $12 million.

Faced with high costs, the high-pressure demand for immediate and visible results, the most common approach is to trim the first box. Suppose it was trimmed by one-half. In theory, the total cost of the first business information system implementation cycle would be reduced to 2.5, that is, 0.5 + (4x 0.5). The total life cycle cost would only be reduced to 15 units, or $6 million. That is a 50% savings. Such savings are however false.

That is because the real requirements are still there. The only thing that has changed is that the analysis to discover the real requirements has been reduced by 50%. In short, 50% of the real requirements are undiscovered.

Now, if the requirements and design efforts are reduced by 50% from 1.0 to 0.5, and given the GAO error allocation of 41% of all information technology errors if this step is underdone or done wrong, the "4x" (the

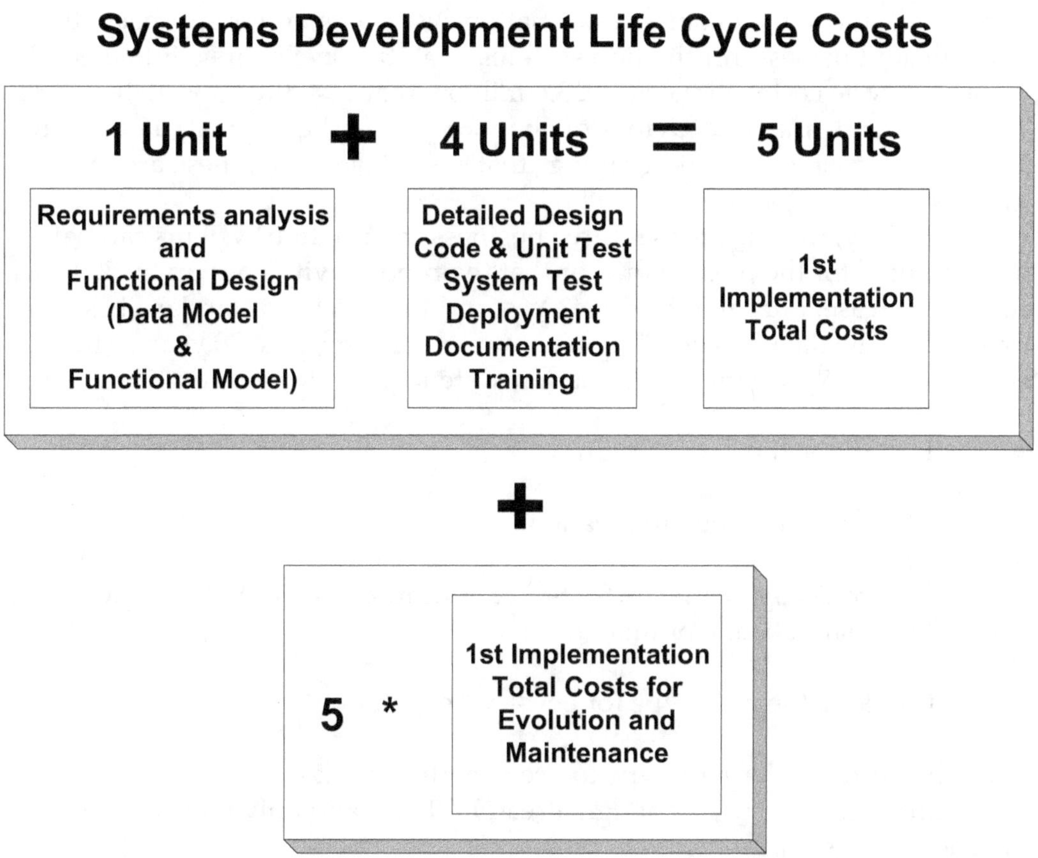

Figure 12. Cost structure of systems development life cycle.

amount to accomplish what is really "required") is likely to be 12x. That is because it is done wrong, it has to be undone and done right. Hence, three times the "4x" number. In that case, the 1st business information system implementation cycle cost is not the original 5 nor the 2.5, but is 6.5 (2.5 (done wrong) + 4.0 (now done right)). The overall life cycle unit quantity becomes 39, that is, 6.5 + (6.5 * 5). This is really a "pay me now, or pay me more later" situation. If a unit costs $400,000, the life cycle cost is now $15.6 million instead of the original $12 million. In short, a requirements analysis and design cut of $200K results in an overall cost increase of $3.6 million. Clearly, these savings are costly.

For a real impact, the savings must be made to the 4x component (that is, Detailed Design, . . . , Training). If the 4x component is reduced to 2x, the life cycle costs are reduced to $7.2 million, that is, 400K * 3 + (400K * 3) * 5. In addition, if due to the business information system generator approach, three of the major business information systems evolution cycles are eliminated, the whole life cycle cost is reduced to $3.6 million, that is, 400K * 3 + (400K * 3) * 2. That represents dramatic savings of about 66% from the original $12 million. Now, in contrast to the $200K savings that cost $3.6 million, these are real savings.

Simply, this means that three business information systems can be implemented for the price of one, or that the productivity goes up by 300%. A compelling case. But can it be done? Yes, and this has been standard best-practice for the data-driven Clarion community for the past 20 years. The steps for a quality business information system generator approach are these:

- Import the database design.
- Generate a first-cut application.
- Electronically prune the first-cut application metadata-based design to the desired behavior model.
- Generate the prototype for end-user evaluation.

After that, present the prototype to users, gather changes, and repeat the steps above until the users say, "enough already." Then, and only then, create the production version.

Once the production version is created, there is the evolution and maintenance cycles. The code-generator approach dramatically affects maintenance as well. There were three reasons cited above that cause maintenance to either remain level or take more time under the traditional approach. In contrast to the traditional maintenance approach, the quality business information system generator approach causes many of the first type of evolution and maintenance problems to be eliminated outright. The second type of evolution and maintenance problem is of course unavoidable, but because 90% or more of any business information system's code is generated, it does not have to be learned. In fact, because the ultimate program's code sort of doesn't really exist, it never has to be addressed at all.

Quality business information system generators operate at a meta-design level. The actual program code is generated from this meta-design level. Changes are made at the meta-design level as well. So, once the changes are made, the ultimate code is regenerated. If it were not necessary to deal with the real-code during the first implementation cycle, there is no need to deal with it now. Finally, the third type of evolution and maintenance problem largely disappears because there are four or more design iterations before the first real implementation. The five cycles of maintenance caused by an immature design are eliminated because the design has matured through prototyping before it is implemented.

It should therefore be quite clear from the business case just set out, that the return on investment (ROI) from effective data semantics management is in two parts: Data, and business information systems. While the ROI for data is significant, the largest ROI is on the development, evolution, and maintenance of business information systems. Simply put, the business case for data semantics management is in four parts: increased productivity, increased quality, reduced costs, and reduced risk. Not accomplishing data semantics management is almost certainly a terminal business decision.

2.4 Error Classes

Chapter 2 set out the first overall framework for addressing the two semantic error classes. The ultimate strategy is to design both these error classes out of existence from the very start. If there are no initial errors, there is no need for follow-up activities. To that end, Semantic Hierarchies support the creation, meaning, and allocation of all the semantic-based words that make up critical elements of fact names.

In all the critical semantic words are known, defined, and employed in a standard manner by all those who create update, and employ data, the first error class will be significantly reduced or eliminated.

Semantically engineered fact names are employed within the Data Element Model, the Specified Data Model, and finally, the two database data models, that is, Implemented Data Model and Operational Data Model. Each of these provides the ability to define fact-based data semantics that control their immediate lower contained levels. Each lower contained level provides the ability to provide a more refined set of semantic definitions and/or restrictions. The parent level of any given semantic definition, for example, concepts within which data element concepts are defined, provide the ability to posit semantic affinities due to common parentage.

If all this is accomplished top-down, there will be a great reduction in semantic errors. The problem will still exist with bottom-up development. That, however, is where most of reality exists. There is no choice then but to have top-down connections to bottom-up developed or imported data models. As this is done, compromises is inevitable. In such cases there at least exists evidence of the mapping decisions. Over time as more and more one-off databases and business information systems are replaced with enterprise-based semantic-driven databases and business information systems, the quantity of semantic errors will be driven out of existence.

The most critical aspect of preventing the two error classes is to have common semantic understanding between the sender and receiver. In that regard there are five layers of unfolding common understandings from the most generalized Data Element Model concept down through to the most detailed DBMS column.

Unless and until there is extensive enterprise-wide data semantics management there cannot be the necessary data interoperability knowledge to support complete transaction semantic awareness. Such knowledge cannot be discovered during a cannot be discovered during a transaction's execution. It must be pre-known. This knowledge can only be developed within the Communities of Interest that develop the shared understanding and metadata mapping.

If COIs have developed sufficient and comprehensive data semantics, there can be dramatic advances in automatic name, data type, and value domain resolution.

If all this is accomplished, the quantity of errors in the two error classes will greatly diminish. The artifacts of the Communities of Interest work are created, sorted, and are able to be employed based on the work accomplished in Chapters 6, 7, and 9 through 15.

2.5 Why Semantics Summary

This chapter addressed the fundamental need for highly engineered semantics as a way to control the error classes cited above. This need was first explained within the context for data management. Key to understanding data management is understanding the role of enterprise-level policies and procedures. Because these need to be enterprise-wide, there must be consistency in their definitions at all levels and across all enterprise organizations.

This chapter set out, at a high level, the diagrams of the various models within enterprise data semantic management. Each of these models had its components identified and described.

The model, database object classes, is probably new to most readers. In today's databases, there can be from hundreds to tens-of-thousands of tables. Clearly not all tables are created equal nor are related to each other. Instead, tables exist in well-defined collections that conform to corporate policy specifications. Data, the by-product of policy executions, are what should be shared within and across corporate organizations. Database object classes are the coherent specifications of shared data. Unique to database object classes are the definitions of the states and state transformations through which a given object proceeds from birth through all life cycle stages until the database object becomes fully mature.

This chapter has as its final section a detailing of cost-benefit value of applying the techniques set out in this book. These costs and benefits are not theoretic. They are proven. Consequently there is no down-side to the prescriptions of this book. Stated again, not accomplishing data semantics management is not only a bad business decision in the short run, it likely is a terminal business decision in the long run. Any executive or business decision maker not embracing these dictums and strategies becomes "the enemy from within."

The next chapter, Chapter 3. Data Semantics Failures, is critical because there has been spent 500 million dollars or more in the attempt to achieve enterprise-wide data semantics, and in virtually all the cases, the result has been failure.

Chapter 3 goes to significant length to thoroughly diagnose the common data semantics failures, including some that are surprising. The lessons learned are clearly understood and have been completely incorporated in the strategies and examples of this book. Everything that needs to be done to engineer and deploy interoperable data environments is set out in this book and in other Whitemarsh books. Additionally, every example from this book is able to be operationally demonstrated via the Metabase System that is downloadable–for free–from the Whitemarsh website.

A key ingredient to the accomplishment of enterprise-wide data semantics management is the establishment of Communities of Interest. The Whitemarsh book, *Data Interoperability Community of Interest Handbook,* sets out the architecture of these groups and the work plans, deliverables, and all necessary policies and procedures for the establishment, operation, and evolution of these groups.

Excuses of not knowing what causes failures, and once known, not knowing what to do have been removed. The time is now, not the tomorrow that never seems to come.

2.6 Questions and Exercises

1. What is the rationale for data management?

2. What is the role of policy in data management? Do you agree with this book's definition of data? Of procedures? What are the interaction between data and process? How does that interaction relate to policy and procedures?

3. What is the definition of enterprise-database? How does this definition differ from a database that you would create with a DBMS? How should the boundaries of a database be set? What's in and what's out?

4. What is meant by "data semantics?" How are data semantic irregularities manifest? What examples can you provide from within your own organization that illustrates these irregularities?

5. Do the definitions associated with the various data semantics components make common sense? Is there value from the progression

from generalized to specialized? Can you see the "define-once use-many-times" approach to these models? What is the value? How would you quantify it?

6. In what way does the Metabase System employ taxonomies and ontologies? Does its use of these components have real value in data semantics management? What is the value? How would you quantify it? The text states that the entire collection of Metabase meta-models are themselves an ontology. Does that make sense?

7. What is the semantic management model? How is it constructed? What is its use when employed with the other data semantic component models? What role does the semantic management model play? What role does each of the contained components play? Do you see any data modeling problems in the model depicted in Figure 3. What are the ambiguities? How would you address them?

8. What are the key components of the Data Element Model? How are they interrelated? What are the benefits from having these levels of abstraction? How is the Data Element Model different from a "normal" data model of say tables, columns and keys. What must be the objective of the differences? Do you see any obvious data modeling errors in this high level model? What are they and what would have to be done to resolve them?

9. What are the key components of the Specified Data Model? How are they related? What are the benefits from this data model of concepts. How is this similar or different from a traditional data model of schema, table, columns and keys? From Figure 3, what is the relationship between an attribute and a data element? What is the benefit of such a relationship? What does the attribute inherit from the data element?

10. What are the key components of the Implemented Data Model? How are they related. How is this data model different from the Specified Data Model and the Data Element Model? From Figure 3, what is the relationship between a column and an attribute? What is the benefit of such a relationship? What does the column inherit from the attribute and in turn from the data element?

11. What are the key components of the Database Object Model? How is this model different from a "data" model? From a "process" model. What role does "state" play in this model? What is the relationship between corporate policy and database object classes? How does your enterprise handle the identification, creation and management of database objects? Formally? Informally? How should business information systems be formed around database object classes? What would be the value? What is the case for making database object classes properly encapsulated within the domain of a DBMS?

12. What are the key components of the Operational Data Model? How are they related? How is this data model different from the Implemented Data Model? What does the DBMS column inherit from the column?

13. What are the key components of the View Data Model? How do the XML and SOA components (see Chapter 14) extend this model? What are the key roles played by a View Data Model in the overall engineering of enterprise-wide data semantics management?

14. After a careful review of the life cycle cost model, do you generally agree with it? Are the factors generally right? Does the purpose and value of iterative prototyping come clear? Does the value of a business information system generator like Clarion become clear? What classes of prototyping problems does a "Clarion" address? Have you attempted iterative prototyping in your enterprise? Has it been of value? If yes, what and how much has it saved in time and money? If it has not been of value, why not? What caused it to fail? What was done given those reasons for failure? Have you been pressured to just release the final prototype into production? What are the advantages and disadvantages in doing so?

15. The Error Classes section outlines the types of errors addressed by this chapter's materials. Do you agree that the error classes are addressed? How? What could be different? Has this chapter's contents and strategy provided an overall justification for its implementation? If yes, why? If now, why not? How would you quantity the benefits from having these error classes resolved?

3

Data Semantics Management Failures

Chapter 2, Why Semantics, addressed the fundamental need for highly engineered semantics, that is, the meaning and rules that govern data. This need was first explained within the context for data management. Key to understanding data management is understanding the role of enterprise-level policies and procedures. Because these need to be enterprise-wide, there must be consistency in their definitions at all levels and across all enterprise organizations.

Another section of Chapter 2 identified the need for database object classes. The specifications of database object classes were set squarely within corporate policy specifications. Data, the by-product of policy executions, are the "what" that should be shared within and across corporate organizations. Database object classes are the coherent specifications of shared data. Unique to database object classes are the definitions of the states and state transformations through which a given object proceeds from birth through all life cycle stages until the database object becomes fully mature.

A key construct in Chapter 2 was the graphical representation of the various layers that start with policy specification and proceeds down through data-based evidences of policy executions. Each layer's data model was graphically depicted and its components listed and described.

Chapter 2 continued with a simple definition of data semantics and illustrated this definition through several examples. Key to the explanation was the simple fact that there is no silver bullet that addressed all the needs of data interoperability across the enterprise. Because there is no silver bullet, the management of data semantics must proceed through several different, but thoroughly interconnected models of metadata. These models addressed: data elements, data structure templates, the employment of these data structure templates within databases, and finally as specifications of semantically complete data exchanges. Each of these metadata models was generally defined in terms of their critical components.

Chapter 2 has a final section that detailed of cost-benefit value of applying the techniques set out in this book. These costs and benefits were not theoretic. They were proven. Consequently there is no down-side to the prescriptions of this book. Again, not accomplishing data semantics management is not only a bad business decision in the short run, it likely is a terminal business decision in the long run. As stated above, any executive or business decision maker not embracing these dictums and strategies becomes "the enemy from within."

This chapter, Chapter 3. Data Semantics Failures, is a very critical chapter. This chapter focuses on the U.S. Department of Defense (DoD) because its data semantics failure was so public and because it occurred over a 12 year period. There were investigations by the U.S. General Accountability Office. The DoD was not alone in these types of failures. These types of failures have been replicated by other sectors such as State and Local Governments, Education, Medical, and industry. Thus, the DoD serves merely as a highly visible "case study." This chapter thus represents a set of very valuable "lessons learned."

The DoD spent more than $500 million attempting to achieve enterprise-wide data semantics and failed. The reaction to this failure, however, was not to understand the causes and try again. Rather it was to abandon the effort because it seemed to be too hard and took too long. The problem with that solution was that the next time data semantics management was tried, and it was, the problem area had grown and the existing ad hoc, stove pipe solutions had become more entrenched.

In the case of the DoD, the reaction to failure was to jump on the next "silver bullet" that whistled through, which was XML. By any real analysis, XML is no silver bullet because in order for the XML sliver bullet to actually work, all the DoD agencies had to first successfully accomplish the very effort that failed, and abandoned. In order to have successful XML exchanges, which is merely a mediation solution for shared data exchanges, there has to be shared semantics in support of the generation of and the decoding of the exchanged XML.

Because data semantics management is a problem that must be addressed before any real enterprise-wide data interoperability can be achieved, it is essential that the failures from the first attempt at enterprise-wide data semantics management be understood and remedied. Thus, this chapter goes to significant length to diagnose the common data semantics failures, including some that may be a surprise. The lessons learned are clearly understood and have been completely incorporated in the strategies and examples of this book. Everything that needs to be done to engineer and deploy interoperable data environments is set out in this book and in other Whitemarsh books. Additionally, every example from this book can be operationally shown via the Metabase System.

A key ingredient to the solution set out in this book is the establishment of communities of interest. The Whitemarsh book, *Data Interoperability Community of Interest Handbook,* sets out the architecture of these groups and the work plans, deliverables, and all necessary policies and procedures for the establishment, operation, and evolution of these groups.

Excuses of not knowing what causes data semantics management failures, and once known, not knowing what to do about them have been removed. The time is now, not the tomorrow that never seems to come.

The reasons for data semantics management failures are legion. Among the most significant are:

- A fundamentally flawed data semantics management model.
- No accommodation for enterprise-wide data architectures.
- Multiple implementation technologies.
- Central standardization maintenance authority.

3.1 Fundamentally Flawed Model

A systems engineering firm, The MITRE Corporation, evaluated the DoD's data semantics management effort in 1994. The summary findings were that the DoD had adopted a data semantics management model that, upon execution, led it further and further away from the ultimate objective, standard data. Despite best intentions and the very hard work of many individuals, the entire $500+ million (2008 dollars) effort was determined to be a failure.

At the very foundation of the DoD's data semantics management model was the DoD's data element. The DoD's data semantics management process, however, forced the data element to take on the granularity of a column within a table. This fatal granularity error was caused by forcing the data element's name to include a prime word (i.e., the name of an entity or table) as the first part of the data element's name. Thus, to the DoD, data elements functioned as columns in database tables, screen cells, fields in a file, columns on a report, or variables in an internal process.

Why this is a terrible consequence is because the quantity of DoD data elements, that is, columns is ever expanding. If a database has a 1,000 tables and each has 15 columns, then there are 15,000 DoD data elements to name and define. If there are only 50,000 databases (a very conservative quantity) there are then 750 million DoD data elements to name and define.

Every new database adds 15,000 new columns, and likely 30,000 staff years of efforts. Clearly this is an ever expanding and impossible to complete task.

In contrast to the DoD's fundamental misapplication, a data element is really a context independent business fact that may be employed in many different contexts. That is, employed from within many different tables, files, screens, reports, or internal process. The DoD's critical methodology flaw caused it to define each data element use, that is, in each different table, file, screen, report, or internal process rather than define the data element once and identify wherever it was used. Because the DoD was in fact defining data at the granularity of columns, the term, column, is used in this chapter to refer to the DoD's data semantics management target rather than the DoD's term, data element. Because of the DoD's fatal error, every staff hour spent standardizing columns was a four-fold waste of time:

- First, there were the time and effort to standardize the column.
- Second, there were the time and effort to recognize the error and have the courage to admit to such a colossal waste of effort.
- Third, there was the time and effort to find all the redundant definitions.
- Fourth, there were the time and effort to delete the redundant (and probably differently defined) definitions and to create the single definition and all the mappings from the defined-once definition to the used-many-times deployments.

When the DoD was informed of this critical flaw, the evaluation report was not allowed to be submitted. What did occur was an *off-the-record* review and acknowledgment by the DoD of the critical error. There were no changes in behavior or direction because the quantity of columns standardized was the success measure rather than the quality, efficiency, and reusability of the result.

The DoD felt that standard data automatically occurs once standard columns were created. Not only is that untrue, it leads to a significant waste of valuable and scarce resources. The converse is true. The standardization of data--through the discipline of information engineering (data models, process models, business rules, et al)--leads to the automatic by-product: standard columns. In short, standardizing columns is a non-productive, ever-expanding effort. MITRE, in support of its summary finding, produced the following findings:

- The DoD's standardization effort concentrated on a level of detail (columns) that is ever expanding, thus prevented completion of the effort.
- Standardization was overly focused on *names.*
- Critical components requiring standardization were being ignored.
- The current effort attempted standardization without critical context and subject matter materials.
- The DoD's repository tool, while adequate for the definition and maintenance of standard columns, was unsuited to the overall task because:
    - ♦ It concentrated on columns, which was a level of detail that was too low and ever expanding,
    - ♦ It lacked critical analysis and synthesis capabilities,
    - ♦ Its meta-model was insufficient for a comprehensive and correct data semantics management effort.
- The existing procedures caused intense standardization of too small an area of the overall data semantics management problem, almost to the exclusion of other critical areas.

A brief explanation behind each of these findings is provided below in separate sections. The recommendations that flowed from the findings were that:

- The main focus of the standardization effort must be on the data element--not the column--so that standardization could reach closure.
- Standardization should be based on a comprehensive meta-model that properly allocated effort all the meta-attributes of all the appropriate meta-entities.
- Standardization efforts must be expended on the identification and standardization of the most common compound and derived data elements, including creating mappings to their atomic and primitive components.
- The standardization effort could only be successful through an outside-in process that gradually, over time, created small islands of standard data that in turn are merged into larger islands, and so on.
- The repository must be thoroughly examined to assess its meta-model. If the meta-model is lacking, a comprehensive meta-model must be built. Once the meta-model is built, an automated system must either be procured or built that is both non-proprietary and can be ported.
- The existing procedures must be designed and built only after a comprehensive meta-model is designed and successfully prototyped. Once prototyped and accepted, the procedures can be quickly and easily built.
- The terminology employed by DoD must be unambiguous and precise. Currently, the only unambiguous and precise terminology is ANSI/SQL.

3.1.1 The Focus of Standardization Was Too Low

The highest and quickest payoffs from data semantics management come from standardization at just the right level. If the level is too high, the subsequent detailed levels of the application of the standards are not consistent. If the level is too low, an excess of time is spent standardizing essentially the same item whenever its context changes even for the most insignificant of reasons.

Within the DoD, the chosen level was too low and thus the standards effort could never reach closure because new tables of columns were being created faster than columns within existing tables were being standardized.

Whenever it was clearly explained to experts in information engineering that the DoD's focus of standardization was the column rather than the data element, the experts were incredulous!

For example, if the data Element, Telephone_Number, is involved in the standardization effort, an essential first step in the DoD's standardization effort was to identify the containing table. The table is the context of its use. For example, Employee, Client, Customer, etc. After every table was identified, every valid column was determined. After each was determined, it was standardized. But, if a data element was represented multiple times in a table, it was first disambiguated by providing a prefix modifier. Thus, if there were four different Telephone_Number data element uses, each different use was specified. For example:

Home_Telephone_Number
Office_Telephone_Number
Boat_Telephone_Number
Airplane_Telephone_Number

If the table was EMPLOYEE, then for EMPLOYEE, they were:

Employee_Home_Telephone_Number
Employee_Office_Telephone_Number
Employee_Boat_Telephone_Number
Employee_Airplane_Telephone_Number

When a new table came along, the standardization effort began anew. For example, if the new table was Customer, the standardization effort reoccurred for:

Customer_Home_telephone_number
Customer_Office_telephone_number
Customer_Boat_telephone_number

Customer_Airplane_telephone_number

If there are 10 more tables, four new sets of definitions must be developed for each. What started as a single data element, TELEPHONE_NUMBER, in this small, trivial example, has burgeoned to forty-eight different column standardization efforts! Such use standardization is clearly not conducive to progress or to closure.

3.1.2 Standardization Was Too Focused on *Names*

A significant portion of the standardization effort was focused on *names,* as if the name, whether it is a column's name or a data element's name, was the preeminent component of the data semantics management effort rather than just one of many meta-attributes that fully define standard data. In the DoD standardization effort, the components of a name were set down in great detail. Each word position was carefully defined, and the allowed words were precisely enumerated. *Naming standards that focus on recording the codification of data semantics into the data's name commonly obscure the needs of the business and the common business meaning of the data.*

When these rules were religiously followed, the results were not understood outside the context in which it was defined. The reason was that the constructed name was often the data element's commonly used name. Instead, the name was a constructed compound, and often derived string. If a data element was supposed to represent a value that was both atomic and primitive, so too should the meta-attributes that describe the data element. For example, when the data element Telephone Number becomes a properly named standard column, it becomes:

Employee_Telephone_Identifier

The reason is that the word Number is missing is because it is a class word, and since telephone numbers can contain non-numerics, it cannot be used. Further, the class word, Identifier, must be used since telephone number is really a form of a unique address of a telephonic instrument.

Another example, Social Security Number, is not really a number. In fact it is the combination of three codes, and, as an identifier its name should really be:

Employee_Social_Security_Identifer_Code

but that would violate the *rules* because only one class word was allowed. Thus, it has to be either:

Employee_Social_Security_Identifer
Employee_Social_Security_Code

And, since both were considered unacceptable, the rules were suspended with the result that it became:

Employee_Social_Security_Number

While that clearly solved the common name issue, the data element was now not selectable either as an identifier or as a code. The problem with the DoD's naming strategies was that it was trying to make the name "say" multiple things. Instead, the name should merely be the common name, and nothing more. The name should be a short-hand string for the data element's business definition, not a compound and derived shorthand for the data element's semantics. All the data elements' semantics, that is, whether the data value set is really a code, or whether it can be used as a primary key, etc. are merely different meta-attributes that contribute to the full description of the standard data.

A byproduct of treating names as merely "one of the gang" of meta-attributes is that there can be synonyms to a name so that different organizations can use the same standard data without having to give-up their current common name traditions. All the meta-attributes of the data elements and columns are known to the Metabase System's database, which, in turn, becomes an integral component of any enterprise data semantics management documentation.

3.1.3 Critical Standardization Efforts Were Ignored

Completely ignored from the DoD standardization effort were two very common classifications of data: derived, and compound. While a clear argument could be made not to have derived data elements within a database, they exist nonetheless for both traditional and performance reasons.

For example, in a set of accounting records, the FINAL_MONTHLY_BALANCE is derived from the series of monthly financial transactions. However to remove FINAL_MONTHLY_BALANCE may be to violate accounting principals. Beyond just retaining the derived data element, also retained must be the formula through which the value is calculated.

Compound data elements, such as the FEDERAL_STOCK_NUMBER is a very common example of a single data element that has contained subordinate data elements. Defining the subordinate data elements without having a mechanism to know of the compound data element is to:

- Discard very important tradition.
- Invite malformed update processing.
- Invite its subsequent reintroduction when it cannot be "discovered" among the many other primitive and atomic data elements.

Some compound data elements and derived data elements have important semantic roles in conveying complete user understanding of the data. It is important therefore that, when necessary, both are allowed and appropriately maintained.

3.1.4 Critical Context and Subject Matter Materials Were Missing

Context material was essential to the correct characterization of semantics surrounding data. One organization within the DoD was interviewed to detect why their submissions of candidate columns were so clean and exact. The answer was simple: their standardization effort was done outside-in. As a system was standardized, all the data facts were first examined, in context, before the encoding of the standardization characteristics.

Once one system was done, another was started. Peer level analysis was undertaken whenever conflicting semantics arose. Slowly and surely the standardization effort was accomplished.

The organization took five elapsed years, spending about 45 staff years to produce about 560 <u>data elements</u> from an original population of about 19,000 application-based <u>columns</u>. That represents a reduction of 380 to 1. The analysis time was about five staff hours per application column. There were between 90 and 100 different applications. When the columns were factored from these 19,000, there were only about 3000. This meant that every data element was used about six different times within each application.

The organization stated that if they only had to standardize the data elements within each applications rather than columns within applications, the staff years expenditure would have been reduced from 45 staff years to just 7.5 staff years. If the organization had restricted itself to just the data elements across all applications, the 7.5 staff years would have been reduced to just 1.6 staff years. At a cost of $180,000 per staff year, the original effort cost $8,100,000. If the organization had worked "smarter" from the very start, the effort would only have cost $288,000. The lesson is simple, if the right paradigm is used from the very beginning, the cost savings are very large.

As of the program's cancellation for failure, the DoD had collected about 70,000 columns from identified systems. None of the critical contextual information essential to resolve semantic ambiguities and conflicts has either been submitted or was planned to be submitted. A brief scan of the submitted columns showed non-standard formats, missing definitions, conflicting semantics, and the like. It was generally believed that the submitted metadata could never be successfully standardized in any reasonable amount of time by any reasonable quantity of staff analysts. A quick workload analysis shows that the required staff years, under the DoD's data standardization architecture would have required 245,000 staff years to complete. That translates to the small sum of $44,100,000,000, which would have paid for eight Nimitz-class supercarriers.

3.1.5 The DoD Metadata Repository System Was Unacceptable

The largest amount of work in standardization is analysis. Currently, as a new column is submitted for standardization, its complete semantics must be studied in detail and compared with all other previously stored columns to decide whether the candidate column is new, or already present. Even if the repository focus was data elements, this analysis would still have to be done.

The DoD's metadata management system contained no tools to help the analyst. Discovery was principally by the brute force method. While such a method was quick at first, it slowed rapidly as both the volume of standardized data enlarged and the quantity of different persons doing data semantics management increased.

The meta-model required to comprehensively support data standardization embraced far more activities and the capture of far more metadata than was possible with the DoD's metadata management system. The repository meta-model scope and accompanying automated assists needed to be completely re-engineered to handle this type of effort.

Finally, three points needed to be made about that DoD's metadata management system. First, it was a proprietary-language application of a single vendor's DBMS. Second, neither its design, nor its programs were portable. Third, its design was not extensible, thus, the DoD's metadata management system was not a long-term, viable tool.

3.1.6 The Existing Procedures Excluded Critical Areas

The effort, as required by the DoD, to accomplish a quality job on data semantics management was enormous. Even if the effort were accomplished, the existing files of now-standardized columns would have, by then, substantively changed, necessitating a new standardization effort.

Missed completely was the incorporation of the standardization results into a comprehensive metadata management system database that was essential to accomplish data migration, ongoing data conversions, and long term data evolutions. Essentially, this would have become an infinite length task being done an infinite quantity of times? Somehow this doesn't seem like a viable strategy.

3.2 No Accommodation for Enterprise-wide Data Architectures

The DoD example is a perfect example of ignoring the down stream costs of not having an enterprise-wide data architecture from the beginning. Apart from blowing up a business and starting over, the following is always true:

> The sooner you start towards an enterprise-wide data architecture, and the smarter you accomplish it, the less enterprise-wide data semantics management efforts will cost.

Strategies do exist. There must also be the recognition that all data are probably localized implementations of commonly derivable definitions.

Without that fundamental understanding, there will always be an insurmountable data semantics communication problem.

Enterprise-wide data architectures must separate out the different concepts from the different implementations and/or styles of the same concept. Any data architecture must accommodate:

- Geography areas such as the world, regions, and countries.
- Subject areas such as human resources, finance, marketing, manufacturing, sales and marketing, and research and development.
- Languages, where the same concept is represented differently.
- Cultural/legal, where it either is traditional, required by law, or is disallowed by law to collect and store certain information.
- Local laws that may, for example address sales accounting, finance such as accounts payable, receivable, and general ledger, when business is "booked," taxation, and payroll and compensation.

The data architecture must handle the similarities at the highest level possible and the push the differences down to the lowest level where they can be screened/blocked by technology. When different representations of essentially the same concept are allowed to flourish, they become the "sacred tradition" that caused the 19,000 columns that had become an almost total and complete obstacle to standard data. These columns were just different uses of about 600 data elements.

An area completely unaddressed in the DoD data semantics management effort evaluation was data semantics reuse with respect to data model templates. Here, a data model template is a collection of attributes within a collection of entities from a set of subjects that are commonly employed across the enterprise. Examples abound within subject areas such as human resources, logistics, inventory, transportation, order-entry, invoicing, and finance. These subjects can all be categorized as data-based specializations of the six interrogatories of who, what, when, where, why, and how.

The DoD, beyond creating the massive storehouse of columns, also created standardized data models. These were called the Defense Data Architecture and consisted of about 10,000 entities across about 50 subject areas. These were created because it was assumed that every DoD agency would just simply adopt these common data models and immediately retrofit all their business information systems to conform to these models. In the DoD at the time there were well more than 500,000 discrete business information systems.

While wholesale adoption never occurred, the standard data models potentially had real value. The value was that the existing DoD database and business information systems could be mapped to these "idealized" models. If that had been done, the DoD would have built the Metabase System's Specified Data Models.

A second unrealized benefit from the DoD effort was that the complete collection of attributes from the Defense Data Architecture could have been mined for data elements. If these two steps had been done, the entire DoD data semantics management effort could have been a real success, and there could have been a firm data semantics management foundation for the development of the XML shared data exchanges. It's almost as if the DoD quit their data semantics management marathon race just 100 yards before the finish line.

3.3 Multiple Implementation Technologies

From the very beginning there have been different names' styles, allowed lengths, and rules regarding special characters. Despite all these rules and restrictions, the requirement for commonly understood names continues.

If the names of various meta objects conform to the requirements of their implementation environments, the same concept will take on different name strings across these different environments. A common technique is to create physical names. A physical name is one that conforms to length restrictions. The result however is unacceptable because the physical name is all the programmer sees and all the end-user is likely to employ. Because the same concept can have different physical names in different contexts users may presume that because the names are different, the data meanings are different.

3.4 Central Standardization and Maintenance Authority

A single, centralized effort to accomplish and maintain standard is clearly doomed to failure. Any name must be automatically created through rules, name repositories, and pick lists. The one most appropriate to create a name is the one who is designing and building the database or business information system. If the effort, however, is concentrated in one organization or one person, project schedules and work will clearly suffer. During a project, names are hypothesized and changed. If there is only one opportunity to create a name, either the whole project must go on-hold to await the "God of Standardization" to perform the miracle, or the names must be created once and never changed no matter the changes in reality.

As an example, if an organization has thirty projects underway, and if there is a population of 3000 names (e. g., tables, columns, and keys) for each project, and if it only takes one-half hour to understand, create, define, and store each name, there must be about 9.5 staff months allocated to the naming effort. If a staff year cost of $150,000, the cost of these names is about $120,000. Given that the project's developers have already created all the materials ahead of being used by the naming guru, nearly all the work has been completed before the naming has started. Given thirty projects, the annual cost is about $3.6 million to provide names to concepts that are almost already named! That is 25 staff years that might be considered a total waste.

Suppose that a project is ready for the naming guru and the guru is already working on one project with two others backlogged. Does the naming guru timeshare all four projects causing all four to be delayed, or is the fourth project's team told to just plain wait 10 months? To be sure, they will not wait. The naming will proceed and the organization will be stuck with non-standard names. Soon there will be 19,000 non-standard names!

Maintenance is forever. Thus, there is a continuous need for name maintenance that is essentially automatic and immediate. Given that a project is completed and proceeds into maintenance, and given 20 projects in production that may need some level of maintenance for every project in development, there are about 600 potential maintenance efforts. If only 10% of the projects require maintenance, and only 5% of any project's names change, that's still about 450 data items to maintain. If each item only takes one-half hour to understand and properly record the change, that's still about 250 hours or about $16,250. All these name changes must also be included in the scheduling mix of new name development.

In place of a centralized authority, the solution is the accomplishment of enterprise-wide data semantics management through Communities of Interest. These communities focus on single domains, and once data sharing specification is established within that domain, cross-Communities of Interest can be established to create the inter-community shared data specifications. The Whitemarsh book, *Data Interoperability Community of Interest Handbook*, sets out the architecture of these groups and the work plans, deliverables, and all necessary policies and procedures for the establishment, operation, and evolution of these groups. If these communities of interest employ supports like the Metabase System, the domain-specific shared data specifications can be shared as well.

3.5 Error Classes

The only real value derived from failure is to learn its lessons. Obliterating the occurrence of errors across the two error classes is clearly the objective of any data semantics management effort. If the effort fails, rather than repeat the past, it must learned, re-engineered, and driven towards success. With the U.S. DoD, only the U.S. Navy learned, re-engineered, and persevered to get the "brass ring."

Jumping on the next "silver bullet" almost always results in an even bigger failure because it tends to ignore failures of the past, which cannot be ignored, and the IT efforts, because of the passage of time are now bigger and more costly to fix.

A critical part of learning, re-engineering, and achieving success is the development of end-to-end prototypes of solutions. Prototypes must be tried before launching massive production line re-systemization efforts. Prototyping seems always to cause the discovery of impending production line errors. Prototypes almost always represent well-spent money and time. Once an end-to-end prototype has been created, accomplished, validated, and adjusted, measurements on the errors across the two error classes about avoidance and remediation can be made. For the DoD effort, this was not done. In the case of the U.S. Navy, it was done.

With respect to the two error classes, the failures were essentially these:

- No semantic hierarchies were developed to then aid in the automatic creation and definitions of business facts. This led to facts with very close name with significantly different definitions. It also led to differently named facts with the same definitions.

- The over engineering of the DoD's data semantics management was critically flawed. This in turn led the DoD to attempt the standardization of an indeterminate and ever expanding quantity of facts.

- There was no climate of "lessons-learned" and then advance.

3.6 Data Semantics Management Failures Summary

This chapter presented an analysis of the failures that caused the half-billion dollar DoD data semantics management failure in their attempt to achieve enterprise-wide data semantics.

Excuses of not knowing what causes data semantics management failures, and once known, not knowing what to do about them have been removed. The time is now, not the tomorrow that never seems to come.

The next chapter, Chapter 4, Engineering Data Semantics, is another preparatory chapter. Data semantics management efforts cannot be slapped together and set into motion. The right environment must exist. Ironically, even if all the factors were completely known 20 to 30 years ago, enterprise-wide interoperable data environments were not possible due to immature technologies. If however the dictums and strategies of this book and those from the *Data Interoperability Community of Interest Handbook* were executed, there would have been integrated data across all the stove pipes of database and business information systems. Had this been done when the technology was available, it would have all fallen into place.

It was no accident that one division of a world-wide corporation had virtually a zero Y2K bill while another division of the same corporation spent many millions repairing and re-deploying their ad hoc databases and business information system implementations.

A key set of material in Chapter 4 is the exposition of the evolution of bindings that existed between data, programs, database management systems, and users. There has been a long and gradual trek in the evolution of bindings from the 1950s through to the present day that has enabled, for example, service oriented architectures. It is critical to know, however, that success in service oriented architectures requires careful attention and deployment of the fundamentals of this book. Disregarding the dictums is to court failure.

If the reasons for failure are not addressed and the solutions are not brought to bear, data semantics management efforts either suffer a slow death or are just canceled. For data semantics management effort to be successful, they must:

- Properly focus on the data concepts such that the lowest level column is easily and automatically named and defined.
- Accommodate enterprise-wide data architectures such that data contexts are immediately apparent without compromising data identification, selection, and melding.
- Accommodate multiple implementation technologies such that no matter the rules, data concepts are immediately obvious and relatable to all other data that espouse the same concept.
- Allow front-line project staff to create and maintain their own names under the guidance and work enhancing tools and techniques of a central standardization and maintenance authority.

3.7 Questions and Exercises

1. After reviewing the reasons for data semantics failure, are these reasonable? Which ones are missing?
2. Have you fallen prey to any of these failure reasons? What was the cost? What was the effect? What was the reaction to the failures?
3. Does your enterprise have an overall model for data semantics management? Was this model created and prototyped to ensure validity? How has the model been propagated and accomplished?
4. How does your model compare with that engineered in Chapters 1 and 2? How would these models affect the failures described in this Chapter? How does your model address these failures?

5. How has your organization constructed names? Via the book's approach or by stringing together words? Which is better? More reliable? More repeatable?

6. Do you have a strategy for automatically constructing names? Definitions? Abbreviations? Is this strategy accomplishable across the enterprise or just on an individual project basis? How have you handled "word" changes?

7. Does your strategy accomplish name engineering across the data semantics component levels starting with data elements and preceding through DBMS columns?

8. How does your strategy deal with same-name different meanings and same meaning different name situations? How are they identified and dealt with?

9. Has your organization created a "naming czar?" Has that accelerated work or impeded it? Do projects wait for the czar or do they just proceed?

10. Have you created a metadata management system? Is it available on an enterprise-wide basis? What has been its value? Problems? Is there a return on investment? How has that been calculated?

11. If you had a comprehensive approach to the data semantics failures described in this chapter, what would have been its effect on your development efforts? What would have been the savings and how would you have calculated these?

12 A key problem in the DoD example of this chapter was that it seemed so "Dilbert-esque". It was not one of hypothesis, prototype, feedback and then cycles of re-engineering before proceeding to data semantics management production. Rather it seemed to be one of untested hypothesis, full speed ahead, failure and then coverup and blame others. How different is your culture from that in this case study? If it is the same, how can you put forward a strategy to change it?

13. There were three failure summaries at the end of the Error Classes section. How does your organization compare to these failure reasons? What can you do to prevent or repair them?

14. What approach have you taken to "sell" your approach to management? Was it successful? Was there push-back? How have you responded? What did you modify?

15. How do you counter the push-back that data semantics management takes too long? What are your counter arguments. Have these succeeded? If yes, why? If not, why not?

16. Do you have a tool set that makes creating data semantics across collections of database efforts faster than individual database efforts? What are these? How do they compare to the strategies set forth in this book?

17. If you have been able to effect strategies that are akin to manufacturing data semantics, have you been able to put some or all of those "savings" into other efforts that, in turn, can produce even more savings?

18. The Error Classes section outlines the types of errors addressed by this chapter's materials. Do you agree that the error classes are addressed? How? What could be different? Has this chapter's contents and strategy provided an overall justification for its implementation? If yes, why? If now, why not? How would you quantity the benefits from having these error classes resolved? How would you present the approach, costs, and benefits to management in order to justify funding?

4

Engineering Data Semantics

The last chapter, Data Semantics Failures, was a very critical chapter. The "poster child" case study of the failure was the $500 million the DoD spent attempting to achieve enterprise-wide data semantics. Its reaction to failure was not to understand the causes and try again. Instead, it was to abandon the effort because it seemed too hard or took too long. The problem with that is that the next time data semantics is tried, the problem area will have grown and the existing ad hoc, stove pipe solutions will have become more entrenched.

What the DoD did was to jump on the next "silver bullet" that whistled by, XML, without realizing that the very approach the DoD failed to accomplish was needed as the foundation for successful XML.

Chapter 3 went to significant length to convey the problems associated with common data semantics failures. The lessons from the diagnosis are understood and have been completely incorporated in the strategies and examples of this book. Everything that needs to be done to engineer and deploy interoperable data environments is set out in this book and in other Whitemarsh books. Seeing the solutions is easy, just download the free Metabase System from the Whitemarsh website.

One key reason for the DoD failure was that a one size solution does not fit all problems. Thus, the need for a Community of Interest approach that can have its resultant metadata artifacts federated for all to see and as appropriate use. The Community of Interest approach is described in the Whitemarsh book, Data Interoperability Community of Interest Handbook.

This book sets out the architecture of these groups and the work plans, deliverables, and all necessary policies and procedures for the establishment, operation, and evolution of these groups. Can this book be trusted? Yes, because the vast majority of its content has been modeled after already successful multi-year data interoperability Communities of Interest. Thus, the book is not an academic trek into theory, but a handbook of practice.

Because of the material in Chapter 3, and the material from other Whitemarsh books and the Whitemarsh website, the causes of the failures are known and what has to be done has been set out.

This chapter, Chapter 4, Engineering Data Semantics, is another preparatory chapter. Data semantics management efforts cannot be slapped together and set into motion. The right environment must exist. Ironically, even if all the factors were completely known 20 to 30 years ago, enterprise-wide interoperable data environments were not possible due to immature computing and network technologies. If the dictums and strategies of this book and the *Data Interoperability Community of Interest Handbook* had been executed, there would have been integrated data across all the stove pipes of database and business information systems today. That is because as computers and networks became sophisticated, the inter-operable infra-structures would have already existed and could have been employed. As stated previously, it was no accident that one division of a world-wide corporation had virtually a zero Y2K bill while another division of the same corporation spent many millions repairing and re-deploying their ad hoc databases and business information system implementations.

A key set of material contained in this chapter is the exposition of the evolution of bindings that existed between data, programs, database management systems, and users. There has been a long and gradual trek in the evolution of bindings from the 1950s through to the present day. Because of this, service oriented architectures are now possible. Success in service oriented architectures requires careful attention and deployment of the data semantics management fundamentals from this book. That is because this book sets out the strategies that can be applied given the bindings' evolutions.

As the restrictive and encapsulating bindings presented in this chapter fell away starting with those of the 1950s and continuing until today, there became opportunities for data sharing and interoperability. Because of this opportunity, there came an obligation to have semantics homogeneity across the inter-system and inter-database connections.

Without careful attention to the elimination of conflicts between databases and business information systems, costs rise, quality decreases, productivity slows, and risks become critical. Regaining control over these factors requires changes in methods and products produced. The problem lies not in staff nor management, but in the methods and the metadata artifacts that result from knowledge worker effort. Replacing staff, management, or mere exhortations to work harder or with more dedication is not the answer. Working smarter is. Methods and artifacts are as necessary to manage database and information systems development and evolution efforts as are financial transactions necessary to control and manage the business itself. For example:

- If a finance manager wants to know product manufacturing costs, financial records are consulted, reports produced, and trends analyzed.
- Tracking and evaluating cartage companies requires performance records, assessments of delivery records, and timeliness of deliveries.
- Quality manufacturing requires critical research into and assessment of manufacturing techniques, planned and random assessments of manufacturing results, lack of product recalls, and the like.
- Efficient, cost-effective manufacturing requires market demand assessment, the ability to manufacture the right product at the right time, thorough materials, techniques, and resource planning.

Analogous techniques, methods, products, and assessments are required for database and information system design, development and evolution. To do fewer activities, or to spend fewer monies than is proportional to those expended for research and development, manufacturing, inventory, distribution, marketing and sales of a product line itself is to court database and business information systems disaster.

The similarities between the life cycles of a real product and that of a database and business information system are not just accidental, they are essentially the same. The only real difference is the type of process, the techniques employed and the products produced. The encompassing infrastructure of R&D, management, quality estimates, prototypes, production monitoring, and quality assurance, and quality control are the same. If these are necessary for real products they are just as necessary for database and information systems.

There should be no more one-off database and business information systems just as there should be no one-off real products. When today's knowledge of methods and techniques are coupled with the certain productivity improvements from CASE, metadata management systems, and code generators, there should never be a database or business information system that is over budget, behind schedule, or that delivers less than realistically possible.

The main reasons for bad or poor quality databases and business information systems are the same as for bad or poor quality real products. That is, poor research, unrealistic plans, improperly constructed techniques and methods, uncontrolled and unmonitored production and manufacture.

Quality database and business information systems require quality engineered methods, techniques and products, which are broadly divided into data oriented and process oriented methods, techniques and products. These all exist within their three forms, specification, implementation, and operation. Finally, these methods, techniques and products within their three forms are produced from within an infrastructure of R&D, management, quality estimates, prototypes, production monitoring, and quality assurance and quality control.

A complete solution to the evidenced problems requires a complete overhaul in methods and the metadata artifacts that result from knowledge worker effort and in the database and business information systems infrastructure of R&D, management, quality estimates, prototypes, production monitoring, and quality assurance and quality control.

4.1 Data Semantics

Data semantics have to be both specified and implemented. Table 9 lists the semantic component, the meta-object through which it is specified and the meta-object through which it is implemented.

Component	Specification	Implementation
Policy homogeneous business segment	Subject Area	Schema
Well-defined unit of business policy	Database Object	SQL 1999 table
Precise set of data within a business policy	Entity	SQL 1992 table
Discrete evidence of policy adherence, that is, a context based business fact	Attribute	Column
Unique discriminator of policy instance	Primary Key	Primary Key

Component	Specification	Implementation
Relationship mechanism between policy instances of different types	Foreign Key	Foreign Key
Business rule within a business fact	Type of Data Integrity Rule	Column Constraint
Business rule for whole business policy	Type of Data Integrity Rule	Table Constraint, Assertion
Business action consequence when a policy is accomplished or violated	Type of Data Integrity Rule	Trigger or Stored Procedure

Table 9. Proper location of data semantics specifications.

These components are materialized after suitable analysis and design. Each component has a name, definition, description, other appropriate meta-attributes, and relationships to other analysis and design component instances.

4.2 Data Semantics Alternatives

The two alternatives to building Specified, Implemented and Operational Data Models are:

- Standard semantics (data names, data types, data lengths, and data structures) for entities & attributes, and tables & columns. That is, wherever the data semantic concepts are the same they must be represented the same way.
- Semantics mappings from standard, non-redundant, technology and context independent concepts (data names, data types, data lengths, and data structures) to technology and context dependent specified and implemented uses of those concepts.

The first approach is illustrated in Figure 13. In this figure, all the data model meta-objects, from data element through to view columns are named the same. However, this is not possible because:

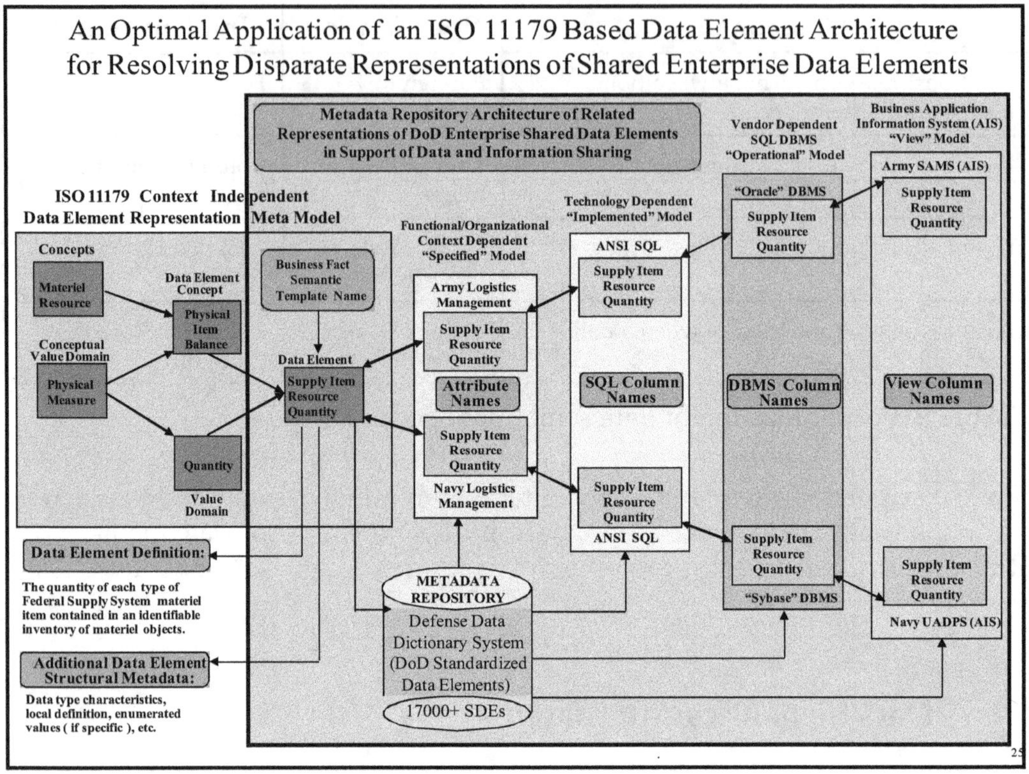

Figure 13. Same named semantics from data elements through to view models.

- The accepted data semantics now in place force most if not all names across all the meta-objects to be different. Thus, concepts that are the same cannot be found because they are at least named differently. In Figure 13, it would be very likely that the view column names, e.g., Supply Item Resource Quantity would be already named years ago when the business information system and database were designed and constructed. The same would be said for the Operational Data Model name, which in Figure 13 is also Supply Item Resource Quantity.

- It requires abandonment of all COTS procurements because the set of semantics within each COTS package are likely different from another package with overlapping database domains. For example, many

COTS application packages all involve semantically different personnel information that addresses the same or essentially similar concepts.

- It requires a complete rewrite of all legacy applications from their non-standard semantics to a set of standard semantics, and the continuous rewrite of these legacy systems every time the semantics standard changes.

Therefore, any proposal that would require the changing of all the names would be rejected out of hand without consideration of any kind. Given that there are only two possible approaches, the semantics mappings' approach has to be explored. This second approach is illustrated in Figure 14. What is different in Figure 14 is that the names of all the meta-objects can be different.

A full treatment of data semantics requires addressing:

- Subject Area, and schemas, which are policy homogeneous business segments.
- Database objects, which are well-defined unit of business policies.
- Entities and tables, which are precise sets of data within a business policy.
- Attributes and columns, which are discrete evidences of policy adherence, that is, a context-based business fact.
- Primary keys, which are the uniqueness discriminator of policy instances.
- Foreign keys, which are the relationship mechanisms between policy instances of different types.
- Data element, attribute, or column fact-based data integrity rules, which are the data-based rules that govern business fact integrity.
- Entities or table data integrity rules, which are the specific assertion data-based rules that affect a whole business policy instance.

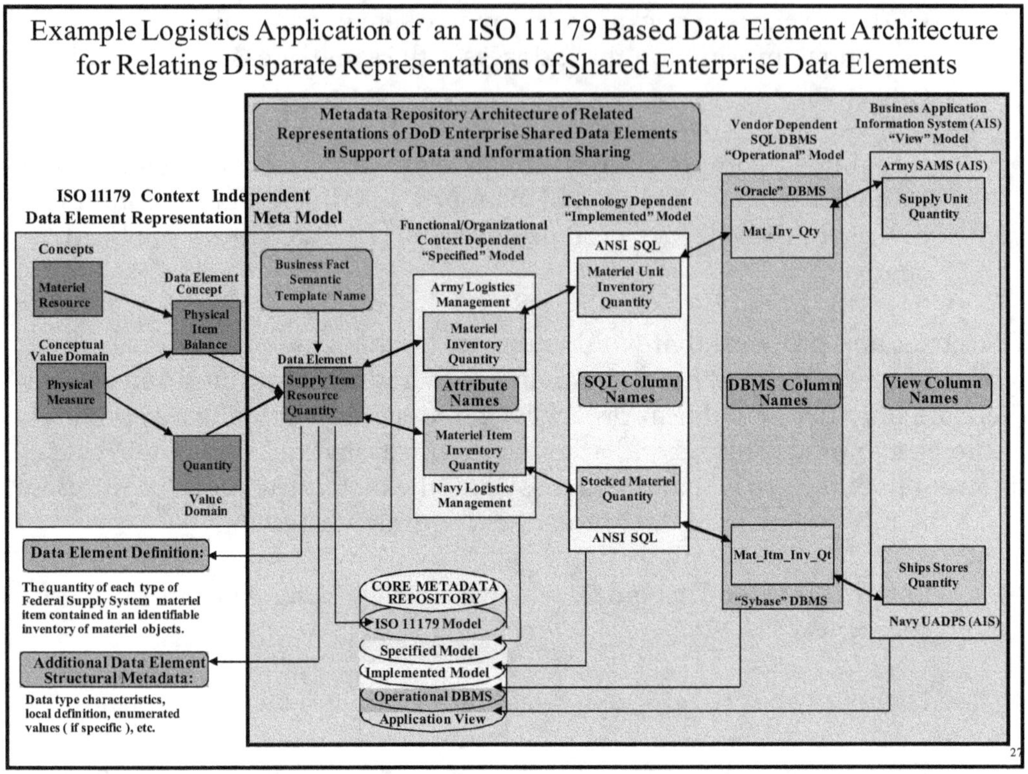

Figure 14. Differently named semantics from data elements through to view models.

- State-based data integrity rules, triggers or stored procedures, which are the business action consequences when a policy-based state is accomplished or violated.

This chapter only addresses attributes and columns because they represent the discrete evidences of contexts deployed business policy facts. Data elements, that is, business policy fact specifications are comprehensively addressed in Chapter 9, Data Elements. The complete specification of the semantics of a business policy fact (e.g., data element) consists of its:

- Semantic components.
- Policy basis description.
- Computer data type of domain of values.

- Data structure of instances.
- Length of instance.
- Required uniqueness.
- Relationships.

The semantics of a data element are most commonly used to regulate the attributes of an entity or the columns of a table. Similar uses of data element semantics govern fields within files, cells within screens, report fields on reports, and so on.

An attribute, column, or field is the semantic abstraction of discrete business fact value instances within technology dependent contexts. A data element is the technology independent, context free semantic abstraction of attributes, columns, fields, etc. There is an abstraction hierarchy between data elements and its contained semantics use within attributes, columns, fields, etc.

Each container (entity, table, file, or screen) may also have localized additional semantic conventions. For example, when the data element Person First Name's semantics are employed within an Employee table, its name might be constructed to be Employee First Name. Whenever the context/container's name is employed as a prefix for the attribute or column name, that container name becomes the contained attribute or column's name prime word. In this example, the prime word is Employee because the table name is Employee. Note that in this example, Person is not part of the column's name. That does not mean that its referenced semantic, the data element, Person First Name was changed. To the contrary, the data element's name, Person First Name is mapped to the column Employee First Name. Even though the resultant column name is different, the mapping to its data element remains.

4.3 Effect of Complex Data Architectures

When databases were application-based and narrow, the contexts and users were similarly focused. The databases commonly called application databases usually embraced only one policy area and accomplished application specific original data capture and all the necessary levels of reporting, that is, operational and management.

Database names and contained table names in these "early" databases were usually closely allied to the single policy area represented at the highest level by the database and at the lowest levels by the table. Consequently, the use of a prime word (the name of the container (that is, the name of the table, file, or screen)) was appropriate as the first part of the constructed name for a column, field, or cell.

Despite their narrow focus, each application area database might employ multiple original data capture applications from different COTS vendors that likely have different prime words notwithstanding their common narrow context or policy area. The first figure in Appendix 1 depicts the environment of complex data architectures. A description of each of the data architecture classes is provided in Appendix 1.

Beyond the original data capture data architecture class, there are two additional database architectures: subject area (sometimes called operational data store or ODS), and data warehouse. The subject area (or ODS) database is the melding of the data from the narrow application area databases that accomplished the original data capture. While the prime words in the application area databases are narrow, the prime words in the subject area databases are more broad, even though the policy area addressed by the subject area database may not have changed.

Today's databases extend well beyond original data capture and reporting. There are now distributed networks of original data capture databases and data warehouse databases that exist in two forms, wholesale and retail. Data warehouse databases were not popular until the late 1990s because the database technology that allowed databases with thousands of tables, tens of thousands of columns and billions of rows were either not possible or not practical.

Data warehouse database tables are commonly constructed from columns from many different original data capture databases and tables. Thus, data warehouses address many different policy areas. Even more distant from the original data capture databases are the retail data warehouse databases, which may contain columns from many different wholesale databases and tables.

Common across all four database architectures are references to standard enterprise data including for example, the allowed set of sex codes, country codes, telephonic country codes, postal codes, and other commonly used reference data like facility names and addresses, product classifications and even primitive product data, and finally employee directories with name, address, some biographic, and email names and addresses. These reference databases are used by databases from the four other data architecture classes.

When database first appeared in the late 1960s, each database was a stand-alone island of policy that served the needs of original data capture and all forms of reporting. Today, there are at least four additional major data architectures (original data capture (application and COTS specific), subject area databases (operational data stores (ODS)), warehouse (wholesale (broad audience)) and retail (narrow audience)), and reference. Because of these different database architectures, the old three component scheme of <prime word> <modifier>s and <classword> is too narrow, too context and technology dependent, and often obscures the repeated use of common business concepts.

4.4 Effect of Client/Server

Client/Server has enfranchised small groups of users, sometimes also called a work group, who accomplish their own database and business information systems development. From the earliest days of data processing through the early 1990s database and business information systems have all emanated from a central design and development authority. For reasons beyond the scope of this book smaller and smaller groups are now able to design, develop, implement and maintain databases and business information systems.

Without a central metadata management system to automatically enforce referential integrity between metadata and actual database and information system artifacts, there has been a tremendous rise in undocumented and semantically conflicting databases and business information systems.

This not only prevents the enterprise from sharing data, systems, and staff, it also wastes staff and other resources. A well engineered metadata environment for both data and process semantics both stops the development of conflicting database and business information systems, it also speeds their development, implementation and evolution.

4.5 Effect of Internet

Ultimately, the effect of the Internet is the same as returning to thin clients where all processes and rules operate on the servers. Since client/server environments will remain along with Internet-based environments, it's even more important that there be a comprehensive metadata management system of all data and process semantics that can ultimately be employed to either build and maintain the client/server applications and to build and maintain the server-based Internet applications.

4.6 Effect of Enterprise Data Architecture

To properly deploy database, an enterprise must have a well-designed data architecture. Michael Brackett defines a data architecture as the common context for integrating and sharing data[5]. A data architecture, based and founded on fundamental business policy transcend business style and technology trends. Key benefits that accrue from data architectures[6] include:

- Having a better understanding of business and its state because data is accepted and understood by wider audience.
- Accelerating the incorporation of new data and new uses of old-data because data sources can be quickly understood and are seen through standardized value discriminator.
- Reducing the size of data by eliminating redundant data or data that is different merely for reason of style.
- Freeing up staff time to work on real business problem areas rather than ferreting out the same data hidden under different names.
- Increasing the quality of decision making because data is valid, reliable, and represents discriminating facts about business activities.

The data about a data architecture is a class of metadata. Metadata is collected, stored, maintained, and reported through a metadata management system like the Metabase System.

---

5 Page 31 of Data Sharing, Using a Common Data Architecture. Michael Brackett. John Wiley & Sons, 1994.

6 Ibid, page 34 and 35.

4.7 Effects of Binding Evolutions

During the Summer of 2004 a senior government manager asked what the key implications were from going to "service oriented architectures." After a week of investigation, the implications of Service Oriented Architectures seem to be more of an evolution than a revolution. Essentially, a service oriented architecture is just the next step in "unbinding." A set of overheads were created and "delivered" to the senior executive. Before delivery, the overheads were reviewed with an expert in service oriented architectures. After delivery, the executive seemed satisfied.

Following that session, the overheads were put onto the Yahoo group, DM-Discuss for review. Several responded with good suggestions for modification. The road to SOA seems to have proceeded through these generations:

- 1950s-1960s: Programs and data completely self contained.
- 1960s: Programs accessing independently stored data.
- 1970s: Programs accessing data through database management systems.
- 1980s and 1990s: Programs accessing data through independently created views.
- 2000s: Programs accessing data constructed as XML streams.
- 2006 and beyond: Programs accessing data through SOA Architectures.

Th seven components involved in the evolution from being bound to being unbound are:

- Person connection to computing device for data capture and/or display.
- Computing device transfer of data to computing network.
- Data transfer between nodes of a network.
- Computer network connection to a computer.
- Accessing of data by a computer program from the network.

- Interaction between a computer program and a data storage device.
- Database management system data interaction with computer program.

Table 10 has three code values, "N," "B," and "U." "N" means that the component did not exist. "B" means that the format is required to be in a specialized format. "U" means a format that is completely transparent and independent of any interacting process.

Component (N = did not exist) (B = exists but bound) (U = exists and unbound)	**Decade**					
	1950s-1960s	**1960s**	**1970s**	**1980s - 1990s**	**2000s**	**2006 and beyond**
Person connection to computing device for data capture and/or display	B	B	B	B	B	U
Computing device transfer of data to computing network	N	B	B	B	B	U
Data transfer between nodes of a network	N	N	N	B	U	U
Computer network connection to a computer	N	N	N	B	U	U
Accessing of data by a computer program from the network	N	N	N	B	U	U
Interaction between a computer program and a data storage device	N	U	U	U	U	U
Database management system data interaction with computer program	N	N	B	U	U	U
Figure Reference	15	16	17	18	19	20

Table 10. Component existence across the decades

Ultimately what has happened over the past 50+ years is a gradual unbinding of all the relationships of components, one to the other. In the early generations, most of these components did not exist. Table 10 illustrates the existence and interconnection among all these components.

4.7.1 1950s-1960s, Programs and Data Completely Self Contained

In the beginning there were only primitive concepts of operating systems and programming languages. The environment is generally illustrated in Figure 15. Most commonly programs and data existed in a set of punch card-trays. There were only primitive tape drives that would hold data. The program's source cards would be read into some type of computer memory, compile and execution would be started. At some point in the execution of the program the data cards would be read into computer memory.

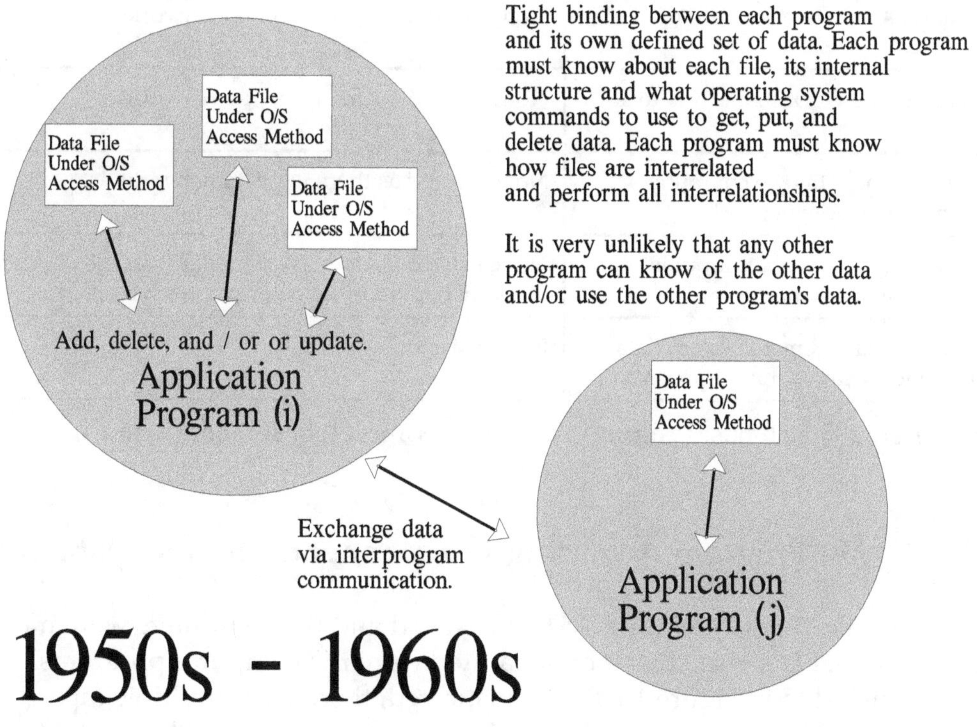

Figure 15. 1950s & 1960s: Complemented embedded computing environment.

Processes would be performed and results printed. The relationship between computer, user, and data was all 1-to-1-to-1. Any possible reuse of data involved duplicating the data card-deck and letting somebody else use it. There was no direct computer-to-user interaction of any kind. Table 11 presents an assessment of the Table 10 components for the 1950s-1960s time period.

1950s-1960s, Programs and Data Completely Self Contained	
Table 10 Component	**Assessment**
Person connection to computing device for data capture and/or display	None existed. You put your cards into a card reading device and pushed a button.
Computing device transfer of data to computing network	None existed as there were no networks.
Data transfer between nodes of a network	None existed as there were no networks.
Computer network connection to a computer	None existed as there were no networks.
Accessing of data by a computer program from the network	None existed as there were no networks.
Interaction between a computer program and a data storage device	None existed as there were no data storage devices other than memory or external punch cards.
Database management system data interaction with computer program	None existed.

Table 11. 1950s-1960s: Programs and data completely self contained

4.7.2 1960s, Program Accessing Independently Stored Data

By the middle 1960s, operating systems existed and the computers were more sophisticated. Data was able to be separated from the computer programs. This is illustrated in Figure 16. By sophisticated, there were multiple tape drives, and some computers even have very large capacity random access disk drives. Does 3 megabytes/words sound like "large capacity?" But still, the overall process was the same. Your program was on a card deck, or was

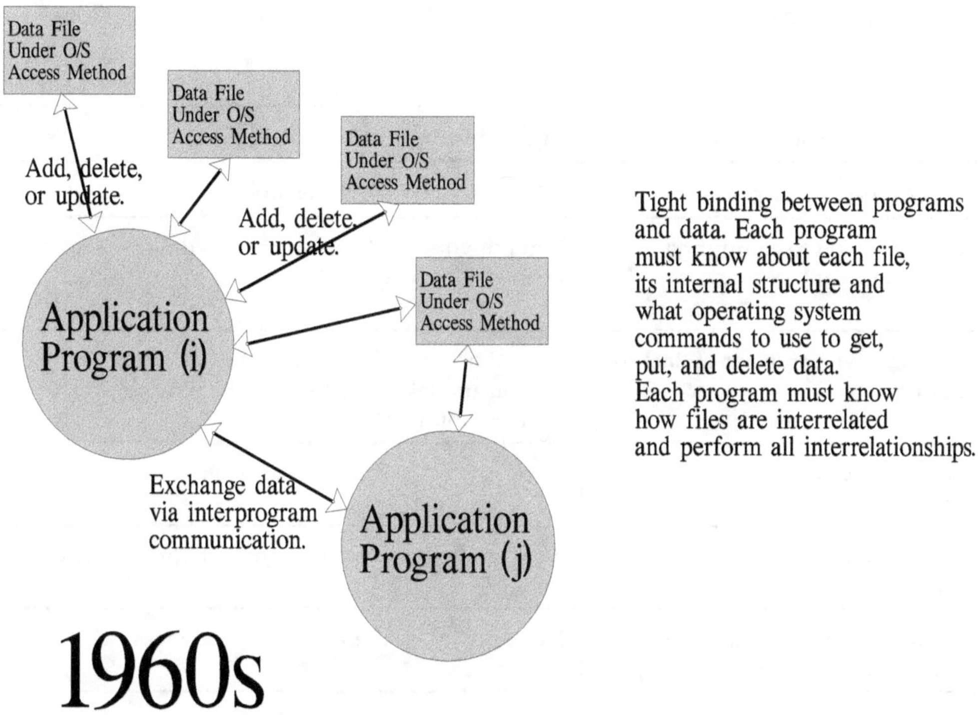

Figure 16. 1960s: Data independently stored from data.

transferred to a tape by a card reader, or in some very sophisticated organizations you could "key punch" your program directly to a tape recording device. Your data was either still on cards, or was copied to tape drives from punched cards readers. The execution sequence was the same. Read in the program, compile it, and execute the program against the stored data.

By the end of the 1960s, the IBM mainframe, "360" was common. This computer had a much more sophisticated operating system (OS 360) and could run multiple "jobs" at the same time. There were also disk drives that existed in banks of eight, for a total quantity of data of about 24 megabytes. The disk drive "cartridges" were actually a "15-18" wide stack of six platters in a sealed case. When computers programs were executing they would access data from one or more of the disk drives. "Databases" could span disk drives if they were very sophisticated.

Still, the relationship between computer, user, and data was all 1-to-1-to 1. For sure there was no direct computer-to-user interaction of any kind.

Table 12 presents an assessment of the Table 10 components for the 1950s-1960s time period.

1960s, Program Accessing Independently Stored Data	
Table 10 Component	**Assessment**
Person connection to computing device for data capture and/or display	You put your cards into a card reading device and pushed a button.
Computing device transfer of data to computing network	Time sharing services enabled multi-site organizational access to central databases and business information systems.
Data transfer between nodes of a network	None existed as there were no networks.
Computer network connection to a computer	None existed as there were no networks.
Accessing of data by a computer program from the network	None existed as there were no networks.
Interaction between a computer program and a data storage device	Computer programs could create punch cards, could read and write computer tapes. There were some very primitive random access devices such as disks or magnetically encoded cards.
Database management system data interaction with computer program	None existed that were publically available. Some existed in the United States Department of Defense and some research agencies. But all were very primitive compared to those of today.

Table 12. 1960s: Program accessing independently stored data

4.7.3 1970s, Programs Accessing Data Through Database Management Systems

In the very late 1960s, there were very early database management systems that could independently define databases and support access from multiple "on-line" users. End-user access to the data on the computers with disk drives was through teletype-like machines. Some DBMSs managed data entirely stored on tape drives. In general, the 1970s was most noted for the growth of

DBMSs that defined databases and managed all data access to and from mass storage devices. Figure 17 illustrates this environment. It was only until the late 1970s that there was any multi-user access of databases. All such end-user access was through paper-based terminals such as the Texas Instruments Silent 700. In a number of in-house mainframe organizations there was access through CRTs. However, there was never any real graphics as both the resolution and throughput was too primitive.

Still in these environments there was no concept of data interoperability among classes of users. The common data access to the database management systems was almost always through direct interfaces. IBM's IMS existed that allowed access through COBOL. COBOL represented the host language for the user's access of data managed by the DBMS. But IMS, at that point in its life, did not allow direct data access of any type of ad hoc query language. Your choice was almost always, direct access via the DBMS to the data, or access to the data via a COBOL program.

In the on-line access, the user had to both connect to the DBMS and

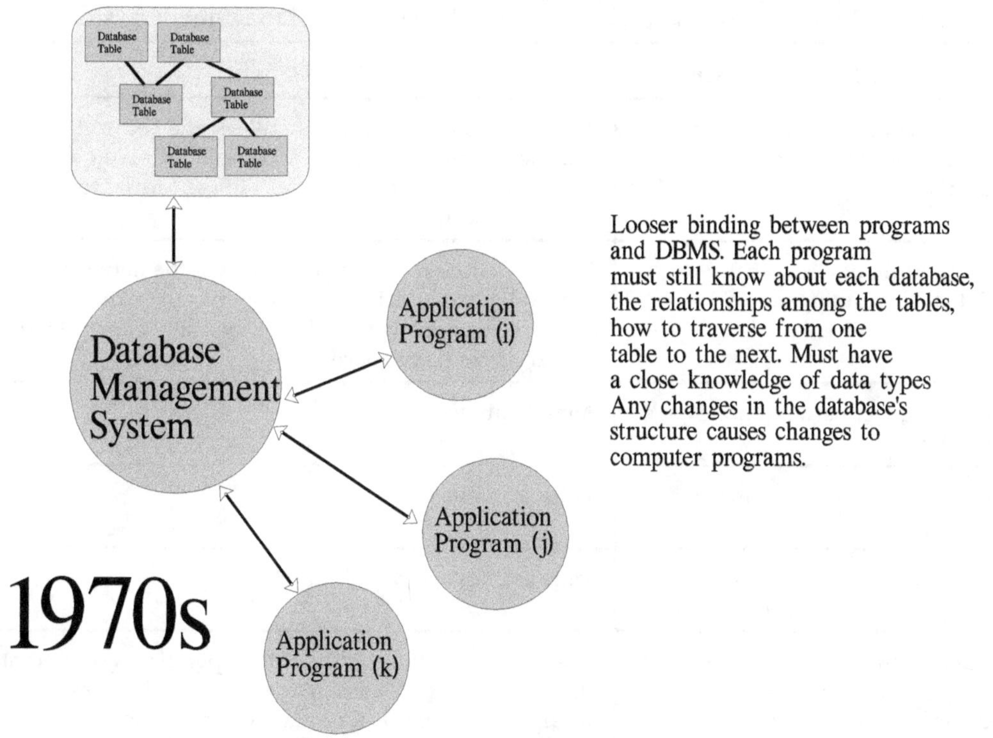

Figure 17. 1970s: Data access through database management systems.

also to the database. There was no way that the user could access more than one database or transfer data from the database via the DBMS to another database. In host language access, the design of the database had to be bound into the computer program. Any changes in the database design required recompiling all computer programs that accessed the database. The database's design had to be incorporated into the application program because database traversal required special commands that climbed a hierarchy, went from one record to the next, selected sets of records, and the like. Any changes in the database's design could have significant effects on the application programs.

The sophistication of query languages and report writers also grew. This enabled users to be more loosely tied to the DBMSs. But query languages or report writer access of multiple databases at the same time was not possible. The benefit to the program was that the program did not have to contain all the database access logic. Table 13 presents an assessment of the Table 10 components for the 1970s DBMS time period.

1970s, Programs Accessing Data Through Database Management Systems	
Table 10 Component	**Assessment**
Person connection to computing device for data capture and/or display	You were able to log into a network and connect to a specific computer program and/or DBMS. You were required to interact with respect to the format for that computer program.
Computing device transfer of data to computing network	This was entirely under the control of the network connections to which you were connected. All networks were private and very specially constructed for very specific purposes.
Data transfer between nodes of a network	Same as above
Computer network connection to a computer	Same as above
Accessing of data by a computer program from the network	Same as above
Interaction between a computer program and a data storage device	This was under the control of the operating system that was supporting the computer program. Libraries and library organizations for data and programs were organized by special languages.

1970s, Programs Accessing Data Through Database Management Systems	
Table 10 Component	**Assessment**
Database management system data interaction with computer program	These were growing in sophistication. Most mainframe DBMSs were of the Network, Hierarchical, and Independent Logical File data models. User interaction with these DBMSs were through carefully crafted forms, screens, and data formats. Each format was very tightly bound to the one specific type of use.

Table 13. 1970s: Programs accessing data through database management systems

4.7.4 1980s and 1990s, Programs Accessing Data Through Views

The key advances during the 1980s and 1990s were the advance in networks, the growth of the SQL language, and views. A view enabled greater separation between the database's design and the application program. Views enabled a very significant part of the database traversal and record selection to be performed by the DBMS before the records were presented to the application program. In SQL:1986 through SQL:1992 DBMSs, a view was a "flat" record that was created within the DBMSs language and included selection and navigation logic that would result in a set of rows. This is illustrated in Figure 18.

Also in the early 1980s, the PC and the mini-computer were born. The mini-computers were very small main-frames and were supported by networks that were either privately owned or leased through telephone companies through "leased lines." Computers, other than for message switching were still all single processing entities. PCs, originally made popular by Apple and Radio Shack were either completely stand alone devices or were employed to capture and construct data into message formats. PCs were able to create and transmit files to networks, and/or to receive and print files to PC controlled printers. When connected to a network the PCs cloned "dumb terminals."

In the middle 1980s, some DBMSs started to appear on PCs. A PC DBMS early one was Information Builder's FOCUS. It was a miniaturized version of their mainframe DBMS.

Also during the 1980s, the great "DBMS Data Model War" was fought. The winner, the relational data model, was heralded by academics and IBM as

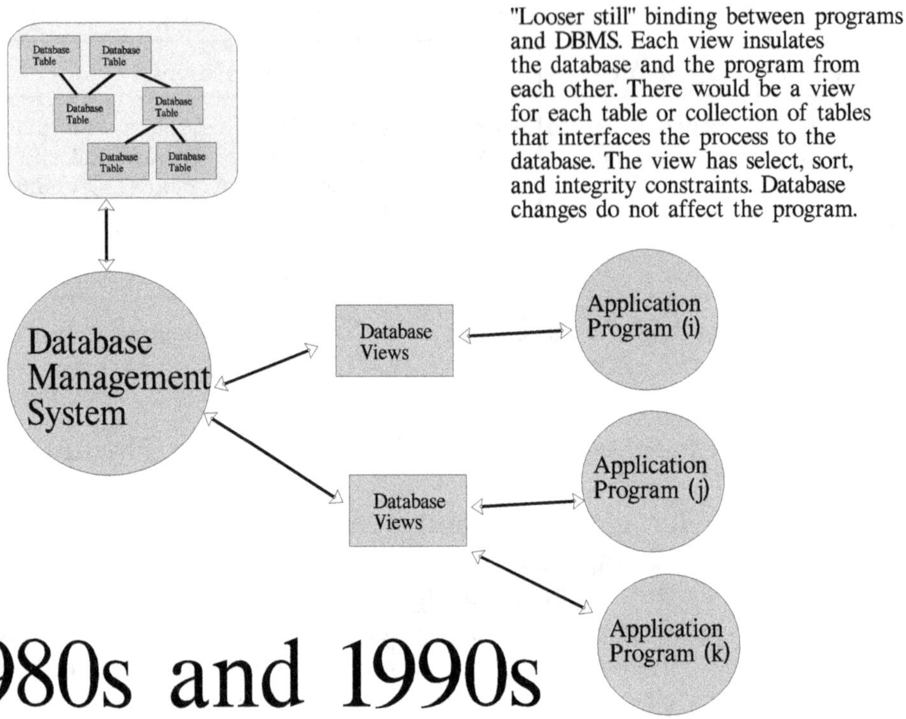

Figure 18. 1980s & 1990s: End user interaction with databases through DBMSs via database views.

the silver bullet for all data management. During the 1980s, the first American National Standards appeared for database languages: Network (NDL) and Relational (SQL). Most of the traditional DBMSs started to fade and were replaced by SQL-based DBMSs. However, many very large production databases such as those required for mortgage processing, insurance claim processing, and those of the DoD Intelligence Community were unable to use SQL based DBMSs because of unacceptable performance. An undesirable side effect of SQL-based DBMSs is that most responsibilities for data selection, navigation, and integrity control was placed squarely on the shoulders of the end-user. This represented a significant step backwards in existing functional capability. The end-user data selection, navigation, and integrity control could, however, be transferred to the server through sophisticated views.

It was not until the late 1990s (i.e., SQL:1999) that SQL based DBMSs, first via standards, started to recover a number of the features and facilities of DBMSs that were the casualties of the DBMS Wars. Recovered mainly were complex data structures and embedded processes that could all be defined and executed on the server side. These new features were collectively known as the SQL Object Oriented features.

Also during the late 1990s, there was a great diversion of computing resources to fix the Y2K problem. It was fixed and life continued. One amusing site, however, was that of the United States Official Clock at the U.S. Naval Observatory. At the stroke of midnight, the website's year was transformed from 1999 to 19100. Oh well.

In all these cases, data formats were prescribed by the computer program (DBMS-based or application system) and were tightly bound. Any end-user variation caused program failure or error. Table 14 presents an assessment of the Table 10 components for the 1980s-1990s time period.

1980s and 1990s, Programs Accessing Data Through Independently Created Views	
Table 10 Component	**Assessment**
Person connection to computing device for data capture and/or display	Either through "green screen" terminals through a specially created software such as IBM's CICS, or through special screen formats designed into PC based computer programs.
Computing device transfer of data to computing network	In the case of CICS terminals the data was directly sent to the receiving CICS programs that were in turn employed by the application systems commonly written in COBOL. In the case of PC captured data, it was formatted into tightly constructed and formatted data streams as required by the application system that processed the data.
Data transfer between nodes of a network	Completely under the control of the proprietary network.
Computer network connection to a computer	Completely under the control of the proprietary network.
Accessing of data by a computer program from the network	Completely under the control of the proprietary network.
Interaction between a computer program and a data storage device	This was under the complete control of the computer's operating system. There was less and less user direction and control over this.

1980s and 1990s, Programs Accessing Data Through Independently Created Views	
Table 10 Component	**Assessment**
Database management system data interaction with computer program	With the existence of the view, more of the selection, navigation and integrity control logic was placed under the control of the DBMSs. Early SQL-DBMSs were lacking in this area and it was not until the early 1990s that some of these problems were solved.

Table 14. 1980s and 1990s: Programs accessing data through independently created views

4.7.5 2000s, Programs Accessing Data Constructed as XML Streams

During the first half of this decade, the emergence of XML as the next silver bullet of all data management problems, arose. Essentially, XML became a standard strategy for embedding metadata with data. XML grew out of SGML, the Standard Generalized Markup Language. This is not a new language. "Conceived notionally in the 1960s - 1970s, the Standard Generalized Markup Language (SGML, ISO 8879:1986) gave birth to a profile/subset called the Extensible Markup Language (XML), published as a W3C Recommendation in 1998." (http://xml.coverpages.org/sgml.html). This is illustrated in Figure 19.

Essentially this is a tag and value language. Key to the usability of XML is the creation of standardized names and definitions of the tags. The tags are metadata. XML moved the issue of the standardized metadata, that exists inherently in data models, into the process of defining XML-based transaction data that is to be created and shipped.

The key value of XML is that it moved the binding of a data's transaction directly into the data itself so that a using process that is quite different from the creating process could know of and decode the data transaction. Not solved, however is whether there is any agreement on what the tag means. For example, one organization might have the value 1 for the male gender and another had 0 for the male gender. Another organization might have Sales defined as net after all taxes and expenses, and another organization has sales defined as gross receipts. These issues, just as before, remain.

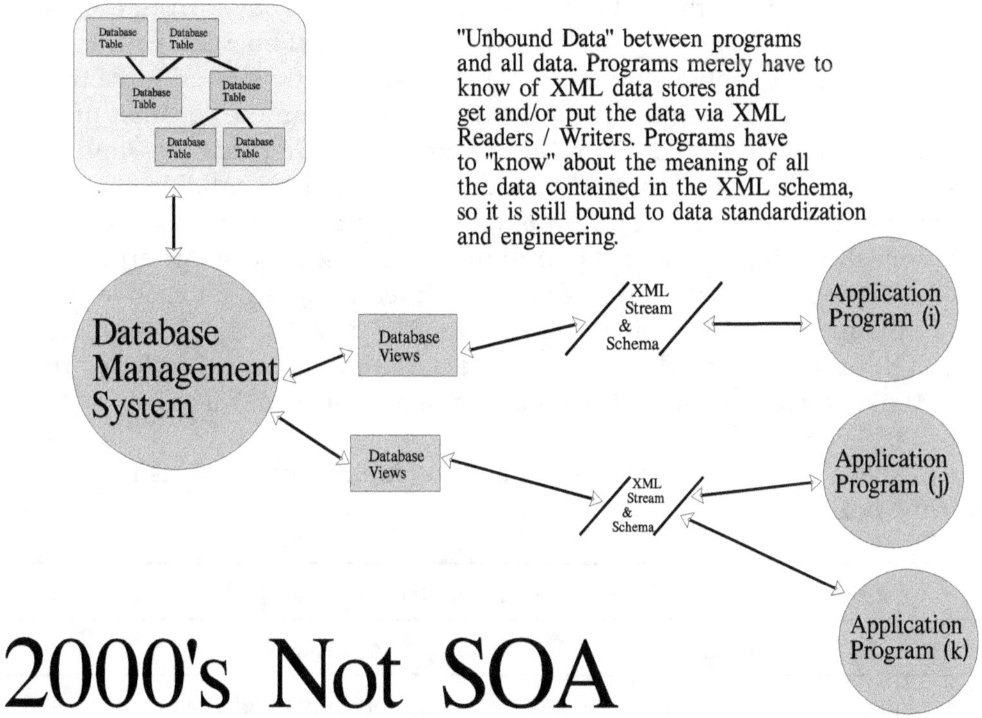

Figure 19. 2000s Not SOA: Programs accessing data constructed as XML streams

The "good news" of XML is that it enables the definition of data exchange specifications in an unbound format. Thus, the data exchange is unbound from either the composing process (and organization) and unbound from the shredding process (and organization). The "bad news" is that all these organizations must generally agree on the most minute detail of the semantics of the composing and shredding processes. Still, what is new? In short, XML is not a silver bullet. Rather, it is a silver bullet opportunity. The key to the velocity and effectiveness of the XML silver bullet is how well, how effectively, and how non-redundantly the XML tags are all discovered and defined. If identified and defined at the transaction level, there will be excessive, wasted, and likely conflicting results. The semantics employed by XML must be defined within the highest levels of metadata that are possible, that is, enterprise-level data elements, and enterprise-level data model templates.

A key issue with the unbinding of the application layer from the data-processing layer (through the DBMS) is the lack of a real business-rule execution interaction between the two. That consequently should force the maximum quantity of business rules to be defined within the confines of the DBMS. That is, within stored procedures, Before and After Triggers, and all manners of constraints both on tables and View. That is essential because there can be, and are, many different language agents that could construct transactions. If business rules are bound into each language agent program unity, the business rules would have to be defined uniquely for each language agent. Both are impossible to keep consistent and are a waste of resources.

SQL vendors, e.g., Microsoft, Oracle, and IBM are now implementing SQL-XML facilities that enable the mapping of XML-based transactions to database table columns.

Table 15 presents an assessment of the Table 9 components for the 2000s time period.

2000s, Programs Accessing Data Constructed as XML Streams	
Table 10 Component	**Assessment**
Person connection to computing device for data capture and/or display	There is now the opportunity to compose data transactions into XML-based formats. This requires recreation of all transaction creation processes. This also requires the movement of business rules to the DBMS layer for consistent execution. If accomplished, a given transaction can be sent to multiple transaction shredding processes not just to one.
Computing device transfer of data to computing network	This can now be accomplished by the Internet by ensuring that the sending transaction knows the URL of the receiving process. Thus, while the format of the transaction is unbound, the "location" of where the transaction will be sent and processed is still somewhat bound.
Data transfer between nodes of a network	Automatically handled by the network.
Computer network connection to a computer	Automatically handled by the network.
Accessing of data by a computer program from the network	This can now be handled by the receiving program if it knows the XML schema of the received transaction. What also must be known for correct processing must

2000s, Programs Accessing Data Constructed as XML Streams	
Table 10 Component	**Assessment**
	also be the semantics of all the tags and the correct valuation (especially in the case of code values) of transaction element values.
Interaction between a computer program and a data storage device	This is entirely handled by the operating system
Database management system data interaction with computer program	This is generally bound because there has to be some declared connection to the DBMS through, for example, Microsoft's ODBC. Other operating system environments have the same class of facility.

Table 15. 2000s: Programs accessing data constructed as XML streams.

4.7.6 2006 and Beyond, Programs Accessing Data Through SOA Architectures

Since this evolution generation is largely about the future and about architectures yet to be accomplished, by institutions such as the U.S. Department of Defense, the descriptions herein are at best conjectures of what is likely. The key point here is that neither silver bullets, nor making "real" that which "does not exist" is possible. The main component that changes in this list is the networking components. The database management systems standards (e.g., ISO/ANSI SQL:2008) all have the necessary contained facilities to accomplish robust server-side processing. What is missing is the creation of the full architecture for a Service Oriented Architecture environment and the recognition that enterprises must get serious about the definitions of transactions. This is illustrated in Figure 20.

Transactions are going to require an extension to traditional metadata management systems to hold not just enterprise architecture, data model, process model, and business information systems metadata but also transaction metadata and network processing metadata. Unless and until all this metadata is captured, integrated and made non-redundant, Service Oriented Architectures will likely be just a far off dream.

In actuality, all these metadata artifacts have to be discovered, collected, loaded and maintained in any event in order for Internet-based

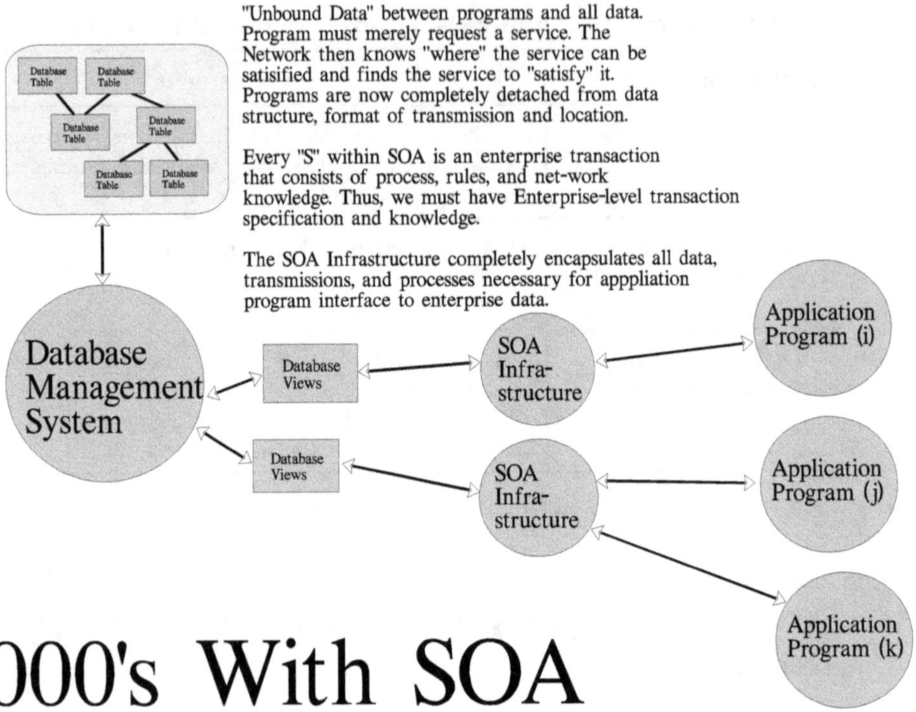

Figure 20. 2000s with SOA: Programs accessing data through SOA architectures

client-server environments to be effective. This is being done today for specific applications and projects. The promise of SOA is that this can be done for the entire enterprise. Table 16 presents an assessment of the Table 10 components for the 2006 and beyond time period.

2006 and Beyond, Programs Accessing Data Through SOA Architectures	
Table 10 Component	**Assessment**
Person connection to computing device for data capture and/or display	This would now be accomplished by a service oriented architecture. What would have to happen is that the transaction would have to include additional information about itself so that the SOA environment could determine what kind of transaction it is, and "where" it can be processed. The transaction composing process would have to encode that information into the

2006 and Beyond, Programs Accessing Data Through SOA Architectures	
Table 10 Component	**Assessment**
	transaction before it is handed off to the Internet. This is different from the table in Section 5 where the transaction would have to know the receiving URL of the application that can process it.
Computing device transfer of data to computing network	Automatically handled by the network.
Data transfer between nodes of a network	Automatically handled by the network. The SOA environment would know how and where to route the transaction for processing. The transaction composing process would have to somehow have enough information to encode in the transaction header to at least get the transaction to the next higher level. Thereafter that layer would possibly have the necessary information to push the transaction onto its next location.
Computer network connection to a computer	The receiving program would have to be discovered by the SOA environment through the information contained in the original transaction. Assuming that the entire transaction is encoded in XML, the header part of the transaction would have to be decoded and processed by some SOA network routing software. Ultimately the receiving program would have to retain sufficient information about the originating location to be able to reroute the answer back.
Accessing of data by a computer program from the network	This is entirely handled by the operating system
Interaction between a computer program and a data storage device	This is entirely handled by the operating system

2006 and Beyond, Programs Accessing Data Through SOA Architectures	
Table 10 Component	**Assessment**
Database management system data interaction with computer program	This is generally bound because there has to be some declared connection to the DBMS through, for example, Microsoft's ODBC. Other operating system environments have the same class of facility.

Table 16. 2000s: Programs accessing data through SOA architectures

4.8 Error Classes

There are two critical elements in data semantics management: domain of interoperability and solution engineering. There is no need to create interoperable data beyond the natural limits of data sharing. There will be not positive ROI. The first error class must be described and enumerated by examples across the boundaries of shared data, which, of course, means shared policy. All data semantics management efforts must therefore focus on these areas first. Within these areas, the shared-data-interfaces need to be examined and their data domains identified and reconciled across their Communities of Interest.

Thereafter, data semantics management efforts need to move outwards, if possible, to the original data capture systems for those shared-data domains. Amend those if possible. Thereafter, proceed inwards. In this way, as the effort proceeds from outside-in, focusing on only shared-data, the two error classes will be designed out of existence.

As shown in the prior section on errors, it is essential to have proper engineering.

The very first lines of defense are of course the operational databases and information systems represented by the Operational Data Models of Chapter 13. Those, if defined in isolation within these models just form islands or stove pipes of semantic consistency. While these are valuable, what is even more valuable are larger areas that would be guided by higher semantic levels. That is, the Implemented Data Models of Chapter 11, and backing those up would be the Specified Data Models of Chapter 10 and finally backing these up would be the Data Element Models set out in Chapter 9.

If these models are implemented, not only are individual database and business information systems more semantically acceptable, but so too are all other related databases and business information systems because they are drawn from common higher semantic levels.

These collective data semantics will be known to all participants and for the data that is shared, the two error classes will largely be designed out. For data that becomes shared in the future, all the shared data semantics will already exist.

4.9 Engineering Data Semantics Summary

This chapter asserts that data semantics management efforts cannot just be slapped together and set into motion because of all the legacy databases and business information systems that have become the tens-of-thousands of data and process stove-pipes in our enterprises.

Many business facts are the same but are named differently, and many business facts are different but are named the same. In addition, many definitions are different when they should be the name, and vice versa. Finally, there are differences related to granularity, precision, and temporal aspects. All this must be addressed and resolved in some satisfactory way before the road to data interoperability can be traversed. Simply put, the right environment must exist.

The ideal vehicles for traveling this road are the Communities of Interest, which are specifically engineered for data interoperability. The Whitemarsh book, Data Interoperability Community of Interest Handbook sets all this out.

Ironically, even if all the factors were completely known 20 to 30 years ago, enterprise-wide interoperable data environments were not possible due to immature technologies.

A key set of material in this chapter was the exposition of the evolution of bindings that existed between data, programs, database management systems, and users. There has been a long and gradual trek in the evolution of bindings from the 1950s through to the present day that has enabled, for example, service oriented architectures. It is critical to know, however that success in service oriented architectures requires careful attention and deployment of the fundamentals of this book. Disregarding the dictums, strategies, and products set out in this book is to court failure.

The very first line of defense is of course the operational databases and information systems that are represented by the Operational Data Models of Chapter 13. Those if defined in isolation form islands or stove pipes of semantic consistency. While these are valuable, what is even more valuable are broader areas across whole business functions and organizations. These would be guided by the higher levels of semantic specifications and would exist within the Implemented Data Models of Chapter 11. Backing those up would be the Specified Data Models of Chapter 10 and finally the Data Element Models set out in Chapter 9. If these models are implemented, not only will individual databases and business information systems be more semantically acceptable, but so will all other related databases and business information systems because they are drawn from common higher semantic levels.

The next chapter, Chapter 5, Semantics of Names, squarely addresses one of the key reasons for data semantics management failure: Names. The point to Chapter 5 is to lay out a formal strategy and infrastructure for the construction of names that does not fall apart when subjected to the slightest of stress.

4.10 Questions and Exercises

1. Table 8 sets out data semantics components in the first column and the difference as to where they exist in specifications and implementations. Are there any semantics components missing? Where should that missing item be specified and implemented?

2. What is the likely effect on enterprise semantics management when purchasing an application package? What strategies do you currently have to accomplish data interoperability when packages are involved?

3. How difficult would it be to implement the data semantics naming strategy shown in Figure 13 as compared to Figure 14? What would be the advantages and disadvantages of each?

4. If Figure 13 from Section 4.2 was declared as the only allowed approach how much would it cost and how long would it take in your organization? What are the cost and time elements? How would you

estimate them? What infrastructure would have to be put into place to carry out this approach?

5. If Figure 14 from Section 4.2 was declared as the only allowed approach how much would it cost and how long would it take in your organization? What are the cost and time elements? How would you estimate them? What infrastructure would have to be put into place to carry out this approach?

6. Sections 4.3 through 4.6 describe the effects resulting from different factors. How have these been dealt within your organization? What steps have you taken to address the different naming strategies that may be imposed by complex data architectures, different data architectures, client-server, internet, and an enterprise data architecture.

7. How does the creation of data architecture classes as set out in Appendix 1 affect the ability to have consistent semantics? What problems have arisen in your organization? How have you dealt with the data semantics across all these data architecture classes? What have you engineered or built? Compare and contrast the efforts and costs associated with having an over-arching data semantics management scheme across these data architecture classes versus ad hoc solutions. Which are you? Which one should you be and why? What would have been your costs and savings?

8. What has been the effect of client-server, where there are many different clients "talking" through ODBC through server-based databases and other systems? Have you created a data semantics management approach across these databases and systems? If yes, what was the costs and benefits? If not, what do you expect were your extra efforts and costs?

9. What has been the effect of Internet-based access between end-user based systems and server-based databases and systems? Has your sophistication gone up or down? Why? What do you do when you need to have complex processing on the end-user side of an Internet-based environment? How have you dealt with the asynchronous precessing challenges? What are the major effects on database and

business information system architecture engineering and implementation as you move from being client-server to being internet-based? What have you gained? What have you lost? How has this affected data semantics management?

10. Has your enterprise attempted an enterprise data architecture? Was it successful? If yes, why? How did the approach compare to the various data semantics models of this book? That is, from Data Element Models down through View models. Did you attempt a Figure 13 approach or a Figure 14 approach? What have been the benefits? If an enterprise data architecture was not successful, why not? What has been the costs?

11. Section 4.7 and its subsections identify and describe the different "generations" of bindings. How have these generations affected the possibility of data interoperability? As each new generation occurred, what type of data semantics management became possible? What strategies could be put into place to accomplish this new level of data interoperability?

12. What are the unbinding factors that surfaced in each of these generations? How did each of these unbinding factors affect the architecture, engineering, implementation, and maintenance of database and business information systems? How did these generations affect data interoperability within a project, function or organization?

13. What benefit would have occurred if additional levels of data interoperability could have been put into place when an unbinding generation occurred? Where would the benefits have accrued? What would the effect have been on the next level of unbinding that would have occurred? Would it have come faster? Seen more clearly? Cost less?

14. The Error Classes section outlines the types of errors addressed by this chapter's materials. Do you agree that the error classes are addressed? How? What could be different? Has this chapter's contents and strategy provided an overall justification for its implementation?

5

Semantics of Names

The last chapter, Engineering Data Semantics, was a preparatory chapter. It made the point that data semantics management efforts cannot just be slapped together and set into motion. That is because the right environment must exist first. The chapter showed that even if all the factors were completely known 20 to 30 years ago, enterprise-wide data interoperable environments were not possible due to immature technologies. Nevertheless, the chapter also showed that if organizations had already properly accomplished best-practice data semantics management, every thing would have just fallen into place when the technologies matured. As evidence of having "best-practice" accomplished when it was needed the most, one division of a world-wide corporation had virtually a zero Y2K bill while another division of the same corporation spent many millions repairing and re-deploying their ad hoc databases and business information system implementations.

A key set of materials from the last chapter set out the evolution of bindings that existed between data, programs, database management systems, and users. There has been a long and gradual trek in the evolution of bindings from the 1950s through to the present day that now enable service oriented architectures. If "best-practice" data semantics management had been done at every generation, not only would IT resources have been saved on existing work, but the work in the next generation would be easier, cheaper, and more effective.

This chapter, Chapter 5, Semantics of Names, squarely addresses one of the sources for data semantics management failure: Names. Take as an example, Social Security Number. According to the traditional naming strategy, that is, a triple of <prime word>, <modifer>s, and <class word>, Social is a prime word, Number is a class word, and therefore Security, the modifier, is the "real part." If the true meaning of a Social Security Number is not instantly conveyed by the string, Security, the point has been made.

In addition, Number, in this case, is a lousy choice for a class word. What is the meaning of an average Social Security Number. The minimum or maximum? What are those dashes, subtraction signs? Finally, Social Security Numbers are often used as identifiers. Thus, the real class word should be Identifier. That would be OK except that a Social Security Number is really a set of three codes. Now, onto Social. If there are so many law firms advertising their services to "fight" the Social Security Administration, they must not be very "Social." As far as security is concerned, the Social Security Trust Funds, which have never really existed, are rushing headlong into insolvency. So much for Security.

Similar examples can be made for Reservation Numbers that seldom contains only numbers or even any numbers at all. Telephone numbers are another example. Are they really the serial number of the telephone instruments? Of course not. The point to all this is that there is a general tendency to overload the semantics and value of names.

There also is an unyielding demand for single definitions for given name strings. At one conference during the early 1990s, the keynote speaker was questioned about the meaning for a given name string. The questioner said that a different definition had been used by members of their discipline for only the previous 2000 years. The speaker immediately responded by stating that they would just have to change. Oh yeah, right. The goal of this chapter, therefore, is to lay out a formal strategy and infrastructure for the construction of names that does not fall apart under even the slightest of scrutiny.

The apex of the business fact hierarchy is the data element. It is the source of semantics for attributes of entities, which, in turn, are the source of semantics for columns of tables, which, in turn are the source of semantics for DBMS columns, and finally, DBMS columns are the source of semantics for view columns. Because of this natural semantic hierarchy, it is critical to get the complete semantics that lead to the engineering of data elements and the construction of data element names straight. In that regard, there have been three long-standing traditions that have greatly diminished the usefulness of the data element:

- Nobody knows, for certain, what a data element is.
- Binding all the data element's semantics into its name.
- Misuse of the semantic components that define the data element.

Sections 2.2.3 through 2.2.8 enumerate the key models of the data semantics solution space. That is:

- Data elements.
- Specified Data Models.
- Implemented Data Models.
- Operational Data Models.
- View Data Models.

These five models are depicted in Figure 3. Tables 2 through 7 identified and described the components within each model. A brief review shows that the data element exists in only one of these models. Therefore, and definitively, a data element is a context independent business fact semantic template. As such it does not exist at the same level of abstraction as attribute, nor column, nor DBMS column. The failures of not placing the data element at the appropriate abstraction level was set out in Chapter 3. This book places the data element at the most productive, define-once, use many-times level.

The second issue above, binding all data element semantics into its name is best illustrated with an example. Suppose the business fact is about an annual sales amount for a particular region of the country. Should the name be:

- Total New England Annual Sales Amount US Dollars.
- Annual New England Sales US Dollars Amount Total.
- Annual US Dollar Sales Total Amount New England.

Are all there of these the same fact? How would you know? What is the core fact? Sales? Total Sales? Where does New England fit here? The point to all of this is that if all these different semantics parts are blended into the fact's name, how do you pick them apart with any rigor? To have consistent semantics that can be constructed in a reliable, repeatable manner the name construction problem has to be addressed in a serious way.

The third issue, misuse of the semantic components that define the data element generally refers to modifiers and class words that prefix and suffixes the core business fact. The most common approach is to have a relatively unlimited set of modifiers (prefix words), and only one class word (suffix word). If the modifiers are drawn from a single pick list, the context of the modifier is lost. The same modifier may exist in different contexts. For example, consider the modifier, Final. Would its definition be the same if it were drawn from money, or from time, from a set of scholastic steps, or the last possible answer? If the same word is used in different contexts, should not all the different contexts remain explorable once the term is used? If there is to be any real progress toward rich and effective semantics, all three of these problems must be addressed.

Chapter 9 focuses exclusively on the first point above, that is, on the nature, scope, engineering, and maintenance of data elements. Chapter 6 focuses on the second and third points, that is, the proper creation and deployment of the semantics associated with business facts and with name construction.

This chapter focuses on the overall problem of creating semantics across all the data model levels that involve the deployment of business facts. That is, data elements within the Data Element Model, attributes within the Specified Data Model, and columns within the Implemented Data Model. Because there is now a comprehensive multi-layer solution, the problems listed in Chapter 3, and the start of this chapter are eliminated.

5.1 Traditional Approach Problems

As stated at the start of this chapter, the traditional approach to constructing names is to bring together the three name parts: prime word, modifier, and class word. Most commonly, the prime word is the entity or table name. The modifier represents an aspect of the entity or table to be represented as the business fact. The class word is an indication of the business fact's value fundamental type. If the entity is Customer, and the property of the customer being represented is Name, and if the type of data is in the form of text, the full name might be Customer Name Text.

This three-part name strategy works OK if there is only a need for single words, and if what is being represented is simple. Data elements, however, are different from attributes of entities or columns of tables. That is because data elements are not encased within a container such as an entity or table. Data elements, in contrast are independent and stand-alone business fact without the encumbrances of any file, entity, or table container. Data elements are thus the ideal mechanisms for engineering enterprise-wide semantics. Because of this, the five main mistaken notions associated with the traditional three part paradigm that do not apply to data elements are that:

- Data elements are synonyms for table columns, screen cells, entity attributes, or report fields.

- Data elements do not have names.

- Prime word, modifer[s] and the class word *are* an intrinsic part of the data element's name.
- Modifiers are from a single homogeneous set.
- There is only a choice of one class word.

The sections that follow address each of these mistaken notions. It is critical to get data elements right at the outset, or alternatively, to iteratively get data elements right given that they are discovered through bottom-up analysis of production class database schemas.

5.1.1 Data Elements Are Synonyms for Table Columns, Screen Cells, Entity Attributes, or Report Fields

Whenever a relational table is reviewed, it is commonly said, "The data elements are: . . . " Ditto when you review a screen, an entity or a report. So, the problem is the inference that the data element is a synonym for a table's column, a screen's cell, an entity's attributes or a report's field. In actuality, a data element is the metadata representation of the semantics for the column, for the cell, for the attribute or for the field. The data element is thus not a synonym for each. Rather, it is a short hand reference for the common and abstracted metadata specification of the column, attribute, etc. That is, a short hand for the inherited semantics.

For example, when you say, "Where are the First Names," you are likely to be answered, in the Person table, or the Biographic screen, the Employee entity, or on the Personnel report. So, if they all *contain* First Name, how can *Person First Name* and *Biographic First Name, Employee First Name,* and *Personnel Report First Name be* the same? Clearly, they cannot. Rather, they all possess "containerized" representations of the First Name's semantics. Hence, data elements are not that which is contained. Instead, they are abstracted common metadata representations drawn from the contained metadata representations of business fact instances.

When you take this "giant leap of common sense," you are immediately freed from the almost impossible, thankless, and worthless job of trying to identify and define every column, cell, or field that has or will ever exist. Because, if you believe that data element is the synonym for columns, cells, or fields, the effort to define these four is exactly that: impossible, thankless, and worthless. If you believe otherwise, you have to totally discount the confession of a senior U.S. Defense Department official who admitted in 1994 to having wasted $55 million on just that effort! That amount was just for the data standardization agency. For the DoD as a whole, it was, in 1994, about $200 million. In essence this U.S. Defense Department agency built a huge landfill of discordant, unintegrated semantics of columns. After 12 years of this completely fruitless work and the total expenditure of over $500 million, the program was cancelled in 2002. The agency's reaction to this colossal failure was not to understand the reason for failure and fix it, but to completely discard the entire set of work products and adopt the complete inverse solution: That is, change from a lock-step, highly structured data standardization process to an "everybody do your own thing" approach. From impossibility to complete anarchy. That is an improvement?

The real solution lies in a very simple change coupled with an elegant data standardization architecture that has as its apex the well-regarded ISO Standard, 11179 for data element metadata. The simple change is this: a data element is the context independent semantic representations of business facts. Data element semantics are thus deployed within many different contexts, e.g., tables, screens, entities, and reports. Data elements "contain," while columns, et al are "contained." Given this REAL definition, the effort to define data elements is finite, manageable, predictable, and cost effective. In general, across an enterprise-wide sea of columns, cells, attributes, and fields, the use-ratio is at least 30 to 1. The broader the area, the larger the ratio. With this new definition, the overall definition effort is cut by 95%, or more.

The concept of prime word is appropriate for columns, cells, attributes, and fields but is inappropriate for data elements. That is because prime word represents the name of the container. That is, the name of the table, screen, entity, or report. To have prime word as part of the contained component's name is to state categorically that the component must be a column, cell, attribute, or field.

What is appropriate for a data element, however is the identification of the business or subject area. This is necessary when we think for example of the many different disciplines that commonly use the data element, *vehicle*, or the data element, *track*. These are valid homonyms and must be disambiguated. Table 17 clearly sets out what metadata is appropriate for a data element and what is appropriate for a column, et al.

Meta attributes for data elements and Context Dependent Business Facts		
Meta attribute or Semantic Part	**Data element**	**Entity Attribute, Table Column, File Field, Screen Cell, etc.**
Business Domain	Yes	No
Common Business Name	Yes	Yes
Prime word	No	Yes
Modifier subclasses	Not recommended	Yes
Class word subclasses	Not recommended	Yes
Generated Policy Basis Description	Yes	Yes
Computer Data Type	No	Yes
Data Structure	No	Yes
Required Uniqueness	No	Yes
Relationship Function	No	Yes

Table 17. Meta attributes appropriate for data elements and contained columns, et al.

5.1.2 Data Elements Do not Have Names

The second problem is the belief that data elements do not have names! In fact, they do. Each has a *common business name.* If however you accept the "traditional" construct for a data element, i.e., *Prime word + Modifer[s] + Class word,* you MUST believe that data elements do not have names. The reason you MUST believe they do not is because under the traditional approach, a data element's name is said to be a concatenated representation of precisely chosen word strings, each of which is chosen to represent a specific part of the overall set of semantics.

Given you believe that to be true, how do you explain, *Employee Social Security Number*? If you examine each word, *Employee* must be the prime word, *Number* must be the class word, and *Social* and *Security* must be two distinct and "nested" modifiers, each of which conveys semantic meaning. One thing you know for sure is that those characterizations are untrue. At the very outset, you must distinguish between the column, *Employee Social Security Number* and the data element, *Social Security Number*. This same forced, delusional analysis would also apply to *Telephone Numbers* and *Part Numbers* where *Telephone* and *Part* are prime words and *Number* is the class word.

So, either you proudly accept that exhibitions of forced, delusional analysis are something to brag about, or you must accept that opposite: that data elements have names that are NOT necessarily characterizations of their semantics. Rather, they are their commonly known names, such as *Social Security Number* (which is a three-part numeric code separated by hyphens) or *Part Number* (that commonly contains multiple embedded codes, some of which contain letters), or *Telephone Number* (which, like Social Security Number, is a multi-part code, but has either hyphens, a set of parentheses and hyphens, or under the new ISO scheme is a four-part numeric sequence separated by periods). How compound and derived data elements are treated is another issue entirely, and are not addressed in this chapter.

5.1.3 Prime Word, Modifer[s] and Class Word *Are* Part of the Data Element's Name

The third mistaken notion is that the prime word, modifer[s] and class word comprise the data element's name. Given that we have shown above that a data element has a common business name which is NOT the prime word + modifer[s] + class word combination, what happens to Prime Word, Modifer[s], and Class word? Well, they are useful and they do describe some important semantics associated with data elements, contained components, or both.

An additional problem arises when the prime word, modifier[s] and class word are melded into the name string: they lose their individual semantic identity and distinguishability. That is especially so when abbreviations are required, or when wholly different words must be used to fit different languages and cultures. Simply put, each is a different meta-attribute within the complete set of meta-attributes that fully define either the data element or the context dependent business fact.

Most importantly is the distinction between that which *contains* and that which *is contained*. Data elements are the former, and columns, cells, attributes and fields are the latter. All these are meta entities within an overall meta model for information technology. It is just that data elements largely stand alone while columns, cells, attributes and field are properly contained in tables, screens, entities, and reports, respectively. Collectively, these latter four are context dependent metadata about business fact value *instances*, while data elements are context independent metadata about context dependent business facts. Because of this difference, each has a slightly different set of meta-attributes. Table 17 presents these differences.

As can be seen, prime word is a proper meta-attribute of a context dependent component but not the context independent component, data element. This same difference holds as well for modifier. Some of the other meta-attributes from Table 17 are common. For example, data structure is now really important for columns since SQL:1999 is no longer relational. An SQL:1999 column within a single row may represent single values, multi-values, groups, repeating groups, nested structures, BLOBS, CLOBS, REF-types to other-site objects, and abstract data types with both data structure and methods.

5.1.4 Modifiers are from a Single Homogeneous Set

It is commonly thought that modifiers are derived from a single homogenous pool. While that may have been true for a single database within a single project that exists within a single functional domain, it is not true for enterprise-wide suites of databases that cross countries and cultures. Table 18 contains a minimum set of modifier classes and examples of the value sets for each. There may be many more classes (columns), and many more instances (rows). Further, both the columns and the rows can be hierarchically organized.

Semantic Modifier Classes and Suggested Contained Values			
Temporal	**Accuracy**	**Geographic**	**Organizational**
Last	Estimated	World	World-wide
First	Projected	Hemisphere	Business Unit
Latest	Revised	North American	Region

Semantic Modifier Classes and Suggested Contained Values			
Temporal	**Accuracy**	**Geographic**	**Organizational**
Earliest	Initial	United States	District
Current	Actual	Mid-Atlantic	Territory
This Year		Maryland	
Last Year		Bowie	

Table 18. Examples of semantic modifiers classes.

5.1.5 That There Is Only a Choice of One Class Word

The final key problem is that class words have always been thought to be drawn from a single choice list. That is, only one class word per component (e.g., data element or column). That, however, has never been true despite our being "forced by tradition" to believe it. For example, suppose there is a Price table, where Price is the primary-key column. Shouldn't Price's class word be Identifier? Yes. But, is not price also an Amount? Yes. Why must we force ourselves to make only one choice of class word? How can Social Security Number be an identifier, but not a code, and certainly not its already existing class word choice, Number?

The way out of this "absurd to be in" situation is to recognize that there are multiple classes of class word, just like there are multiple classes of data use modifiers, a.k.a., Class Words. Table 19 lists suggestions in this area. This categorization and example sets are not complete by any means. For example, the *role* class should probably be divided into a *business* role and a *container* role. A *factor* would be an example of the former and an *identifier component* an example of the latter. The key issue here is that there are multiple classes of class words and that they must be carefully categorized and selected so that the full set of semantics is properly chosen to reflect both the needs of the business and the needs of the role the column performs within the "container."

Data User Modifiers (a.k.a., Class Words) Types and Example Value Sets		
Data Type	**Role**	**Unit**
Date or date component	Identifier component	Day
Code	Factor	Case
Text	Flag	Aisle
Weight	Indicator	Pallet
Dimension	Identifier component	Transaction
Money	Rank	Percent
Integer	Business fact	Inches

Table 19. Examples of data user modifiers (a.k.a., class words)

5.2 A Strategy That Works

When all the meta-attributes from Table 17 are appropriately valued for either a context independent or a context dependent component, a full set of semantics can be printed AS IF THEY WERE the name of the data element or the context dependent component. Figure 21 illustrates just such a scheme. It is critical however NOT to be misled into believing that these brought together strings ARE the component's name. The name may be contrived wholly differently and by another set of rules altogether. For example, the column's name represented in Figure 21 may just be SALARY within the Employee table.

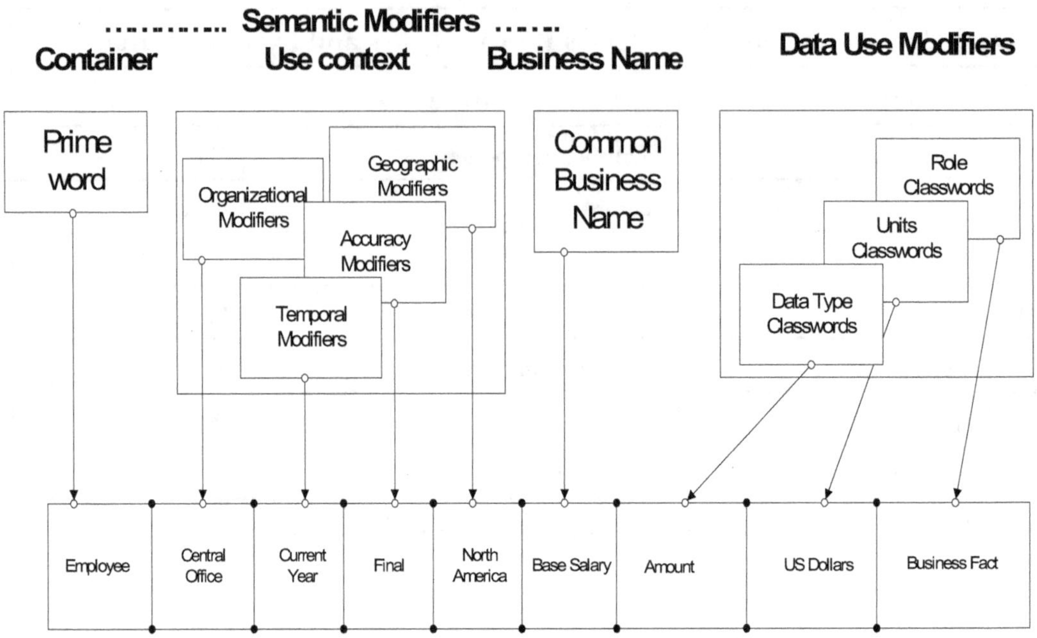

Figure 21. Complete name construction for the business fact:, Salary.

All these value selections (e.g., Central Office) are really just foreign key value instances within their respective meta-attributes of the meta-entity, column. Their originating meta-entities, e.g., Prime Word, or Organizational Modifier, etc., not only have the real values, but also have important information about the real value such as who authorized it, when, why, and definition fragments.

Of especial value are definition fragments. These are brought together to automatically provide definitions for all the context independent and context dependent business facts in the Metabase System. That gives a real boost to the 90% reduction in the quantity of work both to identify context independent and dependent components, and to define them.

The pay off from this approach is really compelling. In a reflective study of a very large data standardization effort, the following techniques comparison, shown in Table 20, was developed. Which cost item do you want to take to your management to fund?

This strategy of having a high quality "driving" metadata management system like the Metabase System coupled with an end-user supportive CASE, code generators and database project methodology is critical if we are ever to empower the end-user database designer.

Data Standardization Alternative	Final Quantity of defined components	Cost
Traditional (prime + modifier + classword) across all systems	19,000 (at the columns, cells, fields, and attributes level)	$6.75 million
Accomplished by standardizing closely named columns and fields	3,000 (still at the columns, cells, fields, and attributes)	$1.06 million
Accomplished through standardization techniques in this book.	560 (at the data element level and generation of automatically defined columns, fields and cells	$450,000

Table 20. Data element standardization techniques comparison.

Today, it seems that everybody should go through data administration (DA) or database administration (DBA) in the construction of data architectures. While that may "appear" to provide data administrators or database administrators feelings of importance and criticality, such feelings will be short lived and ultimately suicidal because they are often overwhelmed with work, and they are seen as preventing business information system development in a timely manner. The "glory" can be true only when it is reflected "glory." That is, when data administrators or database administrators empower others to use facilities that engineer data semantics management success. Then, and only then will DA/DBA success be real.

The data of a data architecture is called metadata. Metadata is collected, stored, maintained, and reported through a metadata management system. The high level specifications for the data models portions are depicted in Figure 22.

Significantly more detailed versions of these high level models is provided Chapters 9 through 13. A current set of Metabase System user guides, and the Metabase System itself can be downloaded from the Whitemarsh website.

A more detailed graphic representation of the business fact components of the data semantics meta model is provided in Figure 23. This figure shows the basic hierarchy in terms of four life cycle steps, and the proper placement of business fact detailing and specification within each of the major life cycle steps. While it would be preferable to create business facts in the life cycle sequence, this is almost never possible. What is most common is that the apex of the life cycle, the data element, is the last built because it is build through bottom-up analysis of existing and operational production class databases.

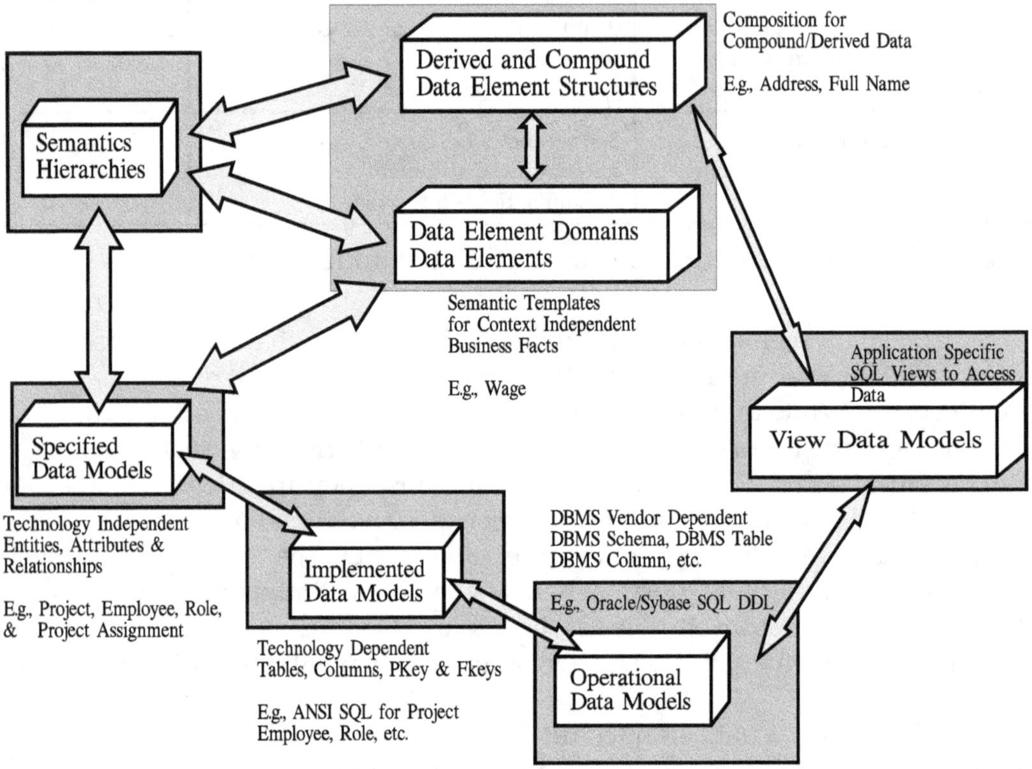

Figure 22. Metadata architecture for data semantics management.

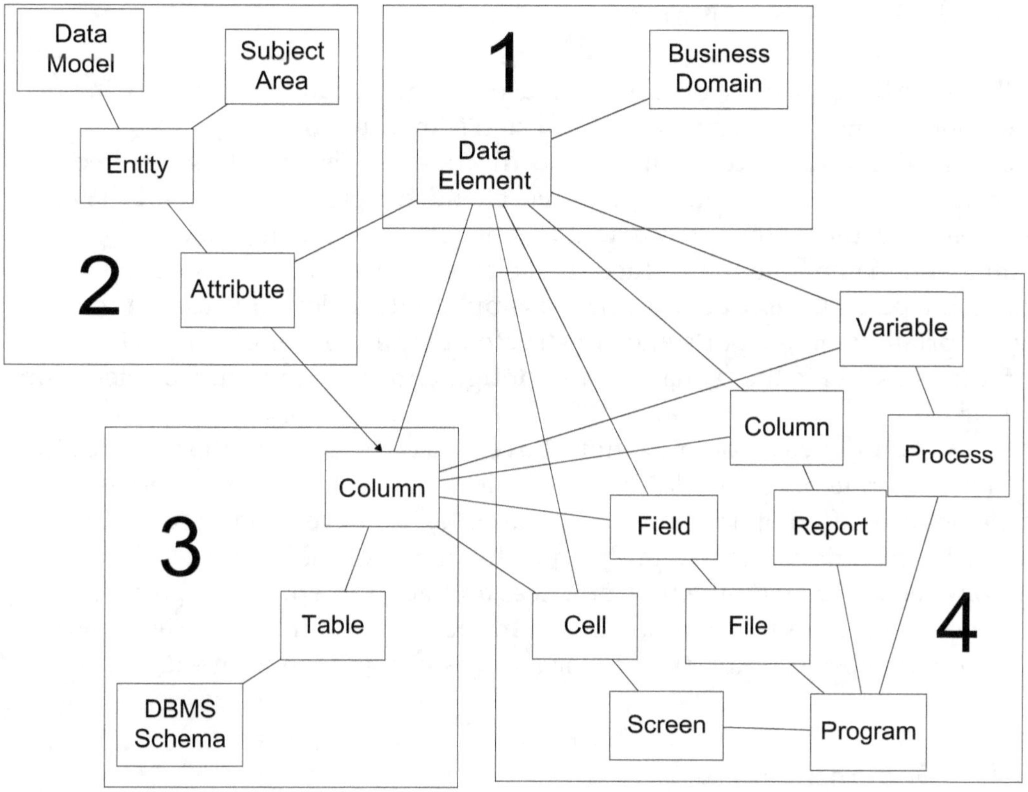

Figure 23. Levels of business fact specification and binding.

The next two sections, 5.3 Single String Names, and 5.4 Semantic Name Parts delve much deeper into the details of properly constructing names.

5.3 Single String Names

Whenever the data element's name space becomes a single melded string of the concatenated name parts of its selected semantics, the ability to distinguish the selected semantic parts disappears. This melding resulted from the use of a data element naming scheme developed in the early 1970s. At that time, there were no metadata management systems that could be employed to generate the melded data element name from the selected semantic parts. Consequently, analysts applied the rules for selecting the appropriate semantic parts and constructed the single name string. The only product to survive these analysis and design efforts was the constructed name string.

Over the years organizations have concocted all sorts of strategies to visibly separate prime works from modifiers from class words. Many of the schemes are different, mixed with one another, and even confused within the same organization. The real problem is that many of the schemes are not automated and rely on the best-practices and good intentions of participates. In the best of times this is problematic. In the worst of times such as crisis or death-march projects, a lack of consistency is virtually guaranteed.

5.4 Semantic Parts

The second problem is an antiquated, heterogeneous set of semantic parts used to define a data element. The three parts are:

<prime word>
<modifier>s
<class word>

An example of a constructed name string from these three parts is <employee> <first> <name>. This component set was also developed in late 1960s, almost before DBMS. The first wide-scale use of the three component data element naming scheme was to name fields within COBOL files. Every file addressed a single policy area and had hierarchically contained sets of data fields. Computer programs that manipulated these fields needed to have unique names for every field to have an unambiguous name space. Prefixing the name of the file onto each field was easy, precise, and solved the ambiguity problem.

When commercial DBMSs came along in the early 1970s (for example, IMS, IDMS, Adabas, Total, and Datacom/DB) the COBOL file's three-part field naming convention just continued because the databases were application specific and had to be accessed through COBOL.

Thus, because the application contexts were narrow, because the quantity of tables (40-100), columns (600-1000) and rows (2-10 million) were small, and because COBOL demanded it, the three part component naming strategy was not only acceptable, it was never seen as a problem. Today, however, enterprise-wide suites of databases have thousands of tables, tens of thousands of columns, and billions of rows. Now, the search for the right database-table-column set of metadata as a prelude to selecting data is almost as laborious as acquiring selected rows.

This yesterday versus today situation is analogous to wandering the stacks of the neighborhood library to find a book versus wandering the stacks of the 25 story University of Toronto library. Yesterday's three component name solution just does not fit today's problems. Today, extensive searches must be made of a very large library's card catalogues before selecting and retrieving books. The former activities are really metadata searches and the later activities are book acquisitions that are the natural consequence of the metadata searches.

The antiquated three part component scheme forces common concepts within different contexts to be named differently and obscured for three reasons: first, every old scheme-constructed name is required to have a prime word, that is, the name of the context within which the name exists, such as Employee, or Inventory Item Detail.

Second, while modifiers, the second old scheme component, permit the concatenation of multiple words, these words come from a single list rather than from multiple modifier subtype lists. For example, the modifiers: Canada, Current, Estimated, First, Home, Last, Prior, United States, World, and Year-to-Date all come from a single list. To make modifiers relevant to today's enterprise-wide, heterogeneous database and information system environments, they must be subtyped into, for example, the following subclasses: geographic, organizational, accuracy, and temporal.

Finally, only a single class word is permitted even when more than one is either necessary or required. For example, identifier and code, both class words are not mutually exclusive. Yet, only one is allowed to be chosen. An examination of a typical class word list shows that it contains words that relate to a data element's role within a container, granularity of the data value, or its business data type.

Not allowed at all within the three component scheme are the common business names used in everyday business situations. For example, to be precise it is wrong to have the constructed column name, Employee Social Security Number. Instead, it would either have to Be Employee Social Security Identifier because Number is a class word and the Social Security Number is really an identifier, or it would have to Be Employee Social Security Code because the Social Security Number is really a coded set of three distinct numbers rather than a "real" number. Simply put, the traditional naming scheme--if enforced--is too simple, obscures the business concept, and requires illogical choices.

A common counter argument is that exceptions to the use of class words are always allowed. But what is the sense in having rules if exceptions are allowed. The issue here is not that humans cannot deal with exceptions, it is that computers cannot. If there is a metadata management system and a query is launched to find all Code columns, Employee Social Security Number would not be found. Or, if the query was launched to find All Identifiers Employee Social Security Number would again not be found.

Under the old scheme, once a designer completes the effort of making the names unique over a database of 300 to 600 columns, the result is often applauded as having been cleverly done. Clever should never be the characteristic. Rather, comprehensive and accurate should be. In summary, the problems with the old three part name scheme are:

- Having prime a word.
- Not having a common business name.
- Heterogeneous modifiers and class words.
- Only allowing one class word.

The sections that follow, while similar provide greater detail to the problems described in Sections 5.1.1 through 5.1.5.

5.4.1 Prime Word Semantic Component

Prime words should exist only as a meta-attribute within the context of a "containerized" use of data element semantics. That is, as a meta-attribute within the attribute, column, field, screen-cell, report-column meta-entities. Examples of a container are provide in Figure 24. Each of these containers should represent a policy-homogeneous expression and the data elements

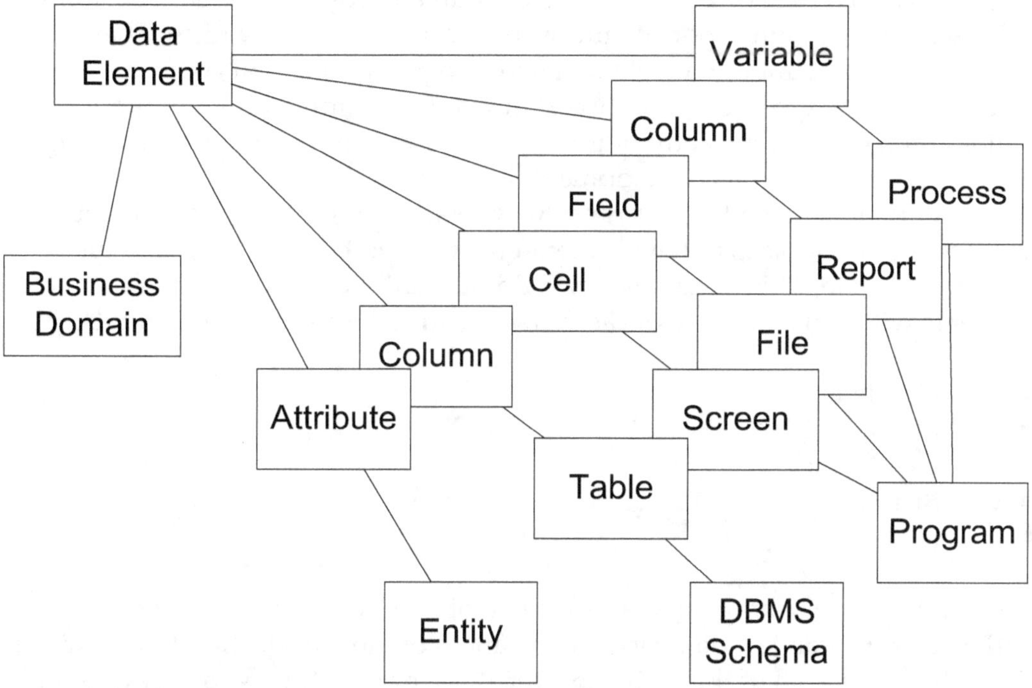

Figure 24. Data element semantics mapping to contained business facts and containers

ascribed to them as columns, fields, screen elements, etc. should serve as the data-based evidence that the policy-homogeneous expression has been instantiated.

Figure 24 also identifies the containers, that is, entity, table, . . . , and process. The figure also identifies, for example, that attributes belong to entities, columns to tables within DBMS schemas, and that cells are within screens, fields within files, [report] columns are within reports, and variables are within processes. The diagram also depicts that screens, files, reports, and processes are all proper contained components of programs. Finally the figure shows that the semantics for attributes, columns, cells, fields, [report] columns, and variables derive their semantics from data elements within specific business domains.

If the prime word is used as a meta-attribute of the data element meta-entity, the granularity of the data element is pushed down to that of the field, screen-cell, report-column. Instead of being able to use and reuse data element semantics an average of 20 to 50 times over an enterprise-wide set of columns, field, screen-cells, and report-columns, they can be used only once. That causes the total count of data elements to be equal to the total count of all columns, fields, screen-cells, report-columns. This completely redundant, tedious, and error prone activity greatly increases the cost of database design and information systems development.

For data elements, the proper way to account for the data element's universe of discourse is to have a meta-entity called business domain, for Example, Person. This meta-entity would have all the Person related data elements within its universe of discourse. Included for example would be:

- First Name.
- Middle Initial.
- Last Name.
- Birth Date.
- etc.

While this looks strikingly like the schema of an HR database, it is not. Types and instances of an HR database are instances of the Application Level Pair of Table 21, while each of the above instances is from the Repository Level Pair column. This table of three sets of type-instance pairs was created by Alan Goldfine of the National Institute of Standards and Technology during the 1980s creation of IRDS-1, the first ISO/ANSI metadata standard.

Each instance of either prime word or business domain must be defined to convey the business purpose whenever and wherever the string (e.g., Person or Employee) is employed. These definitions are critical for the automated definition of data elements or context dependent business facts. The definitions for business domains must be created during the process of installing the business domain that encompasses the set of data elements. The context dependent business facts' prime word is the name of the context (e.g., entity, table, file, etc.) and the prime word's definition is the definition from the context.

5.4.2 Common Business Name Semantic Component

The common business name component of a data element's semantics is the commonly known name for the business fact. It is completely orthogonal from either semantic modifiers and class words. Figure 25 identifies the common business name for a number of business facts. Employing the common business name to imply a class word can be problematic. For example, Social Security Number is the common business name of the coded identifier assigned by the Social Security Administration of the U.S. Department of Health and Human Services no matter whether the "number" has two embedded dashes, is used as an identifier, or is in fact three discrete codes (xxx-yy-zzzz). The same applies for telephone numbers, account numbers, or invoice numbers that contain letters, dashes, or other special characters.

Metadata Repository Fundamental Level			**Types**: entity Relationship attribute
Metadata Definition Level		**Types**: Record type Data item table	**entity Instances**: Record type Data item table
Metadata level	**Types of data items for Employee**: Employee_name Employee_id Department_No.	**Instances**: RT: employee DI: emp_name DI: emp_id DI: depno	
Application level	Data record instance: P. Shaw 1234567 SQL Development		
	Application Level Pair	**Repository Level Pair**	**Repository Definition Pair**

Table 21: Repository levels and pairs

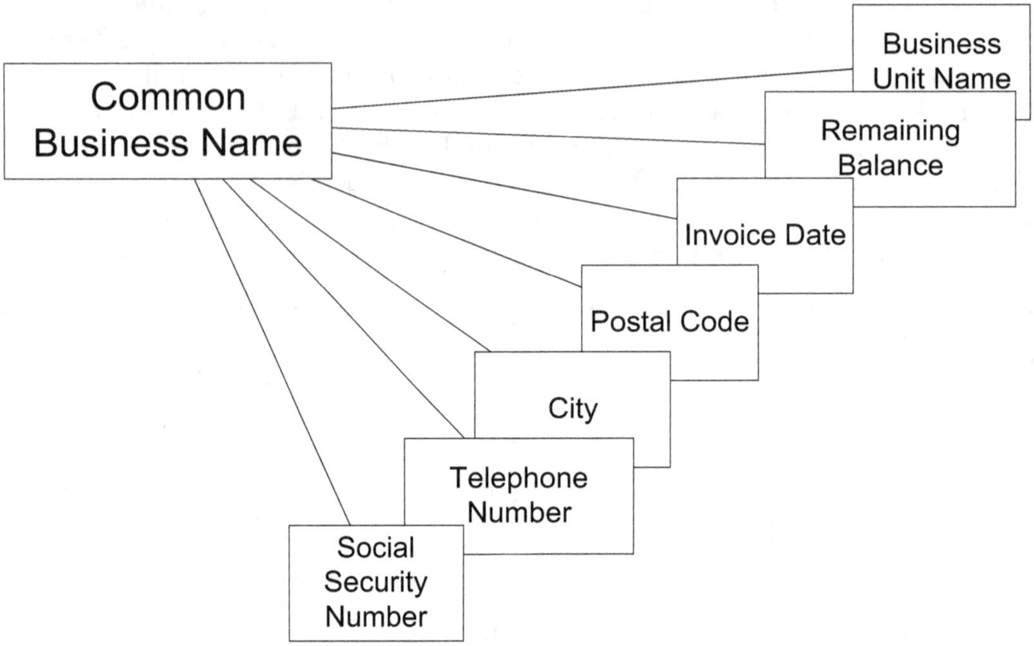

Figure 25. Examples of common business names

While telephone numbers have always fundamentally been numbers, the letters on the phone have not disappeared, and commercial messages such as "Dial 1-800-GoFedEx" appears everywhere. So, what is the telephone ***number***? Other examples from Figure 25 illustrate other classes of problems:

- City—is it the city name? If name a set of letters, what is the spelling of the city, **84, PA**.
- Postal Code—Clearly this one is named correctly, it is a code. Once we removed the data type, is its real name, Postal?
- Invoice Date—is date the data type of values? If so is the real name Invoice?
- Remaining Balance—is balance the domain of values, or a fundamental semantic part of the name?

- Business Unit Name—If Name is the type of data, can "J81" be an allowed business unit's name?

The point here is that these are common business names, and the type and kind of data should not have to be inferred. They are just the common business names, no more, and no less.

The purpose of the common business name is to express a short version of the business fact's essential policy. The common business name is not a synonym for the data's domain of values. Thus, Date is not a common business name because a date is not a business fact. Rather it is specific representation of business fact's value in the form of a date. The business fact common names would be: Invoice Date, Shipping Date, and Order Date, and Backorder Date. In these examples, the data element's name is the name of a value-state that was achieved on the date represented by the contained value.

In this example both the metadata (e.g., Invoice Date) and the date value (01/17/96) are essential to its meaning. In other situations the metadata is less essential to the basic business understanding of the data, for example, Employee Name: Wayne Miller. In all cases both metadata and instances are required to convey full semantics. Additionally, unless the semantics match or are commonly derived, data sharing is at best problematic.

Each common business name must have a definition that conveys its business purpose whenever and wherever the string (e.g., Invoice Date) is employed. These definitions are critical for the automated definition of data elements or context dependent business facts.

5.4.3 Modifiers Subtype Semantic Component

Modifiers, or as they are sometimes called, qualifiers, are essentially semantic modifiers employed whenever the data component is employed, for example, as a column in multiple contexts that would lead to confusion without the modifier phrase. For example, there may be a table for Address. But whose? Company, Employee, Customer, Vendor, etc. While the column in each context is clearly Address, and that Address is certainly the common name, the use of a semantic modifier, such as Company, Employee, Customer, and Vendor is required. Semantic modifiers are most often employed when a column is used multiple times within a table, for example, if the column, Telephone_Number, is employed in a table four different times, a semantic modifier must be employed to separate out the different contextual uses:

<Home>Telephone_Number
<Office>Telephone_Number
<Boat>Telephone_Number
<Airplane>Telephone_Number

Further, if the context is an Employee table, another semantic disambiguator, Employee is implied whenever the "naked" telephone number is seen. The completely disambiguated telephone numbers are thus:

<Employee><Home>Telephone_Number
<Employee><Office>Telephone_Number
<Employee><Boat>Telephone_Number
<Employee><Airplane>Telephone_Number

Note that the strings contained within the "< . . . >" are not part of the column's common name. Rather they are referenced from the column's meta-attributes that reference them. Because of these meta-attribute references, these displays are possible.

These semantic modifiers are called meta category values in the Metabase System. Since they occur before the common business name, these are prefix meta category values. Meta category values and variations are the set of words that provide a further detailing or understanding of the data component's semantics.

After a suitable quantity of data components are identified and analyzed, it is quite likely that a restricted set of meta category value words or phrases can be isolated. Once isolated, these words or phrases can be organized as meta category values, can be defined and made available for data component semantic assignment. Appendix A in Brackett's book, Data Sharing, Using a Common Data Architecture provides a good start. Meta category values are semantic disambiguators: separately identified and mapped to the data components to which they apply.

In summary, a data component is known by all its meta characteristics. These include its subject, common name, and semantic modifiers. The first step on the road of data semantics management is to identify the most commonly seen set of alternatives for each meta characteristic. Each alternative should be defined and installed in the Metabase System. Then, when the database is designed, determining the proper set of meta category values for data components becomes a quick and easy task of understanding the concept and identifying the appropriate set of meta characteristics. Data elements and columns represented through these meta characteristics are immediately linked.

Semantic modifiers are not a monolithic concept. There are many different classes. When the word choices from each of the relevant classes are expressed together, strings such as such that Initial Estimated. Modifiers are thus subclass collections, in which each class has orthogonal choices. For example the modifiers subclasses might be:

- Accuracy (Estimated, Final, Revised, etc.). See for example, Figure 26.
- Geographic region (World, Hemisphere, US, North America, Europe, Asia) See for example, Figure 27.
- Organizational unit (National, Region, District, Territory, etc.). See for example, Figure 28.
- Temporal (Last, First, Latest, Earliest, Current, Future, Last Year, This Year, Next Year, YTD, PTD, etc.). See for example, Figure 29.

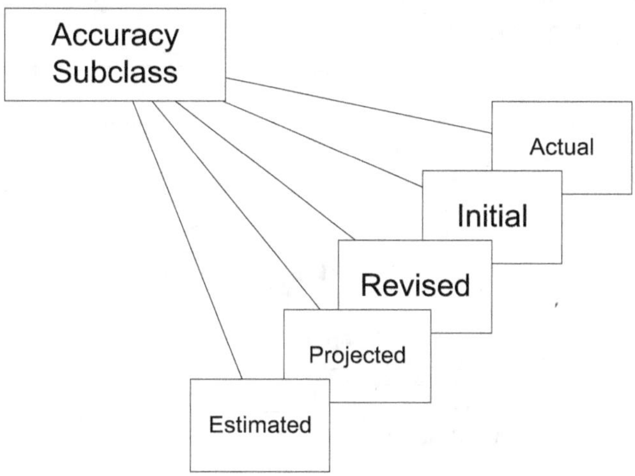

Figure 26. Accuracy modifier example.

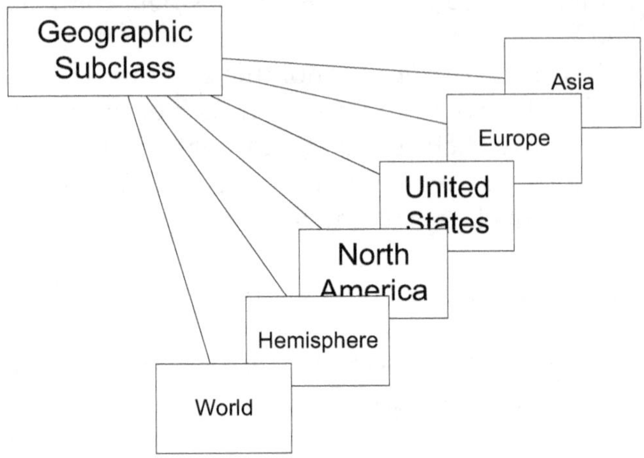

Figure 27. Geography modifier example.

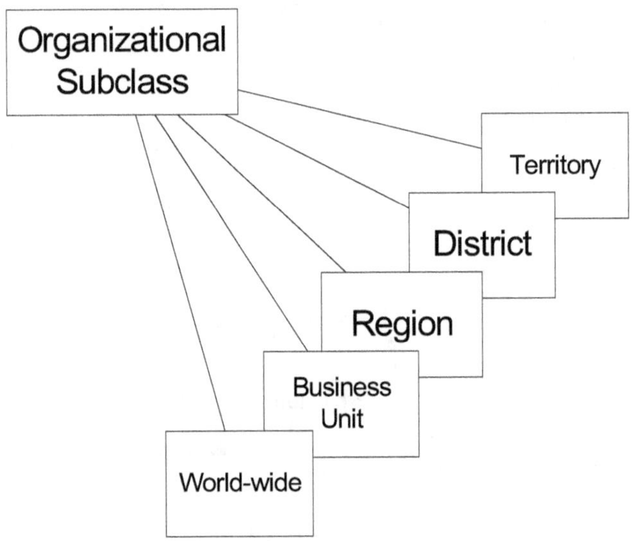

Figure 28. Organizational modifier example.

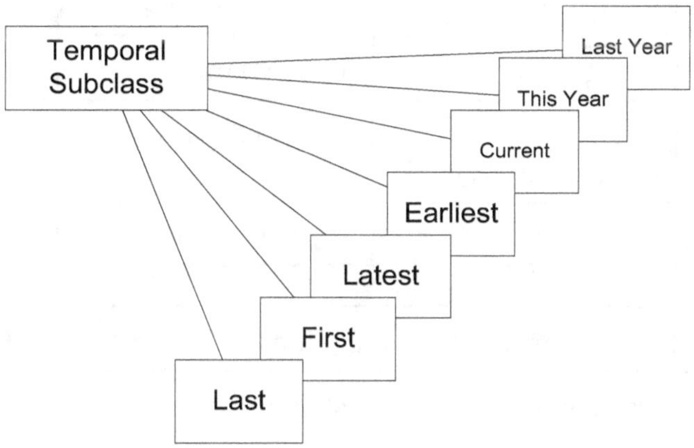

Figure 29. Temporal modifier example.

The quantity of classes and the domain of values each class can only be accomplished after appropriate research. Table 22 presents these presents these four subclasses and suggested contained values.

From this table, combinations of modifiers can be strung together to provide full contextual meanings to the business fact. It can be successfully argued that these modifier name parts could become sibling business fact values to fully define the context of the business fact. Some modifier classes are hierarchical. For example,

(World (Western Hemisphere (North America (Unites States (Mid-Atlantic (Maryland (Bowie))))))).

Regardless of how those arguments are made, the two data structures in table 21 are semantically equivalent>:

Columns	Column Values
Temporal	First
Accuracy	Projected

Columns	Column Values
Geographic	Mid-Atlantic
Organizational	Regional
Sales	$417,000,000

First_Projected_Mid_Atlantic_Regional_Sales	$417,000,000

Table 22. Semantically equivalent data structures.

Each represents an alternative form of expressing the semantics of the data. The choice of one or the other must be made after careful analysis of performance versus flexibility. The issue is not that one is preferred to another, it is that the data semantics management must occur, and after that, the form of database schema implementation becomes the second-order decision process.

Each value within each prefix meta category value subclass must have a definition that conveys the business purpose whenever and wherever the string (e.g., National) is employed. These definitions are critical for the automated definition of the data components to which the semantic modifiers are assigned.

5.4.4 Class Words Subtype Semantic Component

Class words are data use(s) modifiers that indicate the types of uses that mainly befit columns versus data elements. These data use modifiers are also called meta category values in the Metabase System. Since they occur after the common business name, they are suffix meta category values. Some of the uses are oriented towards the column and others for the data element. For example, the use-indicator, Code signifies that the value set of the data element is really a surrogate for an appropriately matched value set of another data element. For example, there is the <state code> such as CA or VA and the corresponding <state name>s California or Virginia.

Other uses, such as identifier, primary key, foreign key are either more appropriate for columns or named collections of columns.

Regardless of whether the data uses apply to data elements or columns, they too are separately identified and mapped to the data elements to which they apply. The class word concept represents at least three different use types:

- Basic data type or form of the data, that is whether the data is money, date, time, decimal, and integer. Figure 30 identifies some of the more common data types that relate to commercial data. These data types are not computer-based data types such as single or double word integers, single or double floating point. Rather they are intended to convey the business purpose of the data regardless of its inherent computer-science data type.

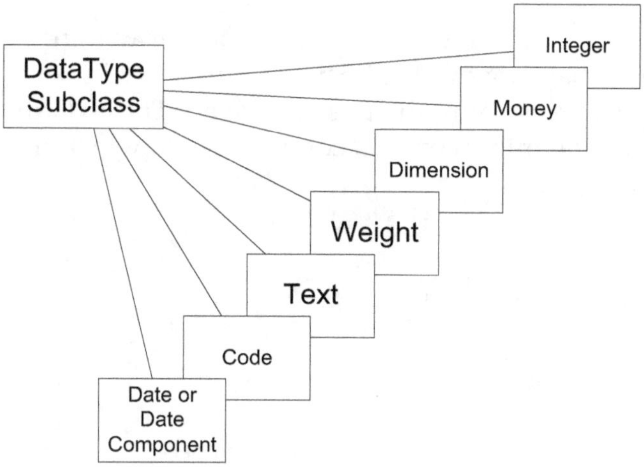

Figure 30. Business data type class word subclass

- Role, that is, the function performed by the column within the table, such as identifier, flag, business fact. Figure 31 identifies the role that this data plays within the context of the container. Note, there is no role for a data element because it is independent of container.

- Units, that is, the granularity of the represented data, such as cases, dollars, and degrees Fahrenheit. Figure 32 identifies the business units that are represented by the data. While the computer-science data type represents the internal form of the data's storage, and the data type represents the business data type indicating that the data is money, weight, or text, for example, this class word subclass indicates the granularity represented by the data.

As with semantic modifiers, each data use modifier within each subclass must have a definition that conveys the business purpose whenever and wherever the string (e.g., Degrees Fahrenheit) is employed. These definitions are critical for the automated definition of data components to which the data use modifiers are assigned.

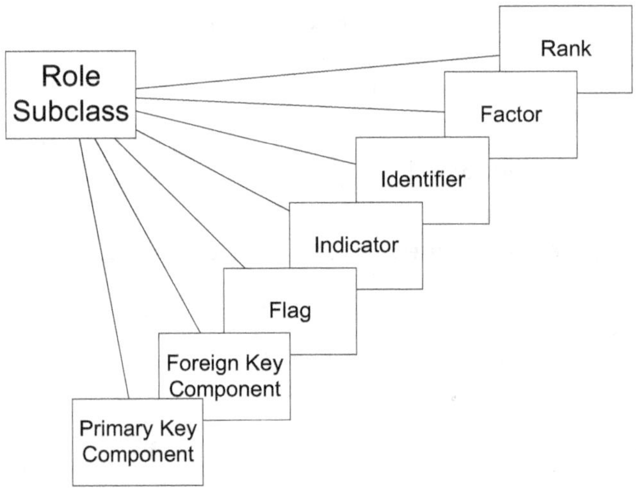

Figure 31. Role class word subclass

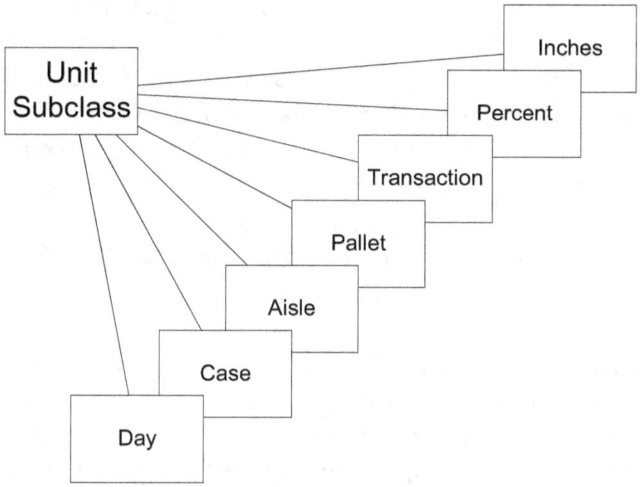

Figure 32. Units class word subclass

5.4.5 Full Name Construction

Full name construction is probably one of the biggest errors committed in the name of data semantics management. Full names must never be constructed from the allocation of all their semantic parts into a single "monster" string and stored in the metadata management system in lieu of all its individualized semantic parts. These "monster" strings may be stored additionally, but never instead. The "monster" strings must always be able to be materialized, on-demand. Full names must, of necessity be created and deployed within DBMSs and computer programs without also containing references to their semantic parts, but never in metadata management systems.

The full set of meta-attributes that represent the semantics of a data element or a context dependent business fact (column, field, etc.) is the result of a process of presenting all the selected names from the different meta-attribute semantic components. Table 17 identifies the meta-attributes employed to represent either the data element name or the context dependent business fact. The containers from Figure 24, include table (column), screen (cell), file(field), report(column), and process(variable). Once the represented meta-attributes are all identified, the resultant set of semantics convey a business fact's purpose.

Figure 21 presents the set of all meta entities involved in fully representing the full meaning of a contained business fact. These meta entities are grouped into container type (prime word), use context (organization, temporal, accuracy, and geographic) modifiers, business name, and data characterization (data type, units, and role) class words. At the bottom of Figure 21 is an example of the complete set of semantics for a contained business fact. The result is, for example,

Employee Central Office Current Year Final North American Base Salary Money Dollars

It is easy to see why the full name must never be materialized and stored exclusively in the metadata management system in lieu of the referenced semantic parts. The distinguishability of the individual meta-attribute values have been melded into a single string. Thus, under the "monster" string approach, it would be almost impossible to recognize that the common business name was Base Salary.

Within the metadata management system, the actual values of the meta-attributes associated with either the data element or the contained business fact do not actually exist. Rather the foreign keys to these values exist. When full names are materialized for use in DBMSs and computer programs, it is likely that the string produced by concatenation of the complete set of semantic components will exceed either the abilities of a DBMS or a commonly used standard. For example, some prime words are quite long, for example, Inventory Item Detail. If the maximum length of a context dependent business fact, a column name for example, is 18 characters, 21 have already been used (oops!) without even accommodating the other parts, that is, the common business name, semantic modifiers sets and class word sets. It is not uncommon therefore to need 50 to 80 characters for full use of the prime word, common business name, selected modifiers, and selected class words.

To deal with length restrictions, the Metabase System supports the creation of abbreviations within business domain, and the automatic creation of abbreviations through a well accepted and commonly used scheme. Since abbreviations only become valuable in the creation of Operational Data Model schemas, that is the Metabase System's module that supports abbreviation creation.

5.4.6 Semantics Component Summary

Because of the preceding strategies, data element semantics and their use in the formulation of context dependent names can be predictable and rigorous. Whether the context dependent semantics is a direct copy of a data element's semantics or is merely a mapping from the data element to the current nonstandard semantics is immaterial. That is because the data element will have been correctly formed and will represent the mapped context dependent business fact. By whatever semantics the context dependent name is known, its real business semantics will always be retrievable from the data element.

Finally, a data element must never employ a prime word because that forces the granularity of the data element down to the level of a context dependent data representation. To include the context meta-attribute, that is, the prime word within the data element's meta-attribute set is a fatal mistake because it explodes the quantity of data elements to an infinitely expanding set.

At this point in the process of defining the full set of semantics for the data element and for context dependent business facts the following are known:

- Business domain or container's name (prime word).
- Common business name.
- Semantic modifiers that define the business fact's geographic region, organizational unit, accuracy, and temporal characteristics.
- Data user modifiers or class words for its role, units and basic data type.

Additionally known are all the definitions of each of these parts. Lashed together through a sophisticated metadata management system, the data element or context dependent business fact's name can be automatically generated as can its definition.

The payoff is simple. Rather than approximately two hours per column of analysis and definition work, the effort is very likely reduced to 15 minutes or less. That's a savings of about 84%. In addition to this significant payoff, all the metadata is able to be queried from the metadata management system, entire data models can be printed, SQL generated, and tables and columns found because of their specific semantic components. This in turn increases the ability to predict evolution and maintenance time schedules, costs, and significantly reduces risks.

5.5 Error Classes

There is a commonly accepted but mistaken notion that names are the alpha and the omega of data semantics management. Far from it. Names, while clearly the most visible semantic component can be very misleading and can lead to both Type I and Type II errors within the first error class, and can also lead to delays and confusion as second error-class attempts get sorted out.

One U. S. DoD agency spends many millions every year just sorting out errors on logistics transit labels due to reference data errors. These are clearly a second error class activity caused by Type II errors such as sending M1 Abrams battle tanks to Bimini versus Bosnia.

This is not to say that names are unimportant. What is critical is that there is almost automatic strategy for name construction that ensures that when names are the same they have the same meaning, and vice versa. If and when this is done, there will be a great reduction in both error classes.

Having a highly engineered strategy for names is essential so they are constructed in a reliable and repeatable manner. Key to name construction are the semantic hierarchies that provide the domain-sensitive words along with definitions. Once semantics are allocated, names can be automatically constructed by the Metabase System. Definitions are also automatically generated because not only is there a short local definition, there are also all encompassing semantic parts that contribute to an auto-generated contextual definition.

It is critical to have levels starting at the most generalized and proceed down to the most specific so there can be maximum reuse of common name parts. The ultimate and only valid objective are that when names are different, the fact's semantics are too. When names are the same, the fact's semantics are too. When this is accomplished, the quantity of errors across the two error classes will fall away to almost zero.

5.6 Semantics of Names Summary

The goal of this chapter was to squarely address one of the key reasons for data semantics management failure: Names. This chapter set out the five most common problems that plague the construction of names, and provided a clear strategy for a common-sense solution approach.

This chapter further detailed why single string names cause an inability to discern the true semantics of a data element, column, field or other business fact. Over the years, organizations have concocted all sorts of strategies to visibly separate prime works from modifiers from class words. Many of the schemes are different, mixed with one another, and even confused within the same organization. The real problem is that many of the schemes are not automated and rely on the best-practices and good intentions of the participates. In the best of times this is problematic. In the worst of times, such as crisis or death-march projects, a lack of consistency is virtually guaranteed.

In further detailing the problems, a whole section of this chapter was spent on the failed strategy of "prime word, modifier, and class word." While this scheme might have been valid for COBOL file structures, it became outdated and of negative value starting in the middle 1970s. This same section of the chapter showed the key reasons for this three-part naming strategy failure and set forth an alternative approach that has applicability well beyond just a single database to that of an entire enterprise. It is this strategy that is the subject of subsequent chapters in this book.

Chapter 6, Semantic Hierarchies, is the first implementation chapter. That is, it is the first chapter that defines and illustrates the actual establishment of data semantics management. This chapter identifies and provides real screen-shot examples of how to define data semantic hierarchies and how to deploy them in real-world examples. Thus, this is the first "can do" chapter.

The real world examples come from the Metabase System and contained example metadata database.

5.7 Questions and Exercises

1. "What's in a name? That which we call a rose by any other name would smell as sweet." That line out of Shakespeare's Romeo and Juliet would seem to imply that there's more to an object than just its

name. In your enterprise, what are all the semantics that you associate with a business fact? How are they defined, interrelated, managed, stored and reported/queried?

2. Section 5.1 of identifies five types of name construction problems. Which have you had? How have you dealt with them? Are there formal strategies in place, or are there just ad hoc solutions?

3. The traditional approach to business fact naming only has one class word. Is this a realistic approach to the identification and assignment of all the data use type cases? How do you resolve identifiers with data values that can be text, code, numbers, and the like? These are all traditional class words. How can you pick just one?

4. The approach in this chapter asserts that there clearly should be common business names. How many cases in your collections of facts that have modifier and class word only constructed names that would be better labeled if these facts would first start with a common business name to represent the core fact and then the allocation of modifiers and/or a class word?

5. What is your enterprise's definition for a data element? How different is that from a column of a table or an attribute of an entity? How do you keep them separated one from the other? Are these, in your organization, just synonyms for each other or are they really different? What would be the value of their differences?

6. Table 16 identifies a set of meta-attributes for data elements and for attributes et al. Are these all the meta-attributes? What is missing? What is the effect of the missing meta-attributes?

7. This chapter proposes that at the core of every data element is its "common business name." Does that make sense? Why? Review the names you already have and determine which violate the commonly employed <prime word>, <modifier>, <class word> triple that has been employed for decades.

8. The traditional approach to naming is that the prime word, modifiers, and a class word are the essential parts of a business fact's name.

What's the problem with that statement? What are the bad effects of having such an approach. What undesirable effects are caused?

9. Section 5.1.4 asserts that semantic modifiers should not all come from a single homogeneous set. What is that true? Can there be modifiers from different sets with the same name but different meanings? If yes, then what type of infrastructure has to be set into place to deal with this issue?

10. Table 17 enumerates four different modifier classes. What other kinds are there? How are they constructed? Examine your inventory of business facts. Are you employing them? How do you identify, define, and manage them? How do you keep different projects and development groups employing the same words in the same sequence so that same names are the same?

11. Table 18 enumerates three different data use modifier (class word) sets. What other kinds are there? How are they constructed? Examine your inventory of business facts. Are you employing them? How do you identify, define, and manage them? How do you keep different projects and development groups employing the same words in the same sequence so that same names are the same?

12. The traditional approach to naming asserts that there can be only one class word for each business fact. Provide examples where this cannot possibly be true. What did you do about such exceptions?

13. Figure 21 illustrates a strategy that works. How is this strategy different from just stringing words together? What would be the benefits from employing this strategy within your enterprise? What infrastructure would need to exist to accomplish it?

14. If the strategy illustrated in Figure 21 is employed, how should it be employed in the cascading semantic models: Data Element, Specified, Implemented, and Operational? What would you have to create to prevent duplications and conflicts?

15. Figure 21 illustrates the Metabase System's engineering that enables the discovery of business facts, that is, data element concepts, data

elements, attributes, and columns by the allocation of semantic modifiers and data use modifiers. How would this strategy help in a given project, an organization, a function and the enterprise as a whole? How would this help in the discovery facts by their common business name or by semantic modifiers and data use modifiers? What value would exist? How would you quantify the benefit? What strategy would you take to convince your management to fund and support such strategies?

16. Table 19 shows the cost advantage from applying the right approach to business fact name construction. After a cursory examination of the inventory of business fact names in your enterprise, how many semantic duplicates would exist that have different names? Would it be practical or possible to change all the names? If not, what approach should be undertaken?

17. The Error Classes section outlines the types of errors addressed by this chapter's materials. Do you agree that the error classes are addressed? How? What could be different? Has this chapter's contents and strategy provided an overall justification for its implementation? If yes, why? If now, why not? How would you quantity the benefits from having these error classes resolved? How would you present the approach, costs, and benefits to management in order to justify funding?

6

Semantic Hierarchies

6.1 Semantic Hierarchies Introduction and Scope

The previous chapter squarely addressed one of the key reasons for data semantics management failure: Names. The chapter presented the traditional strategy for constructing names, that is, <prime word>, <modifier>s, and <class word>. The chapter showed how this strategy, discredited in terms of value starting in the early 1970s was completely misused by industry, and government alike. For example, the U.S. Department of Defense using this scheme not only failed to achieve data semantics management but also wasted more than $500 million.

The last chapter further detailed why single string names make discerning the true semantics of a data element, column, field or other business fact difficult, conflicting, confusing, or just impossible. That was because over the years organizations concocted all sorts of strategies to visibly separate prime works from modifiers from class words. Most of the schemes are different, mixed with one another, and even confused within the same organization.

The real problem is that many schemes are not automated and rely on the best-practices and good intentions of participants. In the best of times this is problematic. In the worst of times such as crisis or death-march projects, a lack of consistency is virtually guaranteed.

The last chapter also presented the necessary detail to support the need for a reliable, repeatable, and auditable approach to the development of names. The chapter set out a name construction strategy that cannot only work on individual databases but can also scale to enterprise-wide efforts.

This chapter, Semantic Hierarchies, is the first implementation chapter. That is, it is the first chapter that defines and illustrates the actual establishment of data semantics management. This chapter identifies and provides real screen-shot examples of how to define data semantic hierarchies and how to deploy them in real-world examples.

The real world examples come from the Metabase System and the example metadata database, Movies.

Semantic hierarchies are a critical element in the development of enterprise-wide semantics because they form the building blocks for semantic engineering and name construction. What is very important about these semantic hierarchies is that they can be allocated to many different business fact components in the data semantics management space including: data element concepts, data elements, attributes, and columns. Not only can semantic elements be allocated to these business fact components, each semantic name string can be unique or duplicated. For example, Length could be a dimension for either an English or Metric measure. When the semantic names are duplicated, each must exist within a unique context, e.g., English or Metric, and thus be able to defined differently.

The Metabase System software ensures that semantic nesting is enforced. Users are not able to allocate to a lower level component, for example, a column of a table, a semantic that is the same as or at a higher level as a semantic that was assigned to an attribute of an entity. That is because attributes are semantically superior to columns. Assigning the same or a lower level semantic to an attribute than is assigned to a column would simply be a mistake. The Metabase System prevents such mistakes.

Finally, not only are the assigned semantics employed to automagically construct names, they are also employed to automagically define data element concepts, data elements, attributes, and columns. Both these capabilities are critical to the efficient and effective management of enterprise-wide data semantics.

6.2 Concepts and Engineering

Business facts, whether context independent or dependent represent a set of semantics established by the enterprise. Semantics are the rules that govern meaning and usage. The semantics are represented through a series of classification hierarchies. Each classification hierarchy represents one complete class of semantics. Each class of semantics relates to a single meta category value type, which itself is organized into a hierarchy. While there is no limit to either the quantity, breadth, or depth of a meta category classification hierarchy, there are two fundamental types:

- Semantic modifiers that provide special contexts for the common business fact meaning. These are the prefix meta category values.

- Data use modifiers that point to the data types, roles performed, and value units. These are the suffix meta category values.

Meta category values can be assigned to any of the following business fact representations:

- Data Element Concepts.
- Data Elements.
- Attributes.
- Columns.

Meta category values are not assignable to concepts because concepts are just that, concepts, and are not to be representations of business facts. Meta category values are also not able to be assigned to DBMS columns because the Metabase System's data model engineering stops the infusion of new semantics with the specification of the Implemented Data Model. The next and more technologically bound data model specification, the Operational Data Model, provides the ability to have different DBMS-specific representations of the same semantics. What can be assigned to DBMS columns are value domains and DBMS-specific data types.

The Semantic Hierarchies Data Model is depicted in Figure 33. This data model contains three meta entities, meta category value type classes, meta category value types, and meta category values. Each of these three meta-entity types is described in sections that follow.

Two of the meta entities are recursive. That means that each can contain hierarchically related rows of data. In general, the meta entity, meta category value type class, divides the allocation of meta category values into two classes: prefix and suffix. The meta category value types provide for hierarchical categorizations of each of the two classes of meta category values. The final meta entity, meta category value, provides the actual words or phrases that are assigned as either prefix modifiers or suffix modifiers.

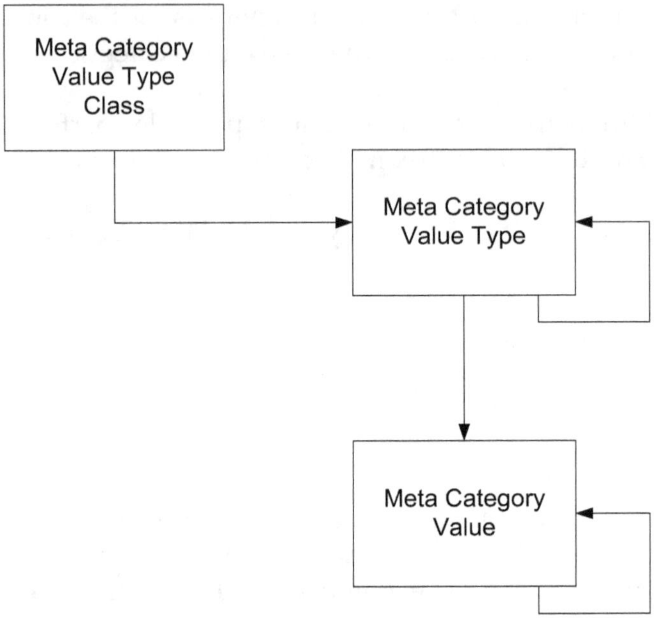

Figure 33. Semantic hierarchy data model.

6.2.1 Meta Category Value Type Classes

The meta category value type class broadly divides the meta category value type and meta category value hierarchies into their class types: prefix and suffix. From Figure 21, and the following column name:

Employee Central Office Current Year Final North American Base Salary Amount US Dollars

The following are prefix modifiers:

- Employee.
- Central Office.
- Current Year.
- Final.
- North American

The following are suffix modifiers:

- Amount.
- US Dollars.

What is left is the actual fact's common business name, Base Salary. How these semantic and data use modifiers are known, discernable, get created, and are assigned is the purpose and content of this chapter.

The only two values allowed for meta category value type class are Prefix and Suffix. When a meta category value type is allocated to prefix meta category value type class, the assigned meta category values appear before the Common Business Name. In the case above, the meta category values would appear before the Common Business Name string, Base Salary.

Conversely, if the meta category value type class is Suffix, the assigned meta category values appear after the Common Business Name string, Base Salary.

This prefix assignment strategy enables the modification of the fundamental semantics of the common business name, while the suffix assignments enable indications of the kind of data, its use role, and business data type.

6.2.2 Meta Category Value Types

Meta category value types represent a way to group collections of the meta category values that are ultimately assigned to, for example, a data element concept. There are three features of meta category value types that are different from meta category value type classes:

- There can be any quantity of meta category value type values.
- The meta category value type values can be hierarchically organized.
- Each meta category value type value has a sequence number.

The rows of data within meta category value types form a taxonomy in which meta category values occur. Because there can be multiple taxonomies, the same term can appear multiple times but with different contextual definitions. For example, take the term, Discipline. From WordNet of Princeton University (http://wordnet.princeton.edu/obtain), discipline is:

- A subject, subject area, subject field, field, field of study, study, bailiwick, branch of knowledge -- (a branch of knowledge; "in what discipline is his doctorate?"; "teachers should be well trained in their subject"; "anthropology is the study of human beings").

- A system of rules of conduct or method of practice; "he quickly learned the discipline of prison routine"; "for such a plan to work requires discipline."
- The trait of being well behaved; "he insisted on discipline among the troops."
- The training to improve strength or self-control.
- The act of punishing; "the offenders deserved the harsh discipline they received."

The question becomes, which one? Must you be restricted to just one, or should you be able to choose which form of "discipline" along with its contextual definition? Of course, with the structure of the Metabase System the answer is, Yes. Examples of these types of choices are provided in Section 6.3 on actual creation and application.

The second characteristic is that meta category value type values can be hierarchically organized into as broad or deep a set of levels as is practical. Thus, any given level of the hierarchy can have as many peer values as necessary, and any given level can have subordinate levels to a depth of 19, which is far deeper than is either practical or reasonable. For example, if a level is Measurement, subordinate levels might be English, and Metric. Within each, there could be Dimension, Weight, Volume, and Distance. Each of these, in turn, represents collections of meta category values such as the English Weight, Pound. The value, Pound, is what is assigned. Knowing what Pound means is something that is reported by the Metabase System. In one case it is weight, and in a different case, it is British money, and in a third case it might mean the intensity of what is being done now, "pound the meaning of meta category value types into the reader's head."

The third characteristic, sequencing, gets back to the opening example. That is, are the following name strings all the same?

- Total New England Annual Sales Amount US Dollars.
- Annual New England Sales US Dollars Amount Total.
- Annual US Dollar Sales Total Amount New England.

Clearly all the words are the same. If the meaning is the same, should not the name be the same? While human analysts might conclude that they are, computers would be hard pressed to come to that conclusion without a very large quantity of rules.

To ensure that the names all appear the same irrelevant of the assignment sequence, each meta category value type has a sequence number. This ensures, for example, that no matter the assignment sequence, the result would always be:

Annual New England Sales US Dollars Amount Total

That is because Annual is from Temporal and has a sequence of 1, and New England is from Geography and has a sequence number of 2 for the prefix assignments. Similarly, US Dollars has the sequence number 1 as a form of monetary units from the United States, Amount has a sequence of 2 as a Quantitative form of Money, and Total has the sequence of 3 as a Computational operation.

The sequence numbers can be modified by the Metabase System user who has access to the data element module. Obviously, great care should be taken to engineer this meta category value type taxonomy as it has great effect on the fundamental process of business fact name construction.

6.2.3 Meta Category Values

Meta category values are the actual values assigned as the semantics of a business fact. In the case above, the common business name of the fact being represented is Sales, and the rest of the words are all modifiers of that fact, where two are prefix semantic modifiers, and three are suffix data use modifiers. The process of assignment is illustrated in Section 6.3.

6.2.3.1 Semantic Modifier Hierarchies

Semantic modifiers represent semantic-based restrictions on both the meaning and the implied value set of a business fact. Semantic modifiers always precede the common business name. The fundamental meaning of the value set is provided by the common business name of the business fact. For example, the common business name, Sales, does not by itself contain contextual restrictions on which Sales are implied. Semantic modifiers provide the semantic contexts. For example, it may be <North American> Sales. Or, it might be <Estimated> <Annual> <North American> Sales. Each semantic modifier conveys a restricted meaning to the value set represented by the business fact, Sales. Each angle-bracketed term is an instance of some meta category value type. A common set of semantic modifier types include:

- Temporal.
- Accuracy.
- Geographic.
- Organizational.

Each type instance may have subtypes and once the type hierarchy is complete, the full meta category value set is provided. There can be at most one semantic modifier from each semantic modifier type hierarchy assigned to the business fact. For example, from above, there can be at most one temporal modifier, one accuracy modifier, one geographic modifier, and one organizational modifier. There can be fewer modifiers than the quantity of first level types, but not more. The data modeler modules of the Metabase System automatically creates names for data elements, attributes of entities, columns of tables, and DBMS columns of DBMS tables.

Every business fact contains three classes of names: Common Business Name, Metabase System constructed name, and User-Set name. The common business name has been described in previous sections, and might be for example, Reservation Number, Telephone Number, or Social Security Number. For these and other similar cases, this instantly resolves the problems caused by the three-part, <prime word>, <modifier>s, and <class word> naming strategy.

The Metabase System constructed names result from the executing of the Metabase System internal process that creates names by stringing together all the prefix assigned meta category values, the Common Business Name, and finally, all the suffix assigned meta category values.

The user set name is a user created arbitrary string, which might be SSN for Social Security Number.

Despite any of the business fact forms, the set of meta category value based semantics inherited through the meta category value types and hierarchies are maintained.

The primary purpose for dividing the meta category value classifications into two distinct hierarchies, that is, one for the meta category value types and another for the meta category values is to ensure that only one value from each meta category value hierarchy becomes a part of any context independent or dependent business fact. In Table 18, which contains a set of semantic modifiers, the first major row has the list of the four meta category value types, that is, Temporal, Accuracy, Geographic, and Organizational. The Temporal column, for example has seven subordinate rows represents the lower level hierarchy. In the case of the Temporal and Accuracy semantic modifier columns, all the entries are siblings. In the case of the Geographic and Organizational columns the entries are hierarchically related.

Each meta category value and meta category value type contains a number of meta-attributes beyond just the value (e.g., North American). Figure 34 presents an instance diagram of the entries of the data contained in the meta category value type and meta category value Hierarchy table, Table 18, for Semantic Modifiers.

In operational databases, semantic modifiers are represented in two different ways depending most generally on the type of data architecture class, and most specifically on the type of data structure in which semantics are implemented.

In the data architecture class databases, semantic modifiers and even some data use modifiers are represented in columns as column name parts so that the columns can be readily recognized. For example, data warehouse and reference data databases, semantic modifiers often become actual values through which data rows are selected, analyzed, summarized, and reported. In the first way, the data structure of the tables and columns is highly specialized as the semantics are bound into the column's name. In the second way, the data structure of the tables is very generalized as the semantics become actual data values. Figure 35 and Table 21 illustrate these two approaches. Because the Metabase System's module, data modeler, approaches data standardization from a database perspective, all data semantics are first and foremost data values that can be used either way. That is, as column names for the specialized data structures, or as column values in rows for generalized data structures.

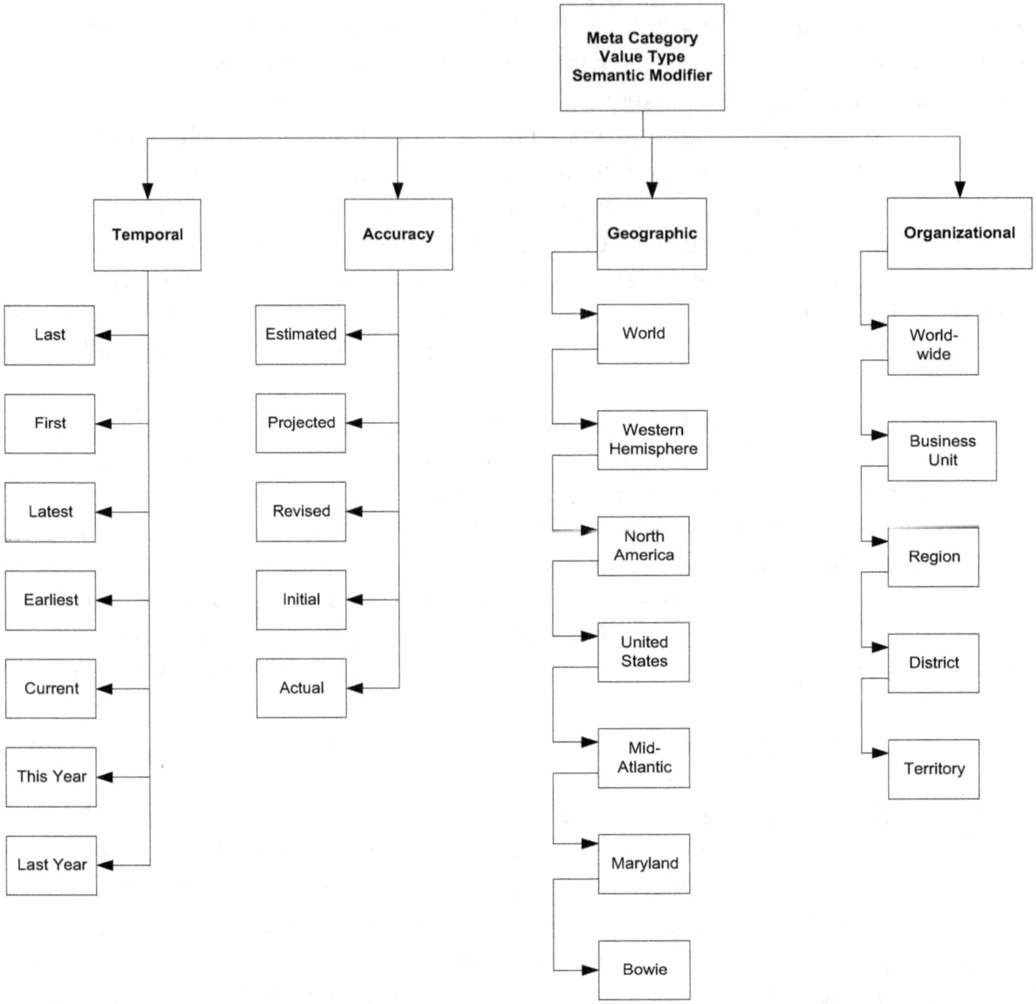

Figure 34. Meta category value types and meta category value hierarchies

Table: Valuables

Columns: Real Estate
Boats
Automobiles
Personal Property

Specialized Data Structure with Semantics Bound into Column Names

Table: Valuables

Columns: Valuable Name
Valuable Value

Generalized Data Structure with Semantics as column values within rows

Figure 35. Two methods of binding semantics into tables and columns of operational databases

6.2.3.2 Data Use Modifier Hierarchies

Data elements have data use modifiers that point, for example, to the data types, roles performed, and value units that serve to describe the use of the data element. Data user modifiers are appended to the end of the common business name. These are properly quite separate from the data element's name as the common business name might misrepresent a data element's allowed use. As stated previously, Social Security Number is not really a number in the mathematical sense of the word. To wit: What is the meaning of the average of a set of 10 Social Security Numbers, or their sum, difference, etc.?

The same problem exists with Telephone Number. It is really a set of joined codes consisting of a country code, an area code, an exchange and finally a four-digit number. To dial a phone number you might enter the numeric string, 1-800-USA-RAIL. That is a number?

The most common three data use modifiers are:

- Data type.
- Role.
- Unit.

Data types represent the fundamental business-use nature of the data values themselves. These are not however computer data types such as CHAR, VARCHAR, INTEGER, and LONG. Rather, they represent the business use of the data value. For example, does the number represent a weight, a code, dimension, volume, money, decimal, or an integer.

The second common type of data use modifier identifies the business role the data element value serves. For example, in the case of a data element called, *Skill Level*, the business data type of a number might be decimal, and the role it plays is that of a *multiplier*. The role represents the "job" the data element does within the context in which it is contained. An example of the types of data use modifiers is provided in the Table 19. In this table, the three broad classes of data use modifiers are: Data Type, Role, and Unit. In the data type category there are seven example instances such as Date, Code, Text, and so on. Figure 36 presents that same information as an instance diagram.

There can be at most one data use modifier from each data use modifier type hierarchy. For example, from above, there can be at most one data type (e.g., decimal, money, integer), one role (e.g., identifier, factor, flag, or indicator), and one unit (e.g., inches, time, and US Dollars). There can be fewer data use modifiers than the quantity of first level types, but not more. As stated above, the data modeler automatically creates names for data elements, attributes of entities, and columns of tables. The Metabase System also automatically creates names for DBMS columns, but just not with the assist of Operational Data Model assigned semantic or data use modifiers. User-names may, of course, be additionally created by the user. Notwithstanding any such changes, the set of semantics inherited through the meta category value types and hierarchies are always maintained.

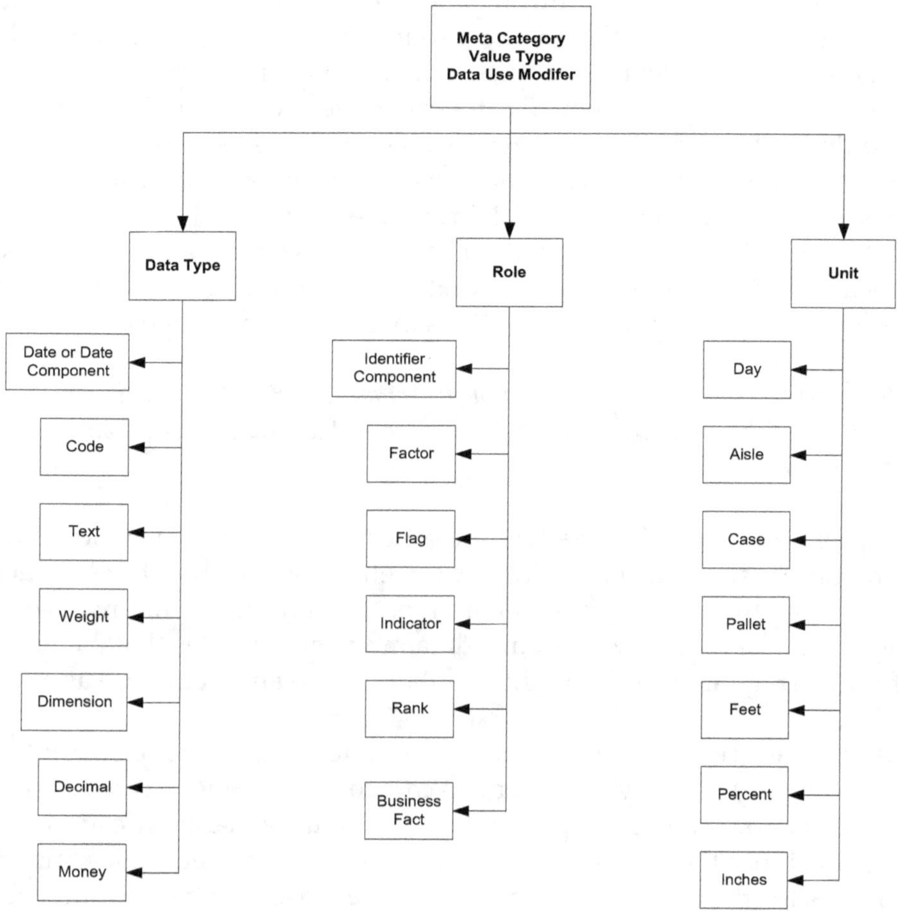

Figure 36. Meta category value type and meta category value hierarchies

6.2.4 Meta Category Value Types and Meta Category Value Hierarchies Summary

Meta category value types and meta category values, along with other meta-attributes such as definition fragments and a defined set of abbreviations greatly enhance and ease the construction process of the semantics of data elements and the semantics for technology dependent contexts such as fields in files, cells in screens, columns in tables, and variables in programs. A most commonly produced meta-attribute is the name. That is, data element name, column name, field name, cell name, or variable name. To assist in repeatable name construction, the Metabase System's data modeler requires that the meta category value types and meta category values be sequenced in an enterprise-defined order. This way, when a name is constructed the name parts are strung together, one with another in a repeatable order.

For example, the definition of North America would be the Continental United States and Canada. The acceptable alternatives might be: N_America[7], and N_A. An easy extension to this architecture is the addition of the values, definitions and abbreviations in different languages.

> *Note well, however, the name is not the semantics. Rather name is just one of the meta-attributes associated with the data element, field, cell, column, or variable.*

The definitions and value alternatives are especially valuable in avoiding the task of creating business fact definitions altogether, which should be no more nor any less than the stylistic concatenation of the definitions fragments of its semantic parts. The Metabase System's data modeler provides that facility. The definition fragments are brought together in the same sequence as are the name parts.

A standard report from the Metabase System's data modeler prints these meta category value types and meta category value hierarchies so they can be reviewed and analyzed against the actual data semantics desired by the enterprise. Since these meta category value types and hierarchies form the essential basis for the most significant of the enterprise's data semantics, great care should be expended in their determination, review and finalization.

6.3 Application of Semantic Hierarchies

Meta category value hierarchies are a key source of the overall set of semantics for data element concepts, data elements, attributes and columns.

[7] Note: computer languages usually "frown" on the use of spaces and periods within names. So, when ever they are encountered in the Operational Data Model, they are replaced with the underscore (_).

The Metabase System has these highly organized such that semantics inconsistencies are prevented. Meta category value hierarchies consist of:

- Meta Category Value Type Class.
- Meta Category Value Type.
- Meta Category Value.

6.3.1 Meta Category Value Type Class

Figure 37 shows the two instances of meta category value type classes. The only two allowed values are prefix and suffix. All meta category value types and contained meta category values are thus grouped to appear either before the common business name or after the common business name. Figure 38

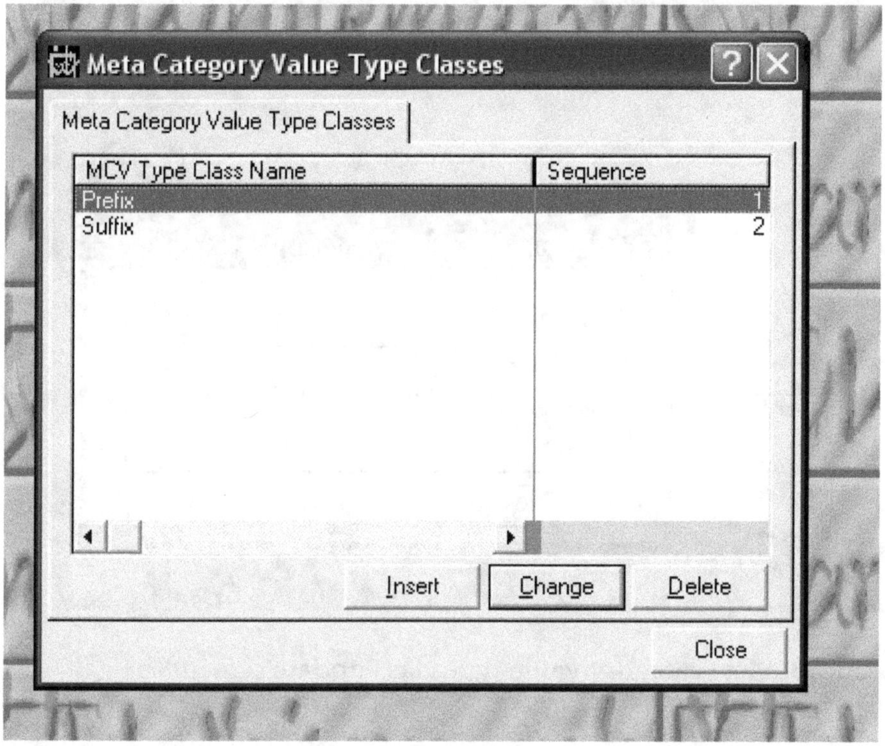

Figure 37. Meta category value type class.

shows the update screen where the actual words that reflect prefix or suffix can be changed.

When meta category value type classes are employed as selectors for meta category value types, screens such as those shown in Figures 39 and 40 result. In Figure 39, Prefix is selected. That causes the meta category value types of Temporal, Accuracy, and Geographic to be visible. Temporal has been selected. Shown in the bottom browse in an exploded form are the hierarchical representations of Temporal, that is, Annual through to Daily.

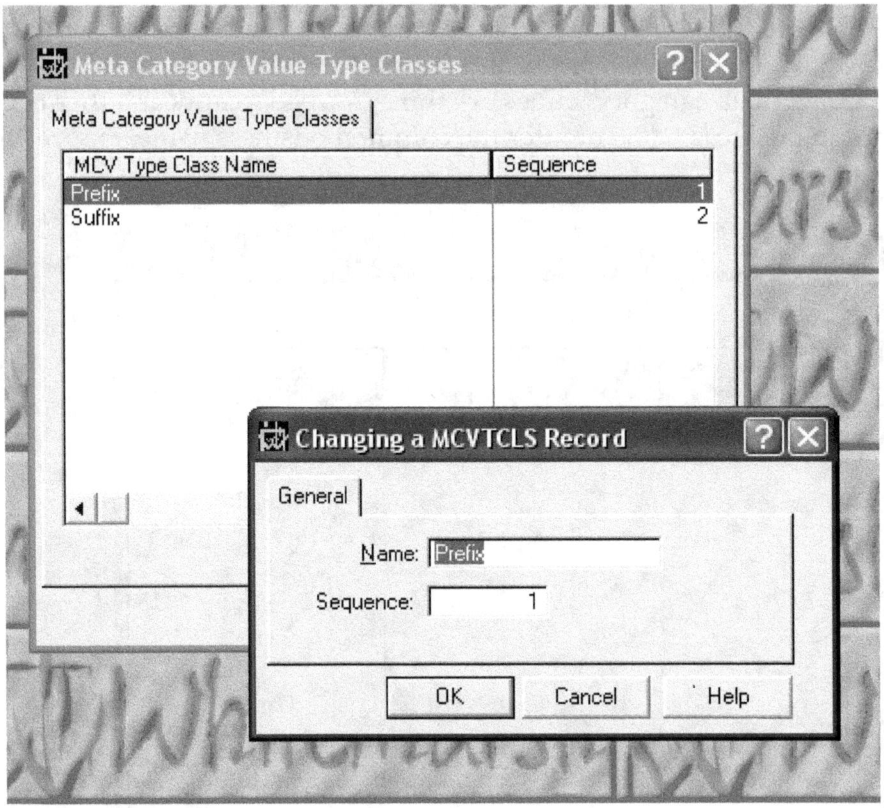

Figure 38. Meta category value type class update screen.

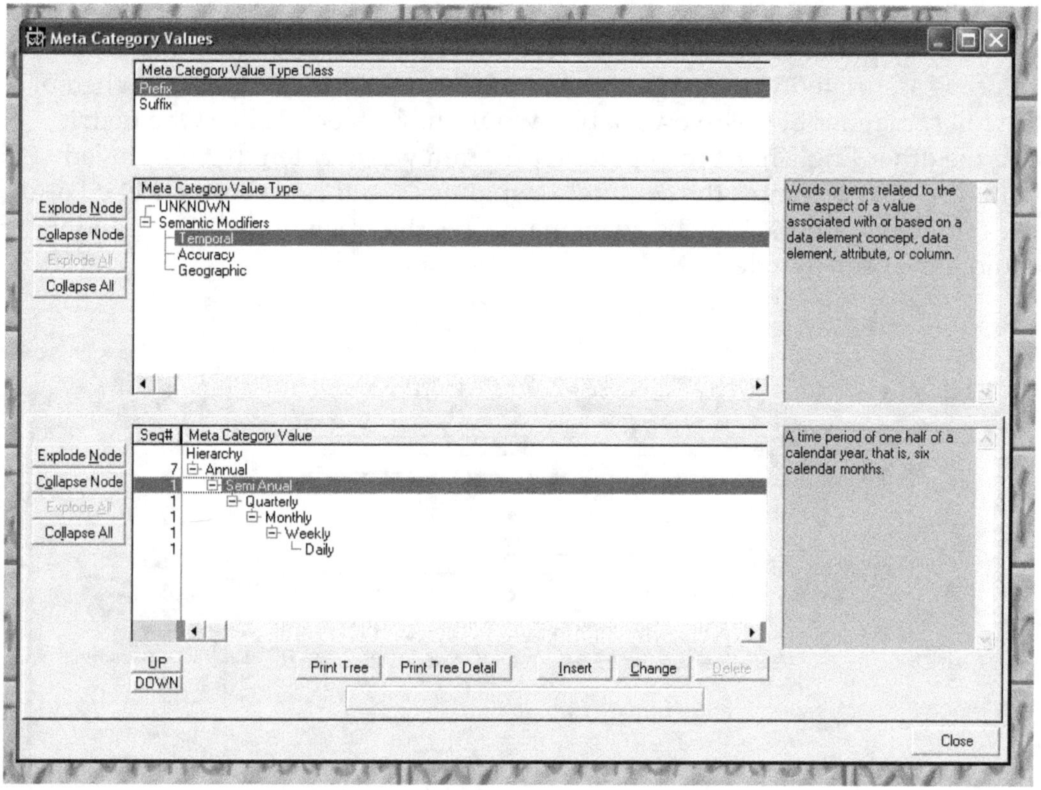

Figure 39. Prefix subseted meta category value types and meta category values.

In contrast, Figure 40 shows that Suffix is selected. The meta category value types that are shown are the Data Use Modifiers. Selected is Dimension below English below Units. Shown in the meta category value window are the three dimensions, Length Width, and Height.

As a practical application of Meta Category Values, if the database columns associated with the NASA Mars Lander had employed this scheme, the full semantics of the two variables would have been visible: One metric and the other English. At the time the program was written that employed both thse data columns, the complete semantics could have been checked and the conflicted notices. May be, just6 maybe be, the crash of the Mars Lander could have been avoided.

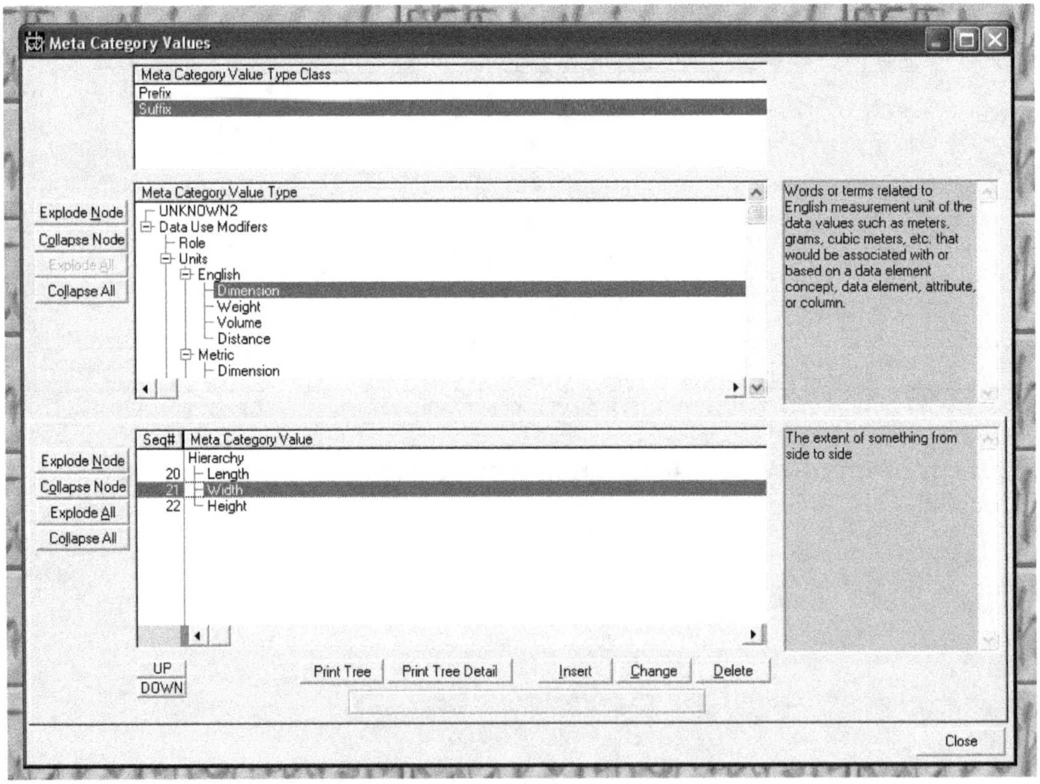

Figure 40. Suffix subseted meta category value types and meta category values.

6.3.2 Meta Category Value Type

Meta category value types are a type of classification hierarchy. Each meta category value type is set within either a Prefix or a Suffix meta category value type class. Meta category value hierarchies exist within specific meta category value type hierarchies. Figure 41 presents a list of meta category value types. From this window there are two super-classes, prefix and suffix. Show are the hierarchy for prefix, that is, semantic modifiers. There may be additional prefix trees. Within this tree the three shown are Accuracy, Geographic, and Temporal. The selected meta category value type, Accuracy, causes the current list of meta category values to be displayed. In Figure 41, the Sequence Number column shows that Accuracy meta category value word choices will appear before geographic category value word choices before temporal meta category value word choices. If the Metabase System administrator wishes to change that sequence, the Up | Down buttons can be used.

Meta category value types can be entered either in a hierarchy or as

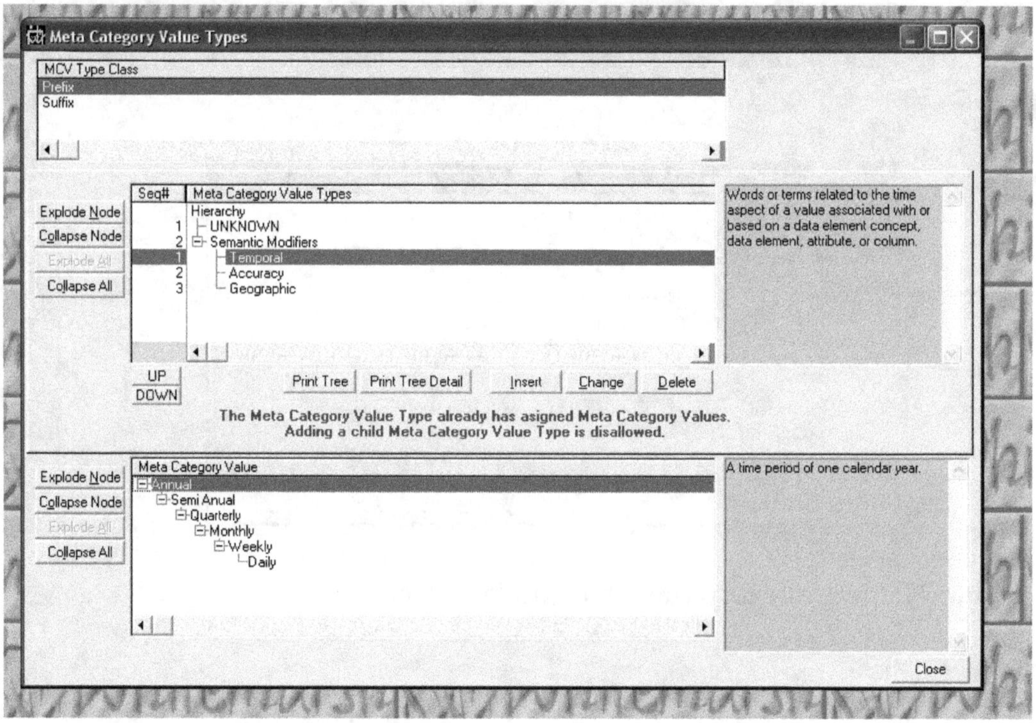

Figure 41. Meta category value types.

siblings. If it is to be a sibling, select the parent meta category value type and press Insert. The screen in Figure 42 is presented. Enter the name and description. Here, whatever is entered will be in a sibling relationship with text, description, date, and definition. The sequence number is automatically generated.

If an additional level of hierarchy is wanted below a meta category value type, select the meta category value type and press Insert. Enter the information as required by Figure 42 and press enter. The new meta category value type and level will appear. If a meta category value type already has assigned meta category values, a lower level new meta category value type cannot be created.

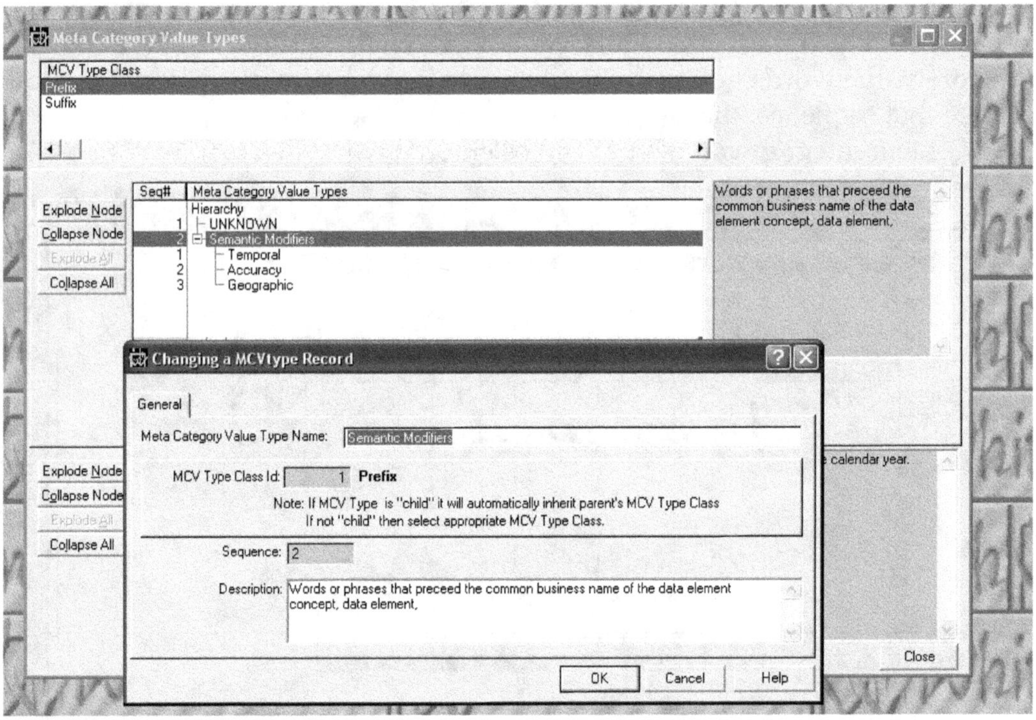

Figure 42. Meta category value type update screen.

The selection of a meta category value Class causes a Subseting of meta category type values. Figures 43, 44, and 45 illustrate that. In Figure 43, meta category value type class, Prefix, is selected. Selected also is meta category value type, Geographic. Note that there is now a message that says that a "child" of Geographic cannot be installed. The reason for that message is that the meta category value words and phrases, for example, United States have been assigned to the Geographic meta category value type. When a meta category value is assigned to a given meta category value type, the ability to install "child" meta category value types is disabled. The logic for disabling the creation of a "child" is that meta category values can be created only at the lowest level of meta category value type. Not allowed is a child of "Geographic." This makes a clean separation between meta category value types and meta category values.

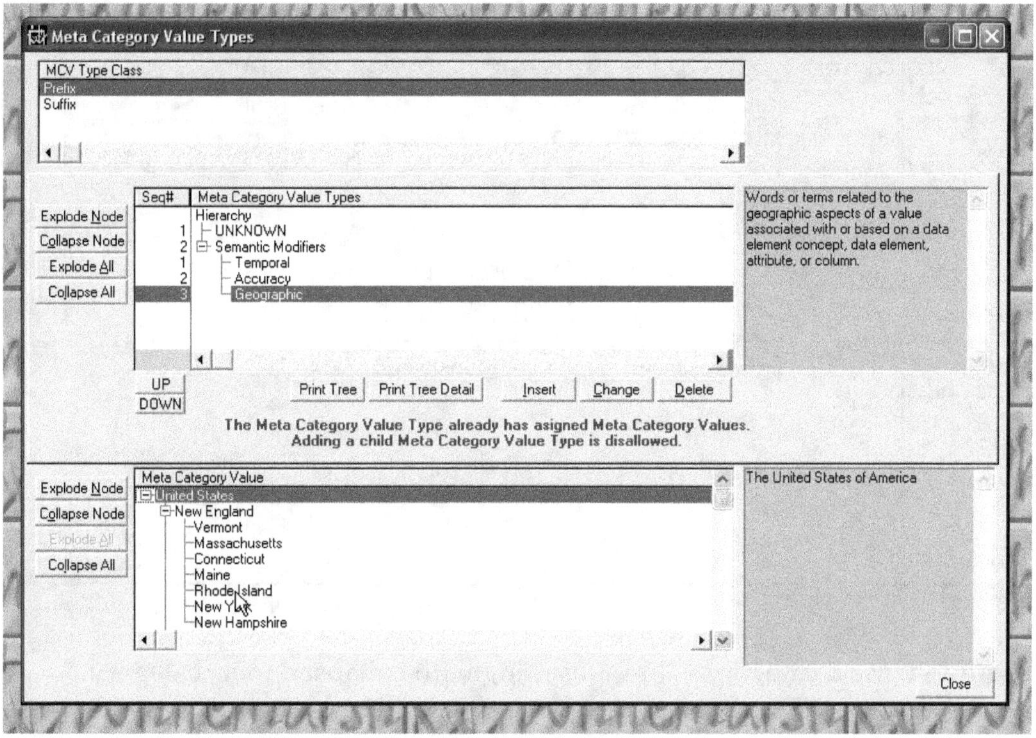

Figure 43. Meta category value type subsetting of meta category values.

Figure 44 also shows a different case for meta category value types. First, the meta category value type class selected is Suffix. Role is the selected Data User Modifier. This shows there are also Units, and within Units, English. What is being shown is that meta category value types can themselves have hierarchies. There is no practical limit to the depth or breadth of the hierarchy. In Figure 44, all the meta category values are un-exploded. There are therefore meta category value breakdowns for Text, Identifier, and Code. Figure 45 is the same as Figure 44 except that the Identifier meta category value is exploded.

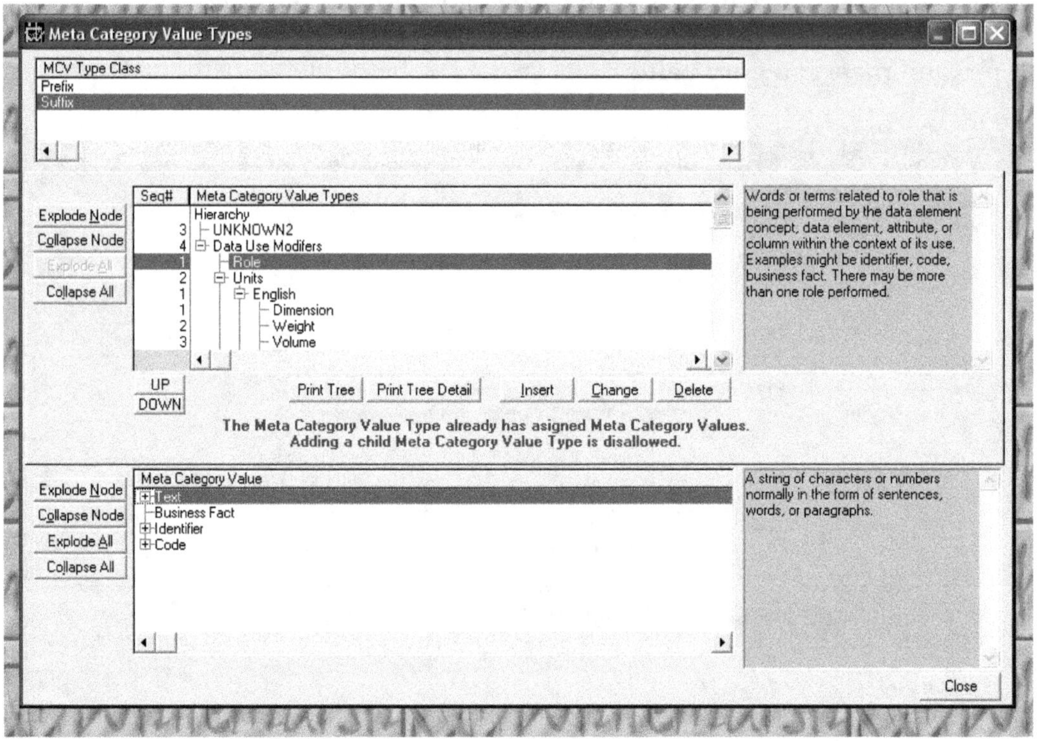

Figure 44. Meta category value subsetting with collapsed meta category values.

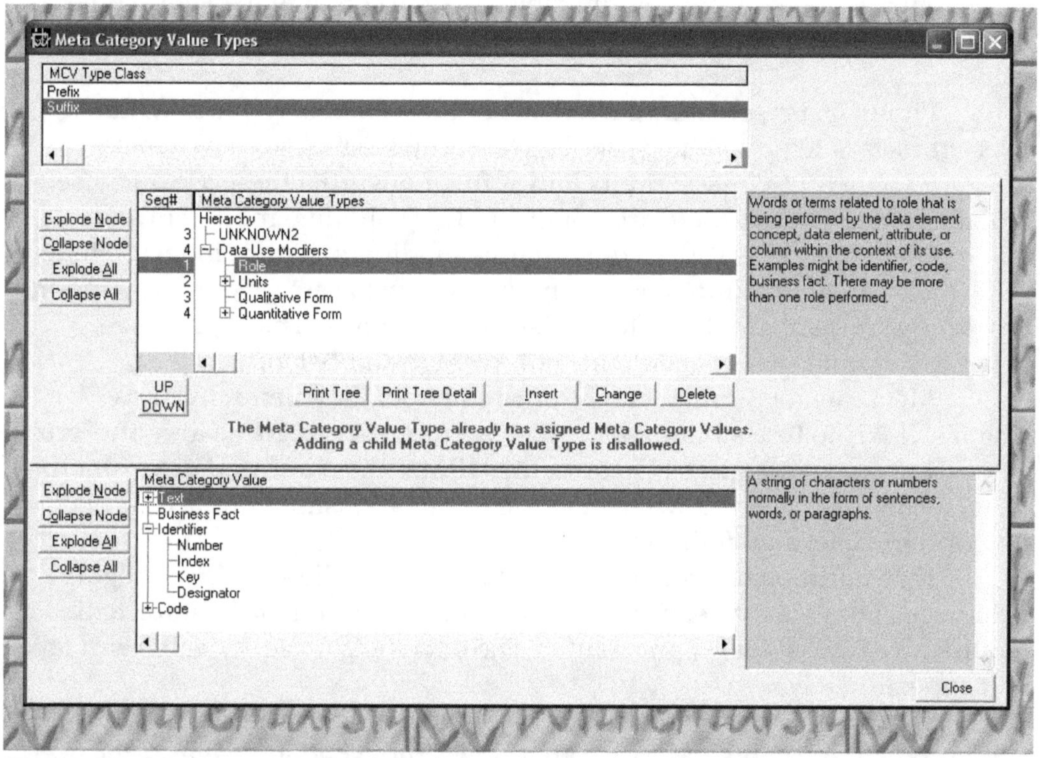

Figure 45. Meta category value type subseting with expanded meta category value.

6.3.3 Meta Category Value

A meta category value is a word or phrase that has a name and a definition. Meta category values can exist within a hierarchy of meta category values. Meta category values exist within a type hierarchy. Finally, there are two classes of meta category value types, prefix and suffix.

It must be remembered that ONLY meta category values are the source for words or phrases that are either prefixed or suffixed to data element concepts, data elements, attributes, or columns. Meta category value hierarchies are employed only to distinguish between sets of meta category values. Usually, data element concepts have few if any assigned meta category values. That is because they are the most general and at the level of a data element concept. Then, come data elements, which are more precise, but still rather general. Finally come context dependent uses, that is, attributes within entities and columns within tables. As the contextual level becomes more precise, there is a tendency to increase the quantity of meta category values.

Figure 46 presents the current set of meta category values within a selected meta category value type. The top browse shows the two meta category value type classes: prefix and suffix. For suffix, these are data use modifiers. Within that, there are, for example, qualitative modifiers and quantitative modifiers. Within quantitative modifiers there is money. Money is the meta category value type that points to a meta category value, Amount, and below Amount are the different kinds of amounts. That is, average, balance, cost, and price. It could also be average and balance.

Meta category values can be entered either as a hierarchy or as siblings. If it is to be a sibling, select the parent meta category value and press enter. The screen in Figure 47 is presented. Enter the name and description. Here, whatever is entered will be in a sibling relationship with text, description, date, and definition.

If an additional level of hierarchy is wanted below a meta category value, select the meta category value and press enter. Enter the information as required by Figure 47 and press enter. The new meta category value and level will appear.

Creating all these meta category value types and values is not simple. It takes much reflection and care. What is accomplished, however is the complete taxonomic classification of all the words that, in collections, define the semantics of all enterprise persistent data. Well done, data becomes semantically interoperable. The alternative is not just semantically unacceptable, but it is very wasteful of human and financial resources. Semantic chaos costs many times more than semantic harmony.

Two cases need to be shown for meta category values. The first is illustrated by Figure 39. It shows that meta category values can be deeply nested. Here, the seven levels go from Annual to Daily. If, for example, Annual is assigned to a data element, Semi-annual can be assigned to an attribute. The opposite is not true. That is, if Annual is assigned to an attribute, Semi-annual cannot be assigned to the attribute's parent data element. This preserves the hierarchical context of the assigned semantics. Figure 40 shows there can be a list of values that are all the same level. That is, length, width, and height. Figure 45 shows that the list can be mixed. That is, some levels are decomposed and others are not.

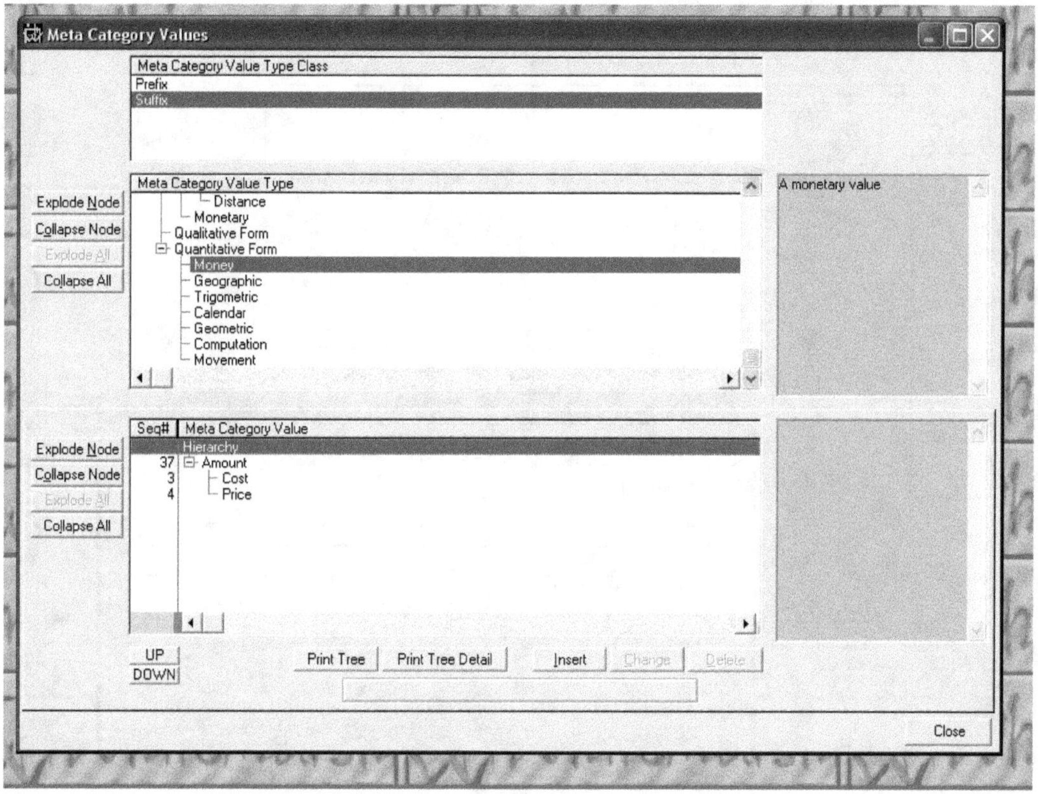

Figure 46. Meta category values.

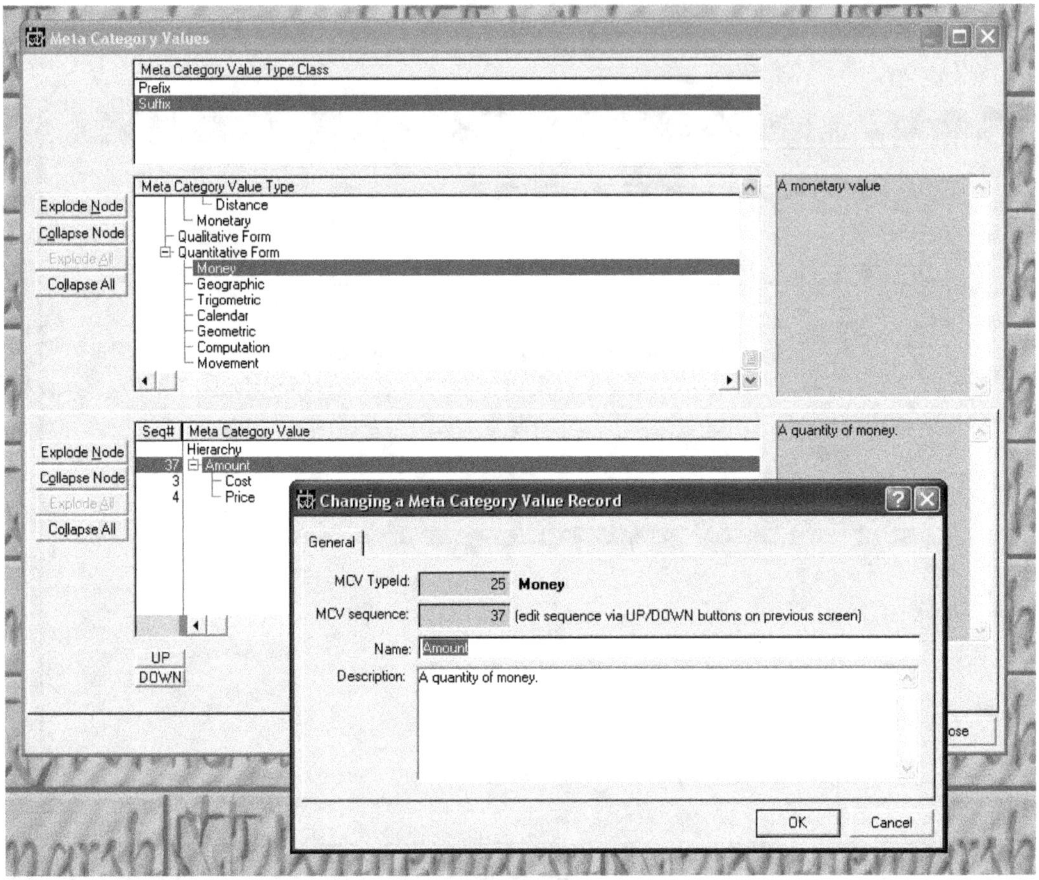

Figure 47. Meta category value update screen.

6.3.4 Meta Category Value Assignments

One or more meta category values can be assigned to any of the following:

- Data element concepts.
- Data elements.
- Attributes.
- Columns.

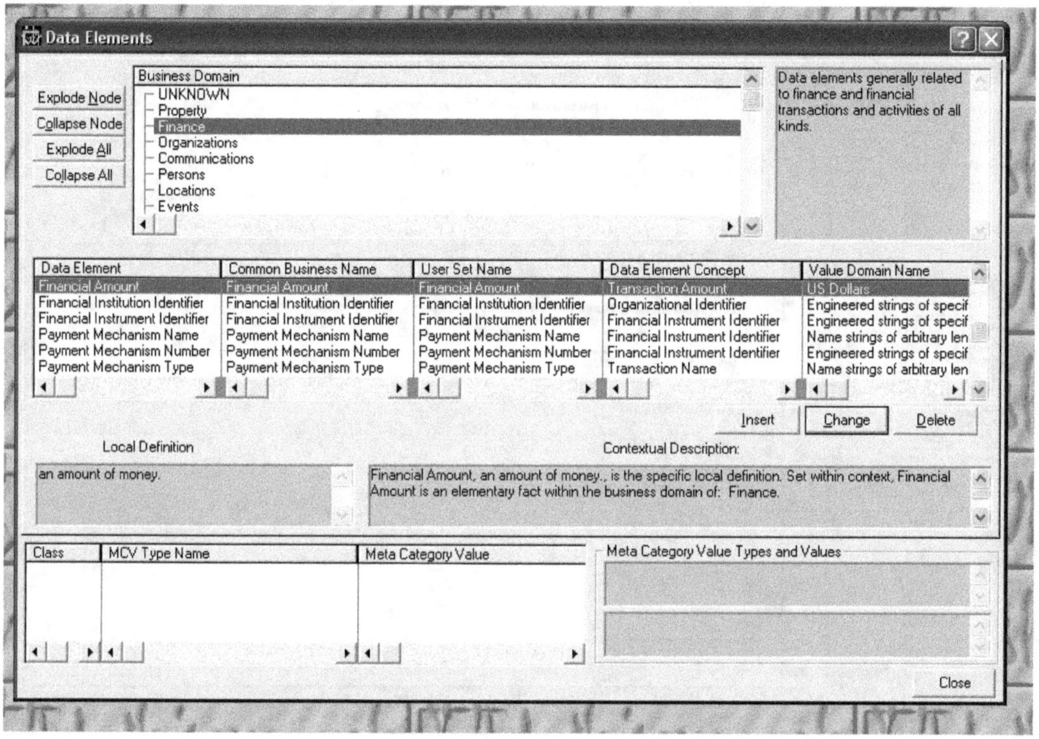

Figure 48. Data element browse screen.

Since the process of assigning one or more meta category values is the same, the process illustrated here is for assigning meta category values to the data element, Salary. This process is illustrated in Figures 48 through 52. In Figure 48, the data element browse screen is shown. In this screen, the business domain, Finance was selected. Shown in the second browse is all the existing Finance data elements. The Insert button is pressed and Figure 49 is presented.

Figure 49 is the data element insert or change screen. In this screen "1" was entered because the data element concept (as populated through a data element concept structure), for this example, is to be Unknown. The Common Business Name default name, Finance Unknown is cleared. Entered is the name, Sales. The data element Name and the User Set Name are cleared out. The Reset button is pressed. The reset process takes all the assigned prefix and suffix meta category values and attaches them before (prefix) and after (suffix) the Common Business Name. In this case the result is just Sales.

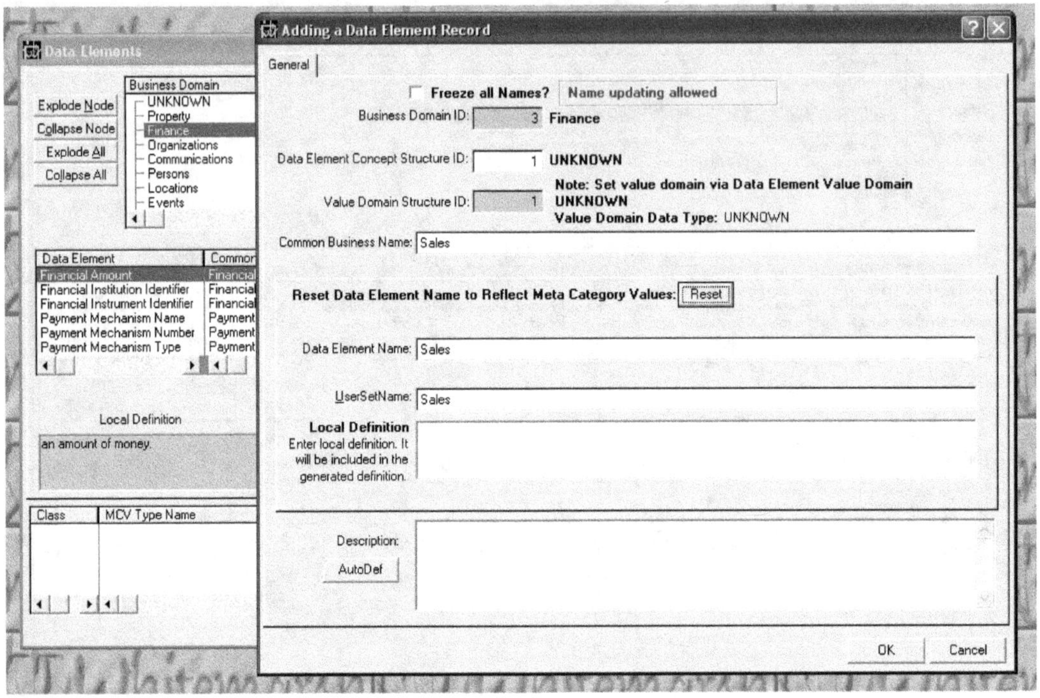

Figure 49. Data element update screen.

The next step is to create some prefix meta category values and some suffix meta category values. The data element built is:

Annual New England Sales US Dollars Amount Total

What needs to be added to the prefix set of meta category values are Annual, and New England. Added to the suffix set of meta category values are US Dollars, Amount, and Total. Figure 50 is an illustration of adding a meta category value. In this Figure, the business domain, Finance is selected. That causes just the data elements within that domain to be shown. The data element, Sales is selected. The meta category value is discovered by first selecting Prefix/Suffix. Finally, the meta category value type, Temporal is selected. At that point, only the Temporal meta category values show. Check the Annual value. Finally, press the Build button. The Metabase System

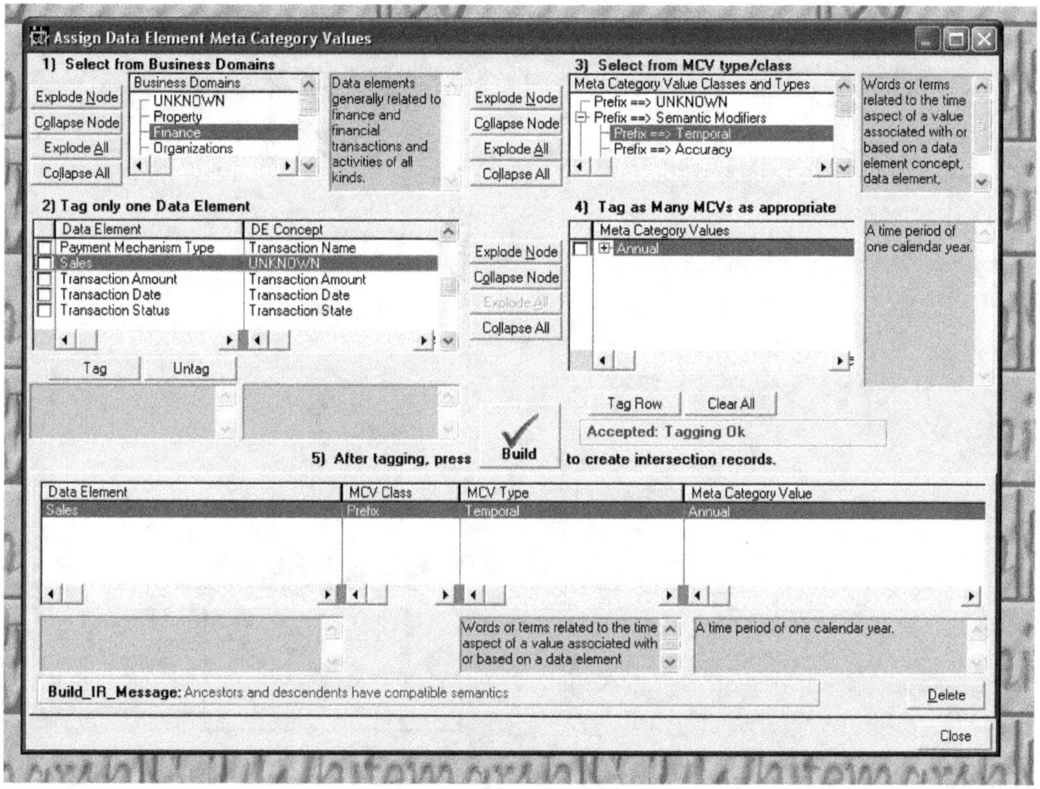

Figure 50. Allocated meta category value, annual to sales data element.

software examines all the choices and displays (at the bottom) a build message, which, here indicates that the semantic affinity of the already assigned meta category values Ancestor and Descendant semantics are OK. The build screen message is shown in Figure 51.

This same process is repeated for the second prefix Meta Category value. The suffix meta category values are allocated. It is not necessary to cycle through this process five different times, that is, Steps 1, 2, 3, 4, and 5. Rather, you can do steps 1 and 2, do steps 3 and 4 five times, and step 5 once. After doing the Build step you will receive five different "Acceptable" or "Unacceptable" messages. Only the "Acceptable" allocations will be shown.

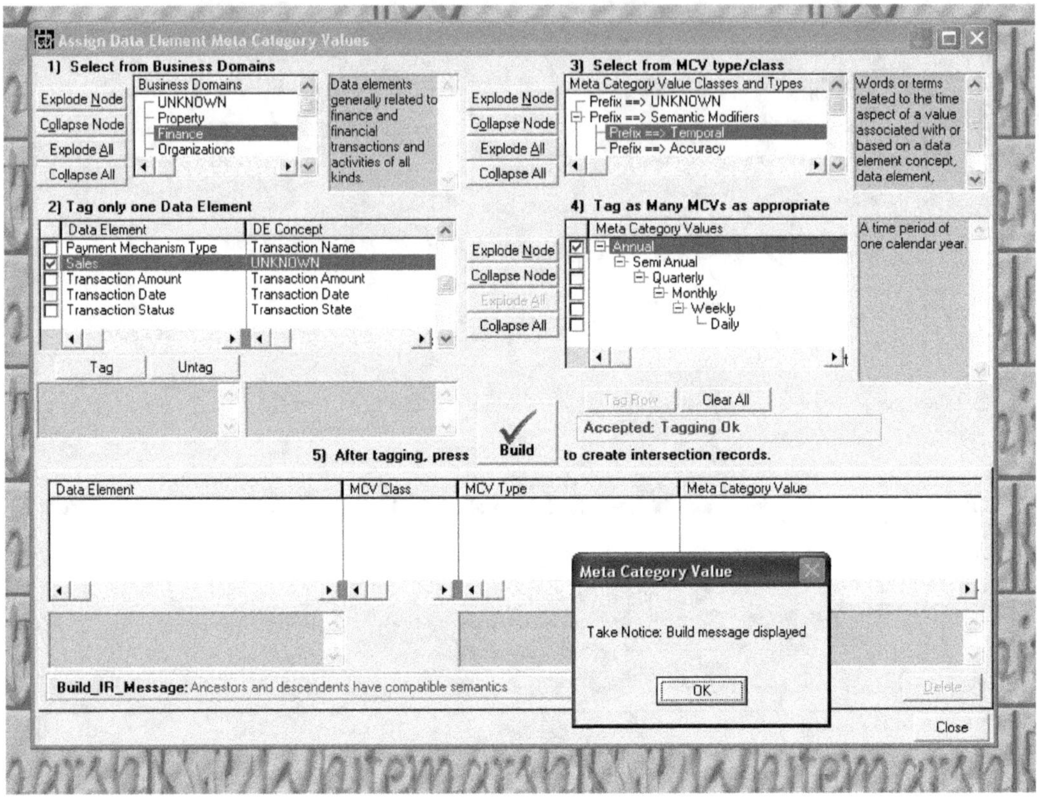

Figure 51. Allocating the prefix meta category value, annual to the sales data element.

Figure 52 shows all the allocated meta category values for the data element, Sales. The next step is to modify the Sales data element name. This is accomplished by pressing the Change button, and then:

- Clear the "Freeze all names" check mark.
- Clear the value, Sales from the data element Name and User Set Name entries.
- Press the Reset button.

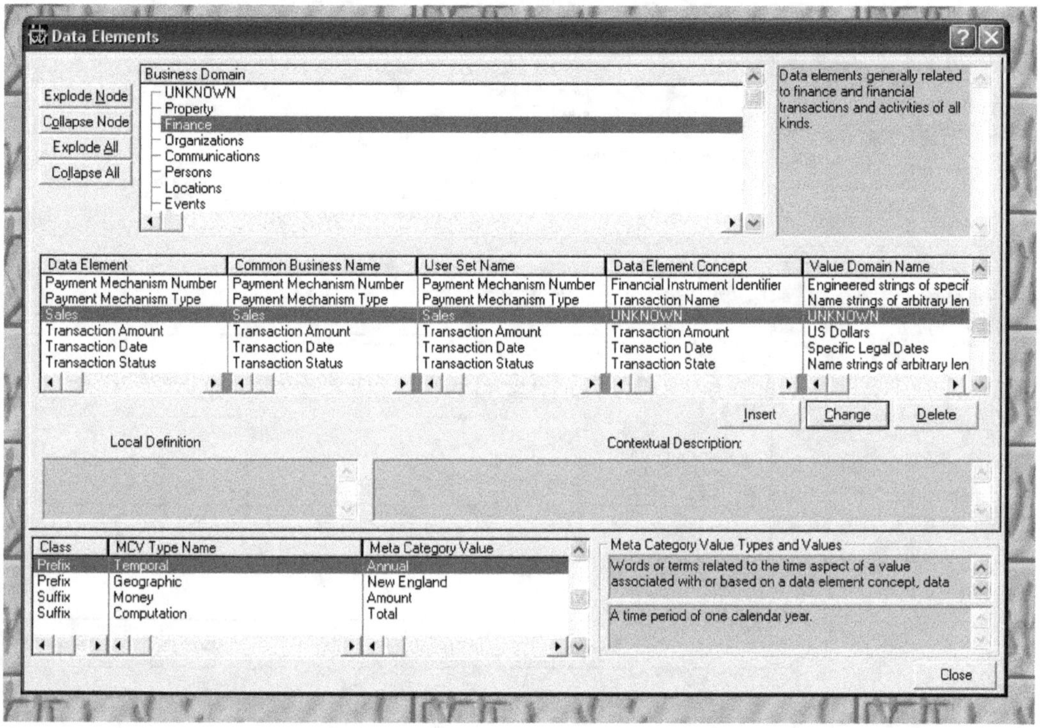

Figure 52. Complete allocation of meta category values for data element, salary.

The Metabase System process accesses all the assigned meta category values and constructs the name. The constructed name is shown in Figure 53.

What is left to accomplish is create a "local definition," and press the AutoDef button that causes the Metabase System to create the contextual definition. The entered local definition is, "money collected because of a set of business transactions between the company and a set of clients." The definition that is created from the AutoDef process is:

> Sales, money collected because of a set of business transactions between the company and a set of clients, is the specific local definition. Set within context, Sales is an elementary fact within the business domain of: Finance.

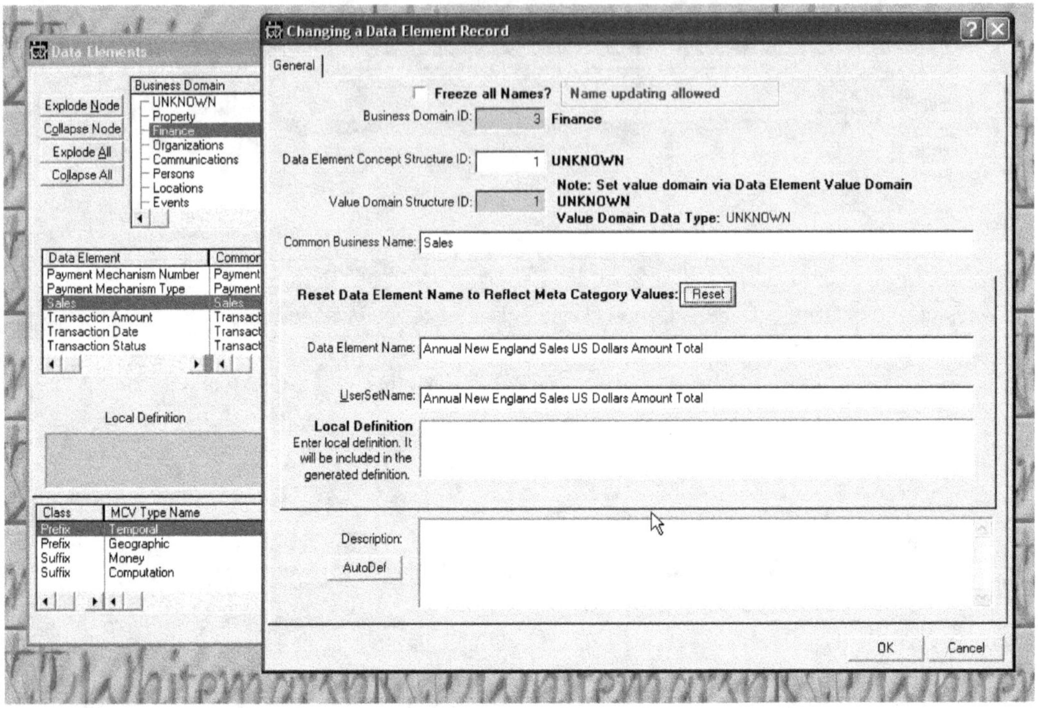

Figure 53. Computer generated data element name for Salary.

The common business name for Sales is: Sales.

Sales is a specialization of the data element concept: UNKNOWN, which is defined as:

Sales is bound by the value domain: UNKNOWN, which is defined as: UNKNOWN It is constrained by the data type UNKNOWN with a scale-allowed indicator of N and a precision-allowed indicator of N

The meaning of Sales is directly restricted by::=

Annual within Temporal. Annual represents: A time period of one calendar year. Temporal represents: Words or terms related to the time aspect of a value associated with or based on a data element concept, data element, attribute, or column.

New England within Geographic. New England represents: A group of states within the North East area of the United States. Geographic represents: Words or terms related to the geographic aspects of a value associated with or based on a data element concept, data element, attribute, or column.

The meaning of Sales is directly restricted by::=

Amount within Money. Amount represents: A quantity of money. Money represents: A monetary value

Total within Computation. Total represents: Arithmetic sum of a set of numbers Computation represents: Forms or kinds of mathematic operations over a set of numbers.

US Dollars within Monetary. US Dollars represents: A money amount expressed in currency from the United Sates. Monetary represents: The names of currency such as dollars, pounds sterling, and the like.

For sure, the computer constructed definition set out above will not win a Pulitzer prize. It does however get accomplished in about 2 or fewer seconds. The "unknowns" above identify the data components for which there are no definitions. Once these are defined their definitions become part of the automated definition.

6.4 Error Classes

The semantic hierarchies chapter is a very critical component because the very words that comprise fact names within an interoperable data environment can be contextually identified, named, and defined through the allocations of specific strings from these hierarchies.

Humans are creatures of commonality. Discourse is based on common phonemes with common meaning within shared domains. The degree to which we are interoperable is directly related to our shared words, phrases, and meanings. We all like to believe that synonyms and nuance differences are valuable and even stylish. These may commonly exist in verbal or written discourse, but when trying to build shared-data business information systems, these differences are the makings of semantic discord. Boring is best; variety is not.

Simply put, words, within their context have single meanings. When definitions are automagically constructed they will always be generated the same way. Hence, they will be reliable and repeatable regardless of the person or organization, or function within which they are generated. What is the same is the meant to be the same, and vice versa. This strategy, if followed will greatly reduce the first error quantity and it naturally follows that the second error class is reduced as well. All these same words and same meanings are carried down through the entire set of Metabase System data models, that is, Data Element, Specified, Implemented, Operational, and View.

Consequently, the words that comprise fact names within an interoperable data environment should be constructed through the allocations of specific strings from within the semantic hierarchies. A very important additional benefit from these semantic hierarchies is the ability of the Metabase System to accomplish automatic name construction and automatic definition of all business facts from data element concepts within the Data Element Model down through to DBMS columns within the Operational Data Model. This enables all uses of databases constructed through these strategies to know the enterprise-wide and consistent names and definitions of all facts. Clearly this addresses both classes of errors.

A very important additional benefit from these semantic hierarchies is the ability of the Metabase System to do automatic name construction and automatic definitions of all business facts from data element concepts within the Data Element Model down through to DBMS columns within the Operational Data Model. This enables all uses of databases constructed through these strategies to know the enterprise-wide and consistent definitions of all facts. Clearly this addresses both classes of errors.

Simply put, when this is accomplished, what is named the same is the same, and what is named differently is different.

6.5 Semantic Hierarchies Summary

This chapter was a "can do" chapter. That is, it is a chapter through which you can define and illustrate the actual establishment of data semantics management. This chapter identified and provided real screen-shot examples of how to define data semantic hierarchies and how to deploy them in real-world examples.

The real world examples come from the Metabase System and the example metadata database, Movies.

Semantic hierarchies are a critical element in the development of enterprise-wide semantics because they form the building blocks for name construction. What is very important about these semantic hierarchies is that they can be allocated to a number of different components in the data semantics management space including: data element concepts, data elements, attributes, and columns. Not only can these be allocated, each semantic name string can be defined singly or in context with its hierarchy.

Further, the Metabase System software ensures that semantic nesting is enforced. Users are not able to allocate to a lower level component, for example, a column of a table, a semantic that is the same as or at a higher level of generalization as an attribute of an entity. Attributes are semantically superior to columns. Thus, assigning the same or a lower level semantic to an attribute than is assigned to a column would simply be a mistake. The Metabase System prevents such mistakes.

Finally, not only are the assigned semantics employed to automagically construct names, but they are also employed to automagically define data element concepts, data elements, attributes, and columns. Both these capabilities are critical to the efficient and effective management of enterprise-wide data semantics.

The next chapter, Chapter 7, value domain Management, is the next "can do" chapter. Increasingly, value domains are important to data semantics management. The Type II error cited above, that is, the ability to specify Gender = 1 as meaning Female in one environment while it means Male in another environment can be a critical source of errors that must be prevented.

6.6 Questions and Exercises

1. Semantic hierarchies are employed to organize word lists that can be assigned to business facts (data element concepts through columns). What are the benefits from having a classes and word-instances separation as is illustrated in Figure 33?

2. Figure 33 shows there are three different sets, Meta Category Value Type Classes, Meta Category Value Types, and Meta Category Values. What is the purpose of each? Why benefits exist because of the ability to have these levels of separation? Does your enterprise have such separation abilities? If so, what is the value? How does it compare to the Figure 33 scheme?

3. What is the purpose of the Meta Category Value Type Classes? Examine several sets of SQL schemas and discover the Common Business names. Identify the Meta Category Values and separate them into prefix and suffix sets. Are any allocated incorrectly? Has that caused confusion? Same word but possibly different meaning? How have you handled that? Wouldn't it have been simpler to have a standard supported by automatic name construction to keep it all straight? How much time and resources could be saved from an automated scheme? Per project? Per functional area? Per organizational unit and for the enterprise?

4. What is the purpose of Meta Category Value Types? How does having the ability to have collections and groups of Meta Category Values help? How have you dealt with same word different meanings? How have you distinguished the different contexts? Has that caused confusion? Same word but possibly different meaning? How have you handled that? Wouldn't it have been simpler to have a standard supported by automatic name construction to keep it all straight? How much time and resources could be saved from an automated scheme? Per project? Per functional area? Per organizational unit and for the enterprise?

5. There are two types of Meta Category Value Type Classes, prefix and suffix. Prefix hierarchical word sets are for semantic modifiers and suffix hierarchical word sets are for data use modifiers. Do you have either of these in your organization? How do you identify, categorize, name, and define them? How do you manage them? Is there a scheme whereby use of these modifier sets are mandatory? What benefit would occur if you had mandatory strategies in your organization? How would you compute the benefit? How would you present the benefits to management such that these efforts would be funded?

6. Can you justify these semantic hierarchy strategies on a one-off project? Would it speed data modeling activities or slow them down? If slow them down, why? Suppose these strategies were alternatively set within an organization, a function, or an enterprise level. Where is the break even point between cost and benefit?

7. Is your organization engineered to see the "big picture" and make productive use of commonly developed materials and problem solutions? If yes, then how do individual projects take advantage of commonly-developed strategies and materials? If not, then how many times do you believe the same problem has been identified and solved by different organizations in your enterprise? What do you estimate the cost to be of not having commonly developed and deployed solutions? How would you quantify the costs? How would you present your findings to management? How would you convince them to fund and support such efforts?

8. If your organization is engineered to identify, create solutions for, and then deploy common solutions to common problems, how are these efforts organized, funded, and maintained? How are these efforts funded? A tax on every project or some sort of project overhead charge? How are benefits from individual projects because of the solutions identified, quantified and reported to management? What has been the support from management? How have these efforts proven their worth?

9. Examine a collection of SQL DDL schemas and identify Meta Category Value Type groups. Across these groups, are there any duplicates? Synonyms. What is the real value of a synonym? Don't they just cause confusion? Why use a different word if there is no difference in meaning? In this case, is not boring better? Why?

10. Examine a collection of SQL DDL schemas and identify the semantic modifiers that are illustrated in Figure 34. Collect these into Meta Category Value Type groups. Across these groups, are there any duplicate semantic modifiers? Synonyms. What is the real value of a synonym? Don't they just cause confusion? Why use a different word if there is no difference in meaning? Now, is not boring better? Why? Wouldn't it have been simpler to have a standard supported by

automatic name construction to keep it all straight? How much time and resources could be saved from an automated scheme? Per project? Per functional area? Per organizational unit and for the enterprise?

11. Perform the same exercise as in Question 6 above but for data use modifiers as illustrated in Figure 36.

12. Request, download, and install the Metabase System. Execute the Data Element module within the Movies metabase database. All the semantic hierarchies are contained in this Module.

13. By means of Figure 41 and supporting text, practice the creation of new Meta Category Value Types. Use the results from Exercise 5 and determine which ones match what you discovered, which are new and which need modifications. Determine your changes and make them to the "Metabase" Metabase System database. Make sure you put them under the appropriate Meta Category Value Type Class.

14. By means of Figure 46 and supporting text, practice the creation of new semantic modifier Meta Category Values for specific Meta Category Value Types. Use the results from Exercise 6 and determine which ones match what you discovered, which are new and which need modifications. Determine your changes and make them to the "Metabase" Metabase System database. Make sure you put them under the appropriate Meta Category Value Type Class, Prefix. Note: you can employ WinSQL's databag feature to export values from a metadata database, for example, Movies and import these values into a different metadata database, for example, Metabase.

15. By means of Figure 46 and supporting text, practice the creation of new data use modifier Meta Category Values for specific Meta Category Value Types. Use the results from Exercise 7 and determine which ones match what you discovered, which are new and which need modifications. Determine your changes and make them to the "Metabase" Metabase System database. Make sure you put them under the appropriate Meta Category Value Type Class, Suffix.

16. Review Figure 53 and supporting text to see the approach for automatic creation of definitions. Is a strategy like this valuable? Are the definitions acceptable for most uses? Given that this automatic definition strategy applies to data element concepts, data elements, attributes, columns, DBMS columns, and to entities, tables, and DBMS tables, will not this greatly ease the tedious, time-consuming, and error prone definition process? Given that this strategy is acceptable, how much time and resources would it save for an average project, over an entire function, and across the enterprise? Note: you can employ WinSQL's databag feature to export values from a metadata database, for example, Movies and import these values into a different metadata database, for example, Metabase.

17. After accomplishing exercise 10, 11, and 12, review your existing data model creation process and compare the types, kinds, and quantities of work. If you had these strategies in place where you could allocate semantics and then automatically create names and definitions, how much time and resources would that have saved? Would it have impacted the "When will it *ever* be done?" syndrom?

18. Given that you agree that this allocation, naming, and definition strategy enables better standardization across databases, projects, functions, and the enterprise, how much faster and easier would projects become? It all this was supported by business information system generators, how much faster still would all these efforts be?

19. How many databases might have been eliminated in your enterprise if it were easier to decide that a given database had the same set of names to support the fact that it essentially had the same data as another database? What would have been the technical infrastructure savings for database creation, data collection and storage, business information system development, and for 10 years of maintenance for all these components. What would have been the savings for all the business information system development? Training?

20. Estimate the quantity of business information systems that exist that are interrelated through ETL (extract-transform-load) facilities that might have been eliminated if they were standardized and/or able to be mapped semantics across these business information systems?

21. The Error Classes section outlines the types of errors addressed by this chapter's materials. Do you agree that the error classes are addressed? How? What could be different? Has this chapter's contents and strategy provided an overall justification for its implementation? If yes, why? If now, why not? How would you quantity the benefits from having these error classes resolved? How would you present the approach, costs, and benefits to management in order to justify funding?

7

Value Domain Management

7.1 Introduction and Scope

Chapter 6, Semantic Hierarchies, was the first implementation chapter. That is, it was the first chapter that both defines and illustrates the actual establishment of data semantics management. That chapter identified and provided real screen-shot examples of how to define data semantic hierarchies and how to deploy them in real-world examples

Semantic hierarchies are a critical element in the development of enterprise-wide semantics because they form the building blocks for name construction. What is very important about these semantic hierarchies is that they can be allocated to a number of different components in the data semantics management space including: data element concepts, data elements, attributes, and columns. Not only can these semantic name strings be allocated, each can be defined individually or in context within its semantic taxonomy.

The Metabase System software ensures that semantic nesting is enforced. Users are not able to allocate to a lower level component, for example, a column of a table, a semantic that is the same as or at a higher level of generalization as assigned to an attribute of an entity. Attributes are semantically superior to columns. Thus, assigning the same or a lower level semantic to an attribute than is assigned to a column would simply be a mistake. The Metabase System prevents such mistakes.

Not only are the assigned semantics employed to automagically construct names, but they are also employed to automagically define data element concepts, data elements, attributes, and columns. Both these capabilities are critical to the efficient and effective management of enterprise-wide data semantics.

This chapter, Chapter 7, value domain Management, is the next "can do" chapter. Increasingly, value domains are important to data semantics management. A Type II error, that is, the ability to specify Gender = 1 as meaning Female in one environment while it means Male in another environment can be a critical source of errors that must be prevented. The first step in prevention is having the ability to know about these possibilities, and being able to control and manage value domain value collections. The strategies set out in this Chapter enable this to take place.

Value domain value sets, that is, reference data, if not properly managed can have very expensive effects as the following example illustrates. The United States Office of Trademarks, many years ago, reorganized itself into "law offices" rather than have a vast space of desks, clerks and lawyers. To make work more efficient, the IT staff created numbered paragraphs that could be used by the trademark analysts to communicate with an applicant. For example, a "24-87-33-68" might mean that a Trademark was granted, while "24-75-46-37" might indicate that a trademark application was turned down.

As the workloads grew in the various offices, some of the administrative work was passed from one office to another. Because there was no central management of the codes and paragraphs, some meanings of the numbered paragraph changed over time. The evitable finally happened. There was a work overload, and when one office asked another office to help, the same-numbered but migrated meaning paragraphs were used. The applicant was issued a trade mark when they should have been turned down. The fault was the migrated meaning contained in a paragraph.

The resolution involved revoking the issued trademark, and compensating the wronged-applicant with several million-dollars so they could research and get a new trademark and conduct a new product marketing campaign.

Another example of mistaken reference data occurred in a different Federal agency. Over the years, manual systems were automated, and finally ended up in a database. All during this data-trek there was no central management of value domains. Eventually there were 9-different values for genders. Nine? How strange. It was eventually resolved, but only after there were embarrassing counts, statistics, and reports to management.

A different example relates to coded values for personnel applications. The Federal agency, Office of Personnel Management seems to be required to preserve the actually entered values that an applicant placed on their application for Federal employment. Over the years, systems have changed from manual to punch-card to tape to disk to database. Maintaining immutable values over many decades in the face of changing requirements, customs, and mores is an impossible requirement. Nonetheless, the requirement seems to exist.

Another example relates to badly constructed and very expensive metadata. In a 2002 Request for Proposal from a DoD Agency, the agency indicated that errors waste about 250 staff-years annually and produce what is known as "frustrated cargo." 250 staff years are worth about $52 million.

To help understand this problem, the DoD agency offered this example: Due to a reference table error, ten $50K items were shipped from Japan to a Maryland Air Force base. They should have gone to an Air Force base in Utah. It was explained that because of "frustrated cargo" problems like this, item managers typically hold more inventory than required just to ensure operations needs are met. As a result of these types of errors, it is estimated that DoD maintains many millions of dollars of excess inventory annually. Clearly, an updated, detailed solution scheme for overall DoD data interoperability is urgently needed. The problem, however, is not only creating reference tables of data for shared use, it is also the question of agreeing on what data to put in the reference tables, and making sure that the data is kept current and synchronized across multiple databases and employing business information systems.

A final example relates to the tallying of the domain of a single type of data field, an Indicator. Traditionally the values are Yes or No. Table 23 tabulates the actually occurring values over about 3400 different indicator fields from a number of different database applications. Clearly, in all these cases, the value domain management function is "challenged."

Tallying of Indicator Values Across 3400 Fields	
Yes | No Freq	**Code values**
1296	1 = yes.....0 = no.....
899	Y = yes.....N = no.....
740	1 = yes..........
75	Y = yes..........
17	0 = yes.....1 = no.....
2	1 = yes.....0 = no.....2= error

Tallying of Indicator Values Across 3400 Fields	
Yes | No Freq	**Code values**
104	Y = yes.....N = no U = unknown
1	Y = yes.....N = no.....D = does not apply
3	0 = false.....1 = true.....
5	Y = yes.....N = no.....N/A = not applicable
90	1 = yes.....2 = no.....
6	1 = yes.....2 = no.....3 = not applicable
1	1 = yes.....2 = no.....N = none
2	1 = yes.....2 = no.....0 = do not use default
88	1 = yes.....0 = no.....99 = unknown
1	1 = yes.....0 = no/negative.....99 = unknown
1	1 = yes.....0 = no.....2 = unknown
24	Y = yes.....N = no.....NA = not applicable
17	Y/N items to be deciphered.....
11	1 = yes.....0 = no.....2 = not applicable
1	1 = yes.....0 = no.....U = undetermined
1	1 = yes.....2 = no.....3 = unknown
2	Y = yes.....N = no.....blank = no
1	0 = yes.....1 = no.....2 = not applicable
1	2 = yes.....1 = no.....
1	1 = yes.....0 = no.....U = unknown
1	Y = yes N = no A = ask
1	Y = yes N = no R = restricted

Tallying of Indicator Values Across 3400 Fields	
Yes | No Freq	**Code values**
1	yes = yes no = no
1	1 = yes 0 = no 2 = ask
3394	Total columns

Table 23. Enumerated code values for a collection of healthcare tables for Yes | No columns.

Demanding that there be completely congruent value domain collections is just not possible, however. Nice, yes. Recommended, absolutely. But absolutely possible? No. This chapter sets out strategies for setting down value domains, associated value collections, and the ability to map one value collection with another. There is, however, no silver bullet here. If one collection has three values for a concept and another five values for a different instances of that concept, there can only be three mappings, and none of these might be exact. This is illustrated in the case of mapping "one-third's based values" to a set of "one-fifth's based values," or academic "letter" grades to precise "numeric" grades. Miracles are not possible. Some domain value mappings are going to be unsatisfactory whenever square pegs are shoved into round holes. At least, this chapter provides a strategy to define the square to round mappings, and a Metabase System solution to know when and where the mappings are occurring.

With respect to the two error classes, initial semantic disconnects, and follow-up semantic synchronization activities, this is also a very important chapter because it provides a very explicit first line of defense in stopping invalid values from entering any database. Value domains and their accompanying values can take on both type and values forms. For example, in a data warehouse database, the column Region might take on the value, New England. While in another database, the column name might be New England Region. It all depends on the data modeling strategy. In either case, the value domains component of the Data Element Model has the values and their definitions from which either the names can be constructed or the values that would become part of reference data can be produced. Thus, there needs to be a good effort to prevent the first class of errors and to therefore prevent the second error class altogether.

7.2 Concepts and Engineering

Restricted value sets are essential for measuring policy adherence. The extent to which these value sets are independent of culture and language is the extent to which data from business units can be exchanged, summarized, and reported.

7.2.1 Detection of Value Sets

Data values do not exist at random. Within society, values fit patters. The patterns are commonly based on established policy intended to maintain order, control, and value-based discrimination. Virtually every business form has a set of directions that contain restrictions on the values that can be employed. For example, you might not be allowed to have numbers in your name, or you must be one of two different sex codes, and in the United States, you are commonly restricted to a discrete few race (ethnic) codes. As to upper and lower limits, some are fixed and others are relative. For example, in the United States, all birth dates for living persons can safely be presumed to be after 1875, but all death dates must be either "Today" or some date in the past.

7.2.2 Process of Establishing Value Sets

The process of standardizing value sets begins with determining if the value set is already standardized. If so, the standard value set must be examined to determine its suitability for the data field under question. Given that there is no value set standard, the following steps are in order:

- Determine the required maximum quantity of distinct occurrences.
- Establish values that are not overlapping in meaning. For example, [O]verhand transportation, [R]ail, [T]rain.

- Establish values that do not contain conflicting semantics. For example, Employee status of [E]mployed, [A]ctive, [P]art time.
- Determine how the value is to be represented. That is, as a code (M or S), and/or as a full value (married or single), and possibly in multiple languages for example, Casado and Soltero.
- Determine if the value is simple or compound. If compound, divide the data field into its constituent data field parts.
- Determine if the value is also complex. If complex, divide the value into its simply contained parts including the definition of additional data fields to fully illuminate its meaning.
- Determine any data value characteristics such as maximums, minimums, allowed to be negative, and null.
- Estimate the maximum quantity of values that may exist in the foreseeable future.
- Create a value set meets the following objectives:
    - ♦ Consists of only simple values.
    - ♦ Can be readily expanded to meet foreseeable value growth.
    - ♦ The value set contains no obvious traps, such as only a two digit year code.
- Determine if the values are to exist, for example, in pairs such as marital status name and marital status code. If so, establish, and map value pairs.

Data set value maintenance is also very important. Whenever there is a need for a new distinct value, for example a marital status value, that is different from either *single* or *married,* a data value maintenance procedure must be followed. The maintenance procedure must produce an impact and proposed change document that contains:

- The values and meanings of the existing set of values.
- The business policy problem that exists because of the existing value set.
- The proposed modification of the existing value set.
- The policy implication of adding, deleting, or modifying the existing value set.
- The information systems technology impact on:
    - ♦ Existing data processing systems
    - ♦ Existing data
    - ♦ Historical data
    - ♦ Supporting documentation and training.
- A proposed time line for modification supported by resource estimates.

7.3 Value Set Enforcement Environments

Value set enforcement environments are different from those that accomplish standard semantics. Value sets can only be enforced through business information systems or schema-based procedures within the DBMSs that the computer programs use. DBMSs should be the first choice. As with data semantics environments, the value set enforcement environments should be both unavoidable and work enhancing.

By unavoidable, all value set restrictions should be encoded within ANSI standard SQL in the form of schema based column and table constraints, assertions and triggers. Referential integrity based schemes can also be used to ensure that specific values can be enforced through the automatic value lookups that are successful only if the value is within the referenced table.

By work enhancing, the value set restriction mechanisms must be specified and implemented through unavoidable core facilities. In the case of persistent data,that means ANSI standard SQL DBMS. If accomplished within DBMS schema facilities, all the various third and fourth generation language programs cannot avoid employing these value set restrictions. This reduces or eliminates all the time and effort for detail design, coding, unit and system testing, documentation, and long term maintenance that would be necessary if these value set restrictions were a component of the 3GL or 4GL business information system environments.

7.3.1 Direct Value Domains

Examples of direct value domains include fields such as Gender, and Birth State. The field essentially represents the purpose, and the values are representative of the intended policy represented by the values. Example values would be M for Male, and F for Female.

If there are changes to the values, for example, a change to the person's gender, the person must be accessed and the field's value updated and the record re-stored. If however the coded value for the value itself must change, say, changing the value "M" for Male to be 1, all the records that have the "M" value must be accessed and changed. Changed as well must all backups of all data records that contain the "M" value. The first type of change is business-based while the second is administrative or IT-based.

There are additional cases of value domains that can exist across two or more fields. For example, in an address, the four fields that commonly have tightly integrated value sets would be:

- Street.
- City.
- State.
- Zip.

Changes to, for example, a person's City would almost always cause a need for a change to Zip code and possibly State. Changes in a person's Street might also cause changes to Zip codes. As a consequence of inter-dependent value sets, greater attention must be paid to the creation and maintenance of the value sets.

Other examples include medical diagnostic codes and the common name for the diagnosis.

A more complex example might be the "Case Code" that would be stamped or recorded on a case that contains a quantity of products. Each case of products contains a "Case Code" that might be a complex string collectively representing the Product Number, Manufacturing Plant,

Manufacturing Line, and Date of Manufacture. If a mistake has been made in the identity of the Manufacturing Line for a particular product, not only might the Manufacturing Line identifier have to change but also the "Case Code" itself.

Other complications occur when a case has multiple, but different products. Like the example above, very sophisticated policies and procedures must be set into place to ensure that the complete set of values are updated in lock-step. Addition complications arise if a manufacturing line identifier value changes, which, in turn, would cause changes to the case codes of newly manufactured products. What happens to existing cases in inventory, or in distribution, or warehouses? In all these examples, reference data value changes must be very carefully examined to fully determine all the ramifications.

7.3.2 Indirect Value Domains

A different class of value domains are those that are indirectly generated or maintained. Included for example are <field name> and <value> pairs. For example, there might be a set of fields for Valuable Name, Valuable Value, and Valuable Status. The Valuable Status codes might well repeat across the different Valuable Names but have different meanings. While such a strategy provides generalization to the database's structure, it complicates end-user understanding as the meaning of the Valuable Status code has to be defined and maintained elsewhere.

Indirect value domains omit the actual encoded value from database records completely. Stored instead is a surrogate key identifier of the encoded value. For example, instead of storing M for Male, what is stored is the value domain value Id value which might be the number 562. When this is looked up in the reference-data section of the database the value might be "M" which would further defined to have the meaning of "Male."

While the value of such an approach might seem obscure, it has the following advantages:

- The explicit value is not stored in the database records thus, the representation of the value could change without having to actually update the database records.

- The value, "562" is a foreign key value in another table that indicates that while the meaning is actually "Male," there could be additional equivalent strings with the same meaning but in different languages, or that might have had different migrating values over a long period of time.

In either of these two cases, there is added flexibility and safety, but the cost is to end-user immediate perception of what the database record is really conveying.

7.3.3 Reference-Data Models

There is a fine line between metadata specifications contained in metadata database systems like those managed by the Metabase System, and the specifications of operational data contained in "real" databases. Metadata existing in Metabase System type systems are mainly specification-based metadata and thus is almost always passive. By passive it's meant that the Metabase System will not automatically enforce any of the contained metadata specifications. By active, the Metabase System would actively interact with an operating database application and allows or disallows certain operations.

Metabase type systems are very difficult to make active. That's because they can be distributed across multiple computers that have different operating systems and DBMSs. Because of this heterogeneous environment, there cannot be any one overarching, ever-present, and always controlling metadata management system. At best there can be distributed data and synchronization supported by good will and quality practices.

One type of solution of the active-passive issue is to have business information systems perform periodic refreshes of metadata to DBMS database catalogues, and/or to classes of reference data resident in database tables. An example of a meta model that could serve active value domain metadata and be periodically refreshed is presented in Figure 54. Certain of these tables belong in the Metabase System database proper, and other tables should reside in the databases that support active-metadata management. The tables and their definitions are set out in Table 24. The tables designated Metabase System have not yet been implemented as of 8/1/2008 in the Metabase System.

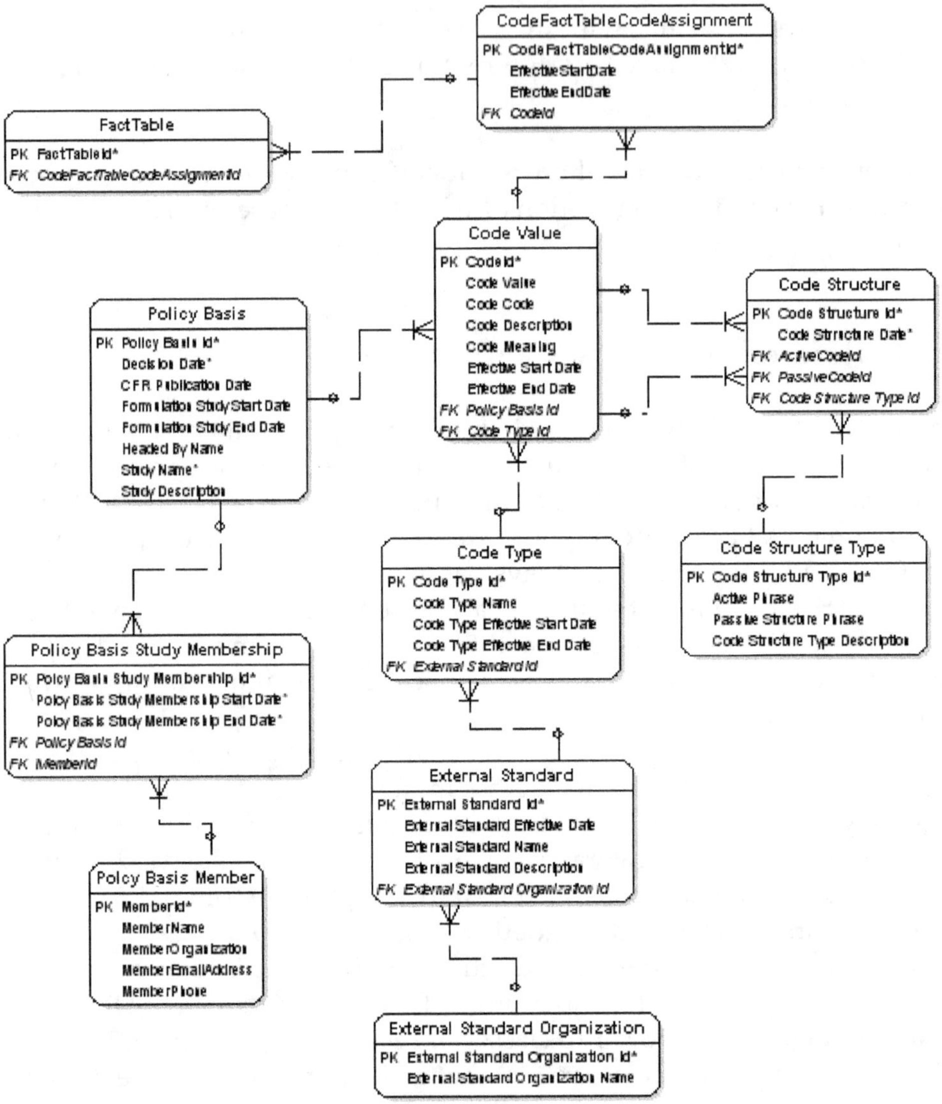

Figure 54. Reference Data meta-data model for active and passive environments.

Tables designated as active are contained in the database's table set. Those tables identified as shared are those that have the rows defined first in the Metabase System, and thereafter, these rows are copied into the active databases. Tables designed as Metabase System are exclusively contained the passive metadata set.

Table Name	Primary State	Location of Metadata	table Description
Code Face table Code Assignment	Active	Database	Contains a surrogate of a code-based value. The primary key value of each row from this table acts as the foreign key value that stored in actual database table rows.
Code Structure	Active	Shared	Contains the mechanisms for relating equivalent code values to each other such as M for Male and 1 for Male.
Code Structure Type	Active	Shared	Contains a classification scheme for the equivalent value-based relationships.
Code Type	Active	Shared	A broad classification of the type of code values that are begin represented, for example, Gender for M and F.
Code Value	Active	Shared	The actual code value that is indirectly represented in the real data rows of the database.
External Standard	Passive	Metabase System	Descriptive information about a reference data standard that is maintained by an outside organization within the enterprise or, for example, ISO.
External Standard Organization	Passive	Metabase System	Descriptive information about the organization that is maintaining the reference data standard.

Table Name	Primary State	Location of Metadata	table Description
Fact table	Active	Database	A "place holder" name for an actual database table and a "place holder" column name that has the row-based value of the Surrogate Id value from the Code Fact table Code Assignment table. For example, the "Fact table" might actually be "Person," and the "CodeFacttableCodeAssignmentId" might actually be "GenderId."
Policy Basis	Active	Shared	A set of descriptive information that provides the policy basis information for the code value including the start and end dates of the policy. This information is "required" solely because of Referential Integrity rules.
Policy Basis Member	Passive	Metabase System	The name of an individual who participated in the determination that the code value is appropriate for this use.
Policy Basis Study Membership	Passive	Metabase System	The membership information for a policy basis member who determined that the code value is a proper part of the code set.

Table 24. Reference data specification data-model coupled with active and passive designations.

7.3.4 Value Set Enforcement Environments

There are five different types of value domain enforcement strategies:

- DBMS schema object bound values.
- Stored procedures, and integrity clauses.
- Foreign key based referential integrity.
- SQL View Clauses.
- Business information system encoding.

The options listed above are characterized by creation and maintenance flexibility, and uniformity of creation and enforcement.

If value clauses are set within SQL schema object clauses, avoiding value-based integrity checks is impossible as it is automatically applied to every insert and update. Changing the set of allowed or disallowed values is difficult and might require "stopping" the DBMS, changing and recompiling the set of schema objects. Various DBMSs also apply changes to the allowed and/or disallowed sets of values differently. For example, some DBMSs will not compile schema changes if there is an existing stored value that should not be allowed.

Stored procedures along with assertions and triggers can also enforce restricted values. These values must be stored outside the DBMS, or they have to be encoded in the computer language of the stored procedure, assertion, or trigger. Similar to DBMS schema clauses, maintenance of these processes may require schema maintenance and recompiling depending on the capabilities of the DBMS.

Foreign key based referential integrity value enforcement is both more flexible and easier to change. That is because the allowed values are represented as data values in specially constructed database tables. Examples include a Genders table, a City-State-Zip table, and the like. Enforcement is through the business systems that force individual users and/or batch data loaders to only store valid values, or to update values from one valid value to another. For collections of valid values, for example, City-State-Zip, the database stored value would be the identifier of a valid City-State-Zip triple. DBMSs, usually have the same sets of Referential Integrity Clauses and Referential Action rules that ensure that only reasonable results occur.

In general, there are two cases that must be addressed when building reference data into separate, stand-alone reference-data tables, and reference data columns within fact tables. A reference data table is for example, City-State-Zip. A reference data column is for example, Gender, and it exists within a fact table such as Employee.

A strategy for addressing both classes can be the same: Factoring into separate reference data table sets. Each reference data table set consists of the following tables:

- <reference data name> Code Assignment.
- <reference data name> Code.
- <reference data name> Code Structure.
- <reference data name> Code Structure Type.

- <reference data name> Code Type.

Figure 54 illustrates a reference metadata data model. This model represents a generalized solution because there is only one set of reference data types. In a specialized solution, each reference data table set would be specially named for the reference data represented. In the list of tables above, an example might be:

- Gender Code Assignment.
- Gender Code.
- Gender Code Structure.
- Gender Code Structure Type.
- Gender Code Type.

Regardless of whether the solution is generalized or specialized, column one of Table 25 lists problems that exist because of un-factored reference data. Columns two through four list the "before" and "after" effect once reference data factoring has occurred for either a reference data column or a reference data table.

	Reference Data Column		**Reference Data Table**	
General Problem	**Before**	**After**	**Before**	**After**
As meanings change, mapping between changes are reflected.	No	Yes	No	Yes
A history of all value and/or meanings is maintained.	No	Yes	Yes	Yes
Massive data value updates to the database are required when values change.	Yes	No	Yes	No
Massive database structure changes along with all database access program changes are required as values change data type.	Yes	No	Yes	No
Valid value lists are visible.	No	Yes	Yes	Yes

Table 25. Reference data problems and approach solution effects.

Regardless of the approach, the general process is:

- Make the reference data table set.
- For a reference data table, replace it with the reference data table set.
- Find the use of the reference data table as a column within the existing set of fact tables and replace the appropriate reference data column with a <reference data> Code Assignment ID column.
- For a reference data column, replace it along with its commonly occurring "effective date" column with the column, a <reference data> Code Assignment ID column. For example, Gender Code Assignment Id.

SQL View clauses are schema objects, but are defined independently from the main set of table and column clauses. An SQL View can have logic and depending on the DBMS can invoke stored procedures to ensure valid data enforcement. SQL Views can be independently defined and maintained without having to stop the DBMS and recompile the schema. SQL views can also be styled to fit the needs of different business information systems, some with common views and others with unique views.

The final strategy is through program language encoding within business information systems. In this strategy, there is maximum flexibility overs what is checked, and the processes that perform the checking. However there can be no guarantee that all inserts and updates across all the business information systems will, be controlled and governed by the very same value-checking process. That is because each process has to be hand-coded into each procedure. Making matters even worse, consistent maintenance ranges from very difficult to impossible.

7.4 Employing Value Domains

Value domains are identified, described, and represent defined collections of values that exist within the context of a conceptual value domain. Specific value domains are assigned to one or more Data elements. If a data element's assigned value domain is to apply to the attributes, columns, or DBMS columns related to a data element, nothing more needs be done as the ultimate ancestor of an attribute, column, or DBMS column is the data element.

If only a subset of values from within the data element's value domain is to be assigned to an attribute, column, or DBMS column, the subsets are defined with value domain structures and are assigned as appropriate to attributes, columns, or DBMS columns.

Employing value domains involves establishing three classes of meta entities. These deal with:

- Value Domain Definition.
- Value Domain Values Definition.
- Value Domain Assignment.

All three collections of meta-entities for value domain management are illustrated in Figure 55. Value domain definition involves two sets of meta entities:

- Conceptual value domain, conceptual value domain structure, and conceptual value domain structure type.

- Value domain, value domain structure, and value domain structure type.

The first set, conceptual value domain, conceptual value domain structure, and conceptual value domain structure type establish the conceptual environment within which value domains exist. Examples of conceptual value-domains include descriptions, numbers, dates, classifications, and the like. These are to represent the concepts that would commonly exist for a collection of value domains.

Numbers as a concept are a collective for real numbers, complex numbers, integers, imaginary numbers and the like. Descriptions, as a collective, could include textual descriptions, names, and definitions. The creation of these concepts is a process suited to those who specialize in classification, generalization, and abstraction schemes.

The process of creating conceptual value domains is iterative and occurs along with the creation of value domains. Conceptual value domain creation will not be perfect nor accomplished on the first pass. Iterations and corrections are common. Consensus is reached over time, and will likely vary with the overarching discipline that is common to the participants.

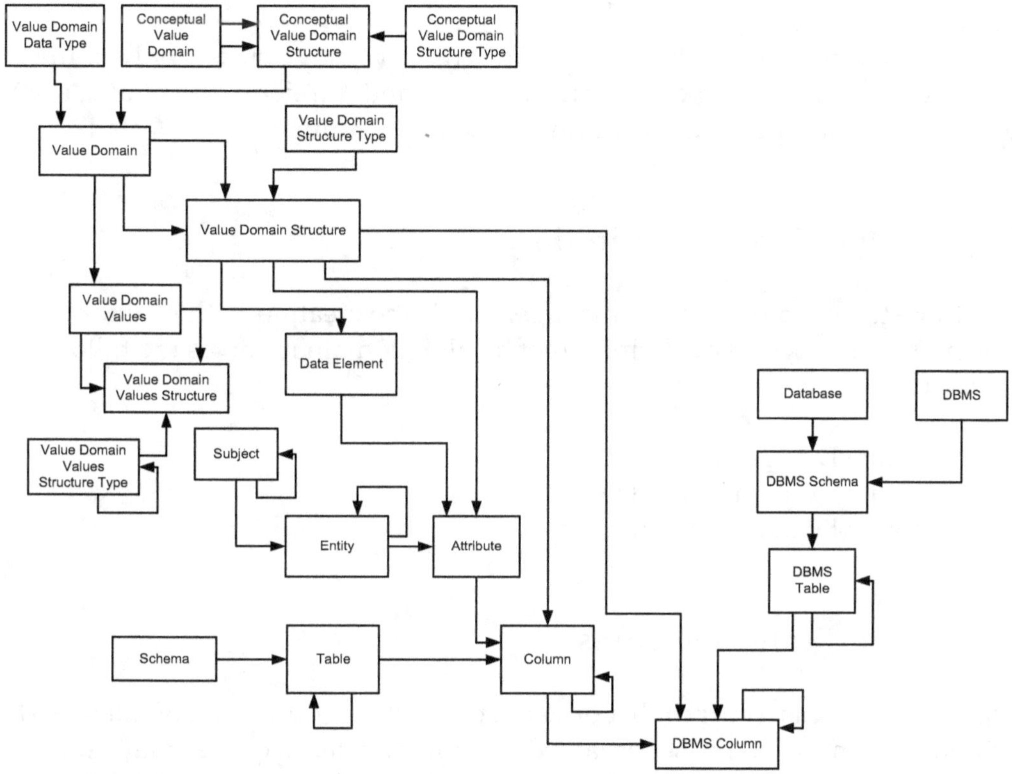

Figure 55. Value Domain management meta model.

The data model shown in Figure 55 shows that conceptual value domains are not simple. They exist in network structures. Creations of conceptual value domain network structures are described in Chapter 9, Data Elements.

The second class of value domain related meta-entities, value domain values definition, consists of the following meta entities:

- Value Domain Values.
- Value Domain Value Structure.
- Value Domain Value Structure Type.

Valuation of these data structures form a network of values and the mapping of values, one to another as may be appropriate when the values are different but semantically equivalent as would be the case between "1" and "M" if both meant Male.

The third class, value domain assignment really consists only of the processes that assign a specific value domain (and implicitly their value sets) to data elements, attributes, columns, or DBMS columns.

7.4.1 Value Domain Definition

The first step in value domain management is the creation of the value domains themselves. The definition of a value domain involves the following meta-entities:

- Value Domains.
- Value Domain Structures.
- Value Domain Structure Types.

7.4.1.1 Value Domains

Figure 56 presents a current list of value domains. Since value domains exist within the context of a conceptual value domain, select a conceptual value domain structure type, and a specific conceptual value domain. This figure also shows that conceptual value domains exist within structures. Once a specific conceptual value domain is selected, the set of associated value domains is presented.

This figure shows a selected conceptual value domain, Integer, from within a containing conceptual value domain, Numbers. Once Integers are selected, the four selected value domains, large integers, medium integers, small integers, and integer numbers are shown. All four of these value domains do not have either precision or scale. Null is not allowed, the value characteristics are both inclusion, and discrete.

Figure 57 presents the update form for adding, changing or deleting value domains. A value domain data type may be numeric. Thus, it can have both decimal places and scale. Also defined is where the set of data values assigned to a value domain can be Null, are inclusive, or exclusive, and finally, represent ranges or discrete values. Figure 57 shows the update screen for medium integers. There is a definition, a zero for length/precision, and scale. Null values are not allowed, and the values are to be inclusive, and all the values are discrete.

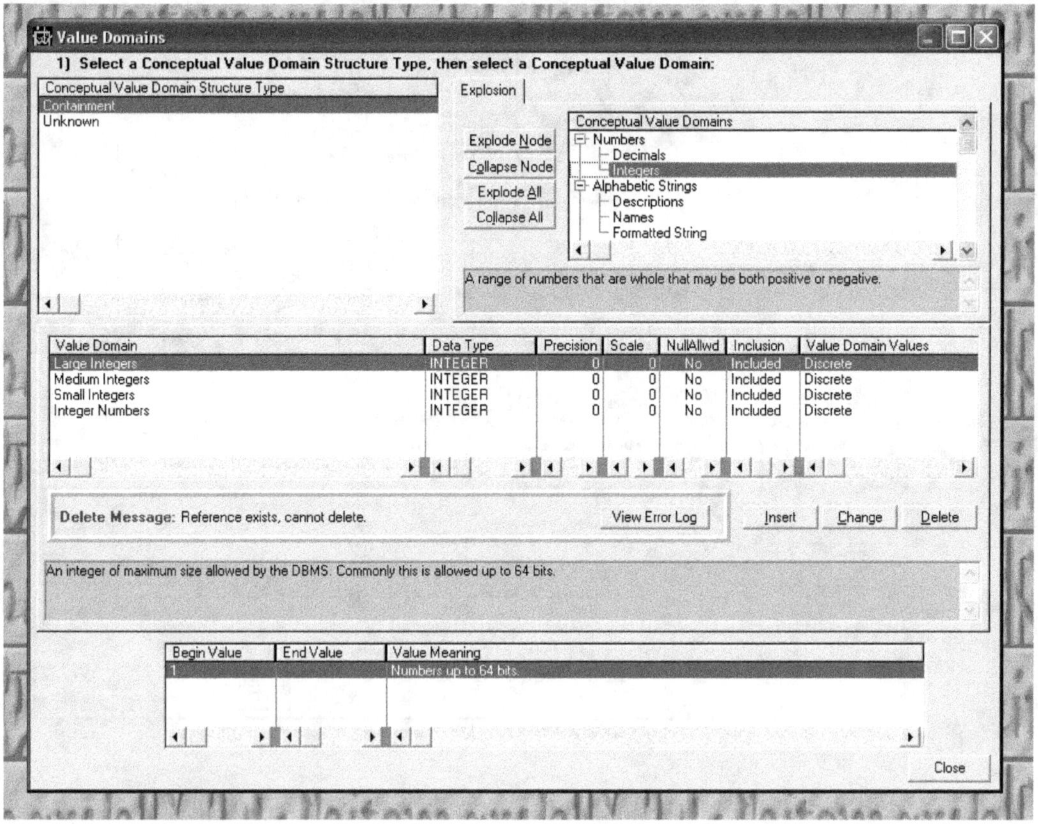

Figure 56. Value domains.

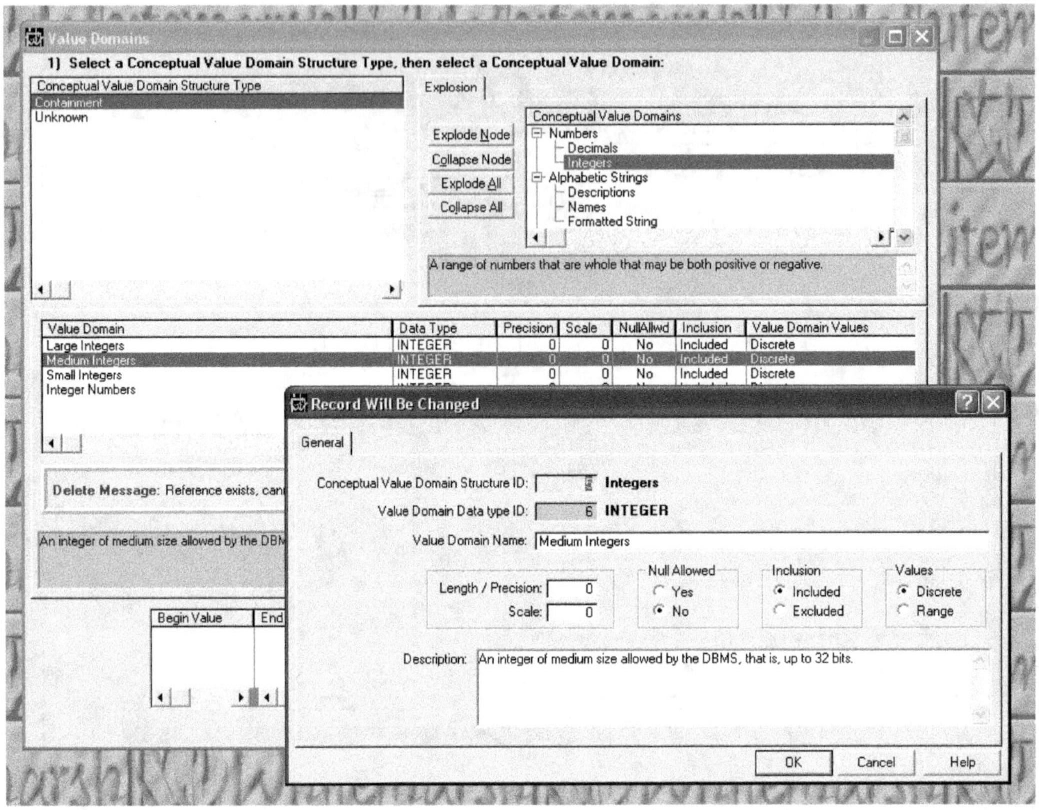

Figure 57. Value domain update screen.

7.4.1.2 Value Domain Structures

Value domains can exist singly, in hierarchies, or networks. In the network case, value domains form a traditional bill-of-materials data structure. Figure 58 shows an existing set of value domain structures.

When a value domain name is selected and the Insert button is pressed, a screen like Figure 59 is presented. This screen enables the selection of an additional value domain that is to be included in the existing set.

Figure 55 shows that the relationship between a value domain and a data element is not a relationship directly between the two. That is, a direct reference to the value domain from within the data element. Rather, the relationship is between a value domain structure instance and a data element. A check of Figure 55 also shows that the relationship between a conceptual value domain and value domain is similarly constructed. That is, there is a relationship line between conceptual value domain structure and value domain. The reason for what would appear to be indirect relationships is straight forward: conceptual value domains, and value domains are

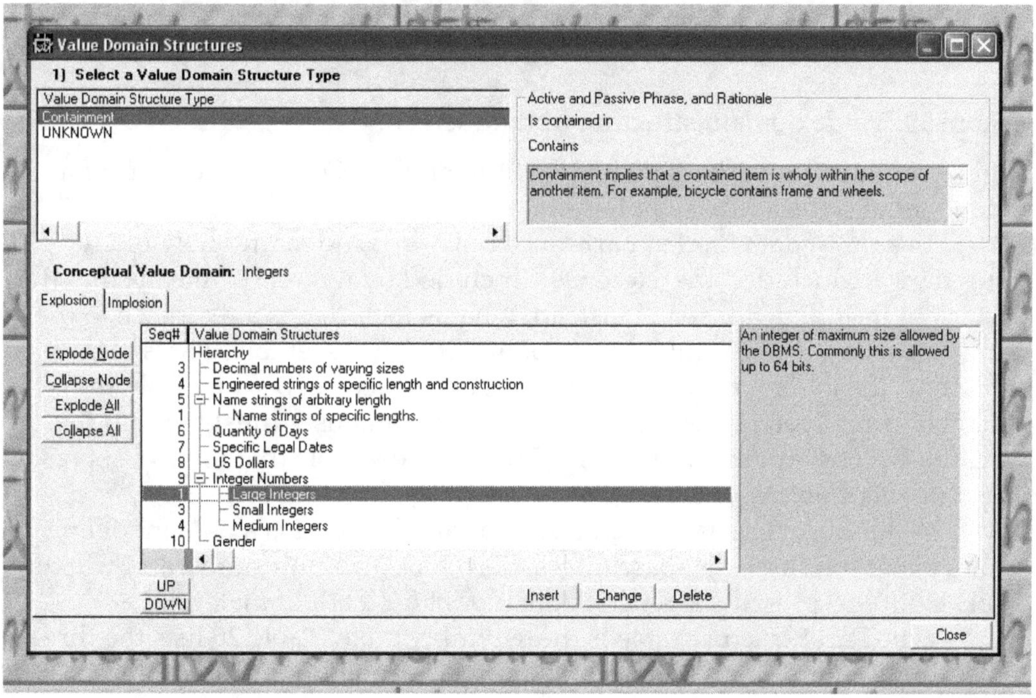

Figure 58. Value domain structures.

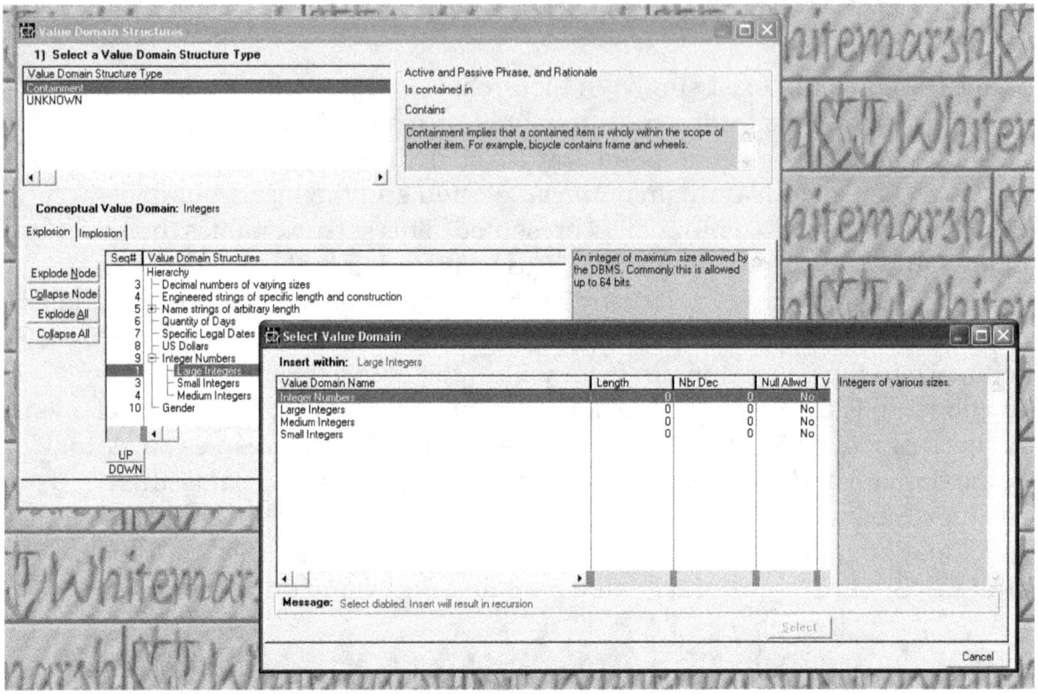

Figure 59. Value domain structure update screen.

engineered into networks called Bills of Materials (BOM) that consists of a unique set of parts and assemblies of related parts.

In a BOM data model, parts exist only once, no matter how many times they are included-by-reference. "Included by reference" means that the part is not included "for real." Instead, what is included is a reference to the part from within the assembly. In actuality, it is the assembly that is real as it represents the joining of two parts into a Part Structure. Because only the assembly is real, references from it to the contextual uses must be from the assembly record, not the part record. The BOM example that follows serves to illustrate these points.

A bill-of-materials, sometimes also called a BOM is a data modeling strategy that has three database tables: Part, Part Structure, and Part Structure Type. Figure 60 presents a generic data model for a bill of materials.

The Part Structure table is the assembly table. Table 26 lists the three tables, the columns in each table, and a brief description for each column. The table naming strategy is this. The subject of the BOM, e.g., concept, data element concept, and the like is the name for the "part" table. Appending the

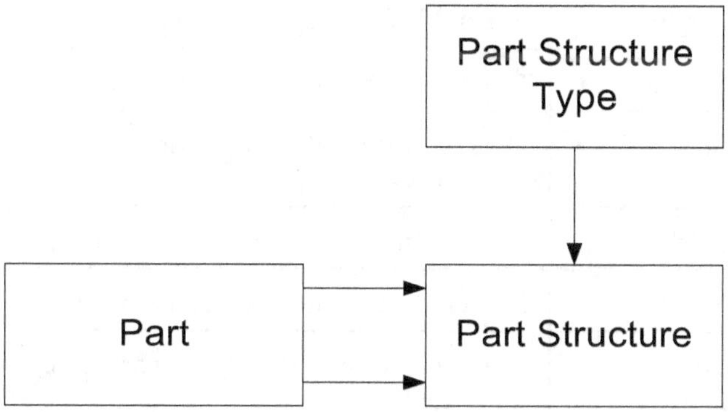

Figure 60. Bill of materials data model.

string "structure" to the name of the part table names the assembly table. Finally, appending the string "type" to the name of the part structure table names the table that distinguishes collections of assemblies.

For the purposes of an example, a very simple bicycle example is used. If this was a real bicycle it would be more complicated. The data records for the three tables is presented in Tables 27, 28, and 29. The Parts table contains one record for each part, which, for this example contains 6 parts. There are 13 Part Structure records that represent the interrelationships between the various Parts. In this simple example there are only four spokes for each wheel, and each is identified as separately numbered Part Structure records as they must be, from front-wheel-spoke-1 through front-wheel-spoke-4, and back-wheel-spoke-1 through back-wheel-spoke-4. Note that Front Wheel, Back Wheel, and Spoke, as parts exist only once. It is their joining in assemblies that repeats. Each set of four spokes has a common Active Id and Passive Id. Each however also has a different Sequence Number. The reason why this set of records forms a network is because the part, Spoke, is reference from eight different Part Structures, that is, Front Wheel and Back Wheel.

Table	Columns	Brief Definition
Part	Part Id	The identifier of the part
	Part Name	The name of the part

Table	Columns	Brief Definition
Part Structure	Part Structure Id	The identifier of the part structure
	Part Structure Type Id	The identifier of the part structure part type
	Active Id	The identifier of the part in the active process
	Passive Id	The identifier of the part in the passive process
	Sequence Number	The sequence number of this part structure in an structure collection
Part Structure Type	Part Structure Type Id	The identifier of the part structure part type
	Part Structure Type Name	The name of the relationship that encompasses a specific sub-collection of part structure instances

Table 26. Tables and columns for bill of materials and single file recursion

This simple example services to illustrate the point of having relationships from the Part Structure table versus the Part table. Suppose there was a need to replace the third spoke in the front wheel. To find it the Part Structure records would be subseted by Front Wheel, and by the part, Spoke. That would produce four records. Then, select the third spoke, that is, the record with Sequence Number 3. If further there was a need to relate that spoke to a database record in another table, the Part Structure Identifier value that would be used would be #12 because the relationship is to the third spoke of the front wheel, not to "spoke" in general. If the relationship was to the Spoke part in the Parts table, there could be no knowledge as to which spoke or either of the two wheels. This example should therefore demonstrate clearly why relationships from within Bill of Material data structures can only correctly be recorded if they come from the Part Structure table.

A quick review of Figure 55 shows that there are three bill-of-materials data structures for value domain management:

- Conceptual value domain, conceptual value domain structure, conceptual value domain structure type.
- Value domain, value domain structure, value domain structure type.

- Value domain values, value domain values structure, value domain values structure type.

There is one record in the Parts Structure Type table for the "contains" relationship. The last column of Table 28 contains the "meaning" of the structure that exists between the two identified parts. That is, <passive-id> <contains> <active-id>. The part-structure-type table is Table 29.

Part Table	
Part Id	**Part Name**
1	Back Wheel
2	Spoke
3	Bike
4	Frame
5	Fork
6	Front Wheel

Table 27. Part table.

Part Structure Table					
Part Structure Id	**Part Structure Type Id**	**Active Id**	**Passive Id**	**Seq-uence**	**"Meaning" of Relationship**
1	1	1	4	2	Frame contains back wheel
2	1	2	1	1	Back wheel contains spoke 1
3	1	3	0	1	Bike
4	1	4	3	1	Bike contains frame
5	1	5	4	1	Frame contains fork

Part Structure Table					
Part Structure Id	**Part Structure Type Id**	**Active Id**	**Passive Id**	**Seq-uence**	**"Meaning" of Relationship**
6	1	6	5	1	Fork contains front wheel
7	1	2	6	1	Front wheel contains spoke 1
8	1	2	6	2	Front wheel contains spoke 2
9	1	2	6	3	Front wheel contains spoke 3
10	1	2	6	4	Front wheel contains spoke 4
11	1	2	1	2	Front wheel contains spoke 2
12	1	2	1	3	Front wheel contains spoke 3
13	1	2	1	4	Front wheel contains spoke 4

Table 28. Part structure table.

Part Structure Type table	
Part Structure Type Id	**Part Structure Type Name**
1	Contains

Table 29. Part structure type table rows.

7.4.1.3 Value Domain Structure Types

The value domain structure type is a way of distinguishing one value domain structure from another. Figure 61 presents the current list. If two different value domain structure hierarchies are interconnected, the intersection can be distinguished from the others by means of a different value domain structure type.

Figure 62 presents the value domain structure type update form. Not only is the name and description of the value domain structure type provided, so too is the active phrase and the passive phrase. The active phrase is employed by the Metabase System when a downward structure is presented. For example, <parent> contains <child 1>, <child 2>, . . . , <child n>. The passive phrase enables the inverse phrases to be presented. That is, <child 2> is contained in <parent>.

7.4.1.4 Value Domain Data Types

Each value set within a value domain must conform to a particular data type. These data types are analogous to those contained in ANSI Standard SQL. Figure 63 presents the list of value domain value data types and Figure 64 presents the update form for a value domain value data type.

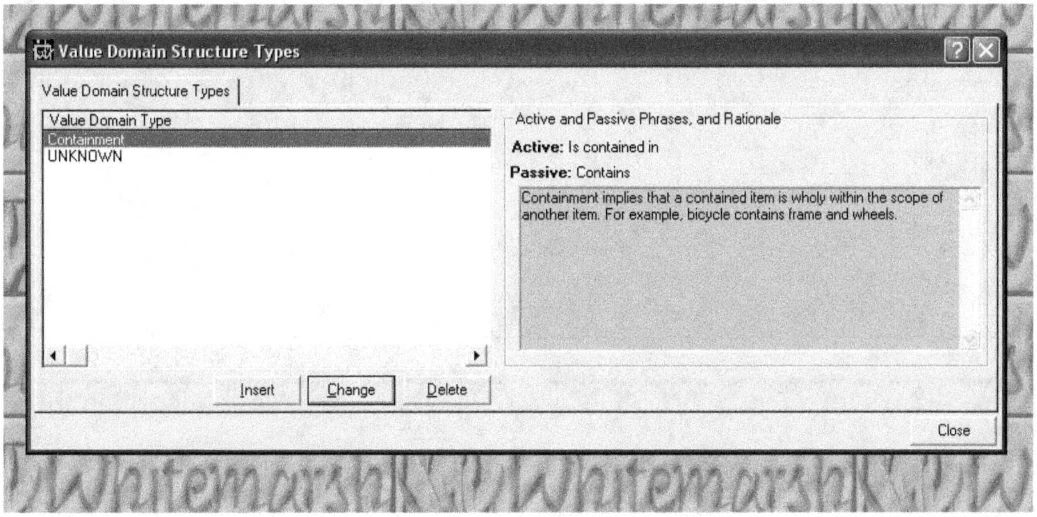

Figure 61. Value domain structure types.

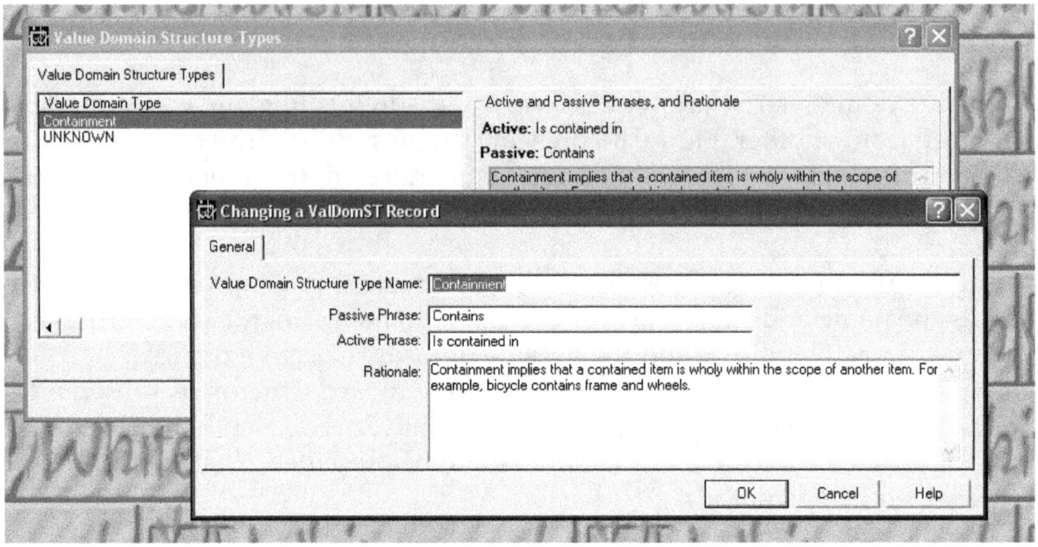

Figure 62. Value Domain structure type update screen.

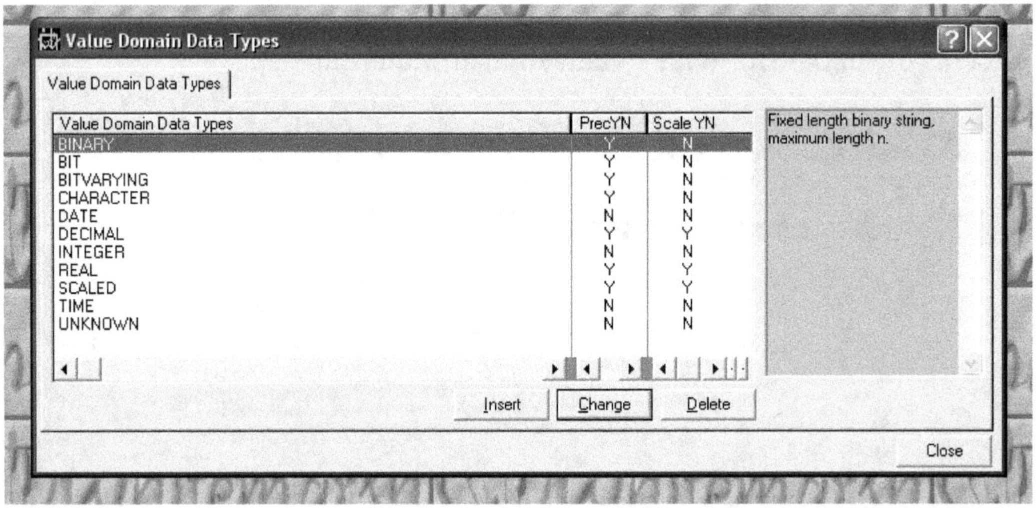

Figure 63. Value domain data types.

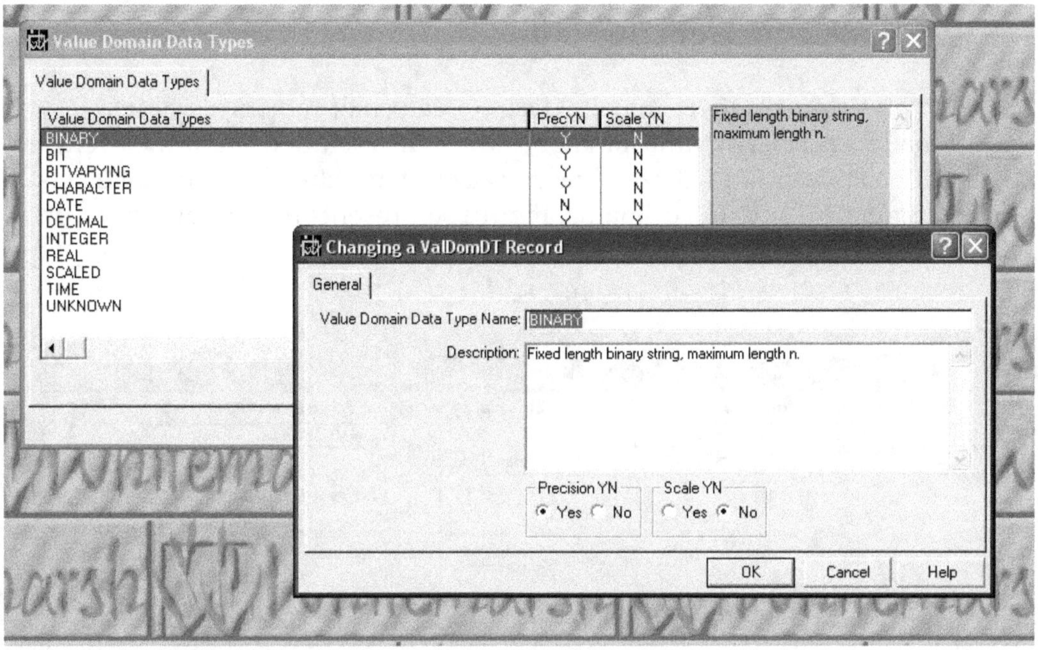

Figure 64. Value domain data type update screen.

7.4.2 Value Domain Value Definition

Value domains are not the values. Value domains are the instance of an existence of a set of values. Management of value domain values include:

- Value Domain Values.
- Value Domain Structures.
- Value Domain Structure Types.

7.4.2.1 Value Domain Values

For any value domain there may be discrete values or ranges of values. Additionally, each set of values may be either the included set or the excluded set. Figure 65 presents a list of values for a selected value domain that exists within a structure of value domains. Figure 66 presents the update form for a value domain value.

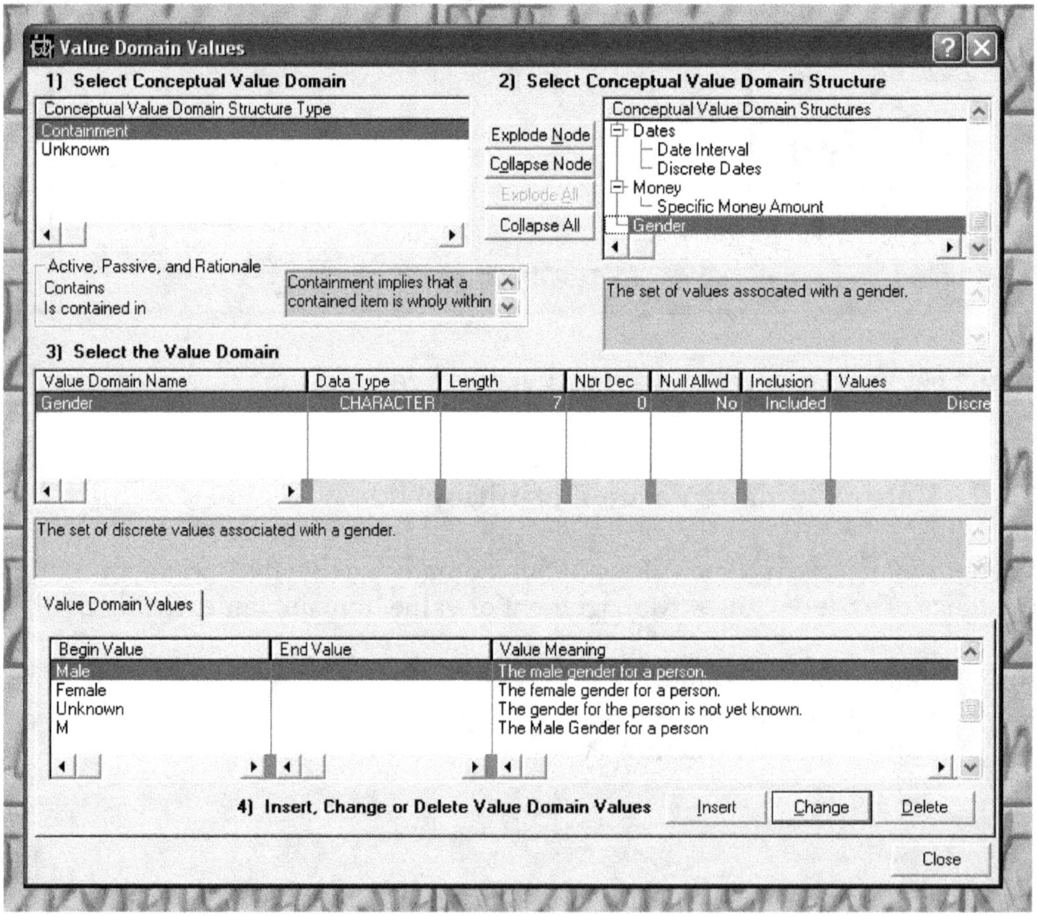

Figure 65. Value domain values.

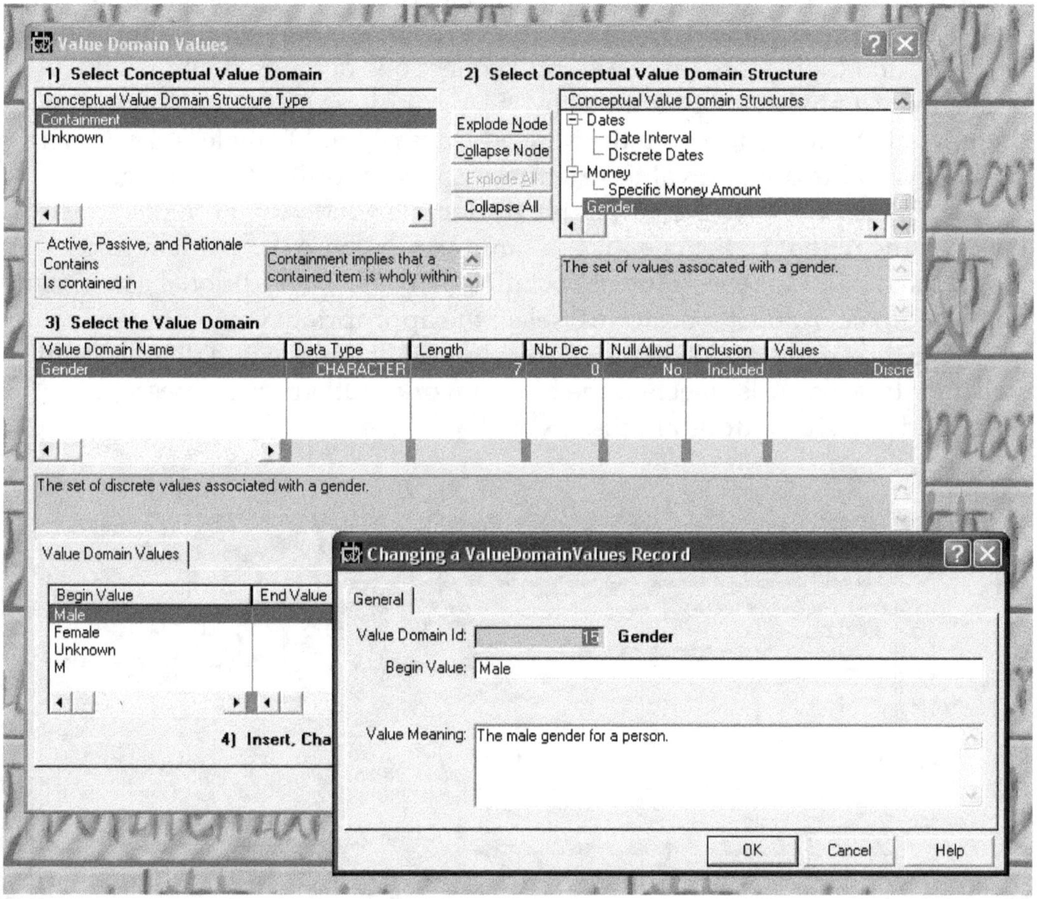

Figure 66. Value domain values update screen.

7.4.2.2 Value Domain Value Structures

Value domain value structures are a way of expressing relationships among values in different value domains. Suppose there was a general value domain, Genders, and there were three different collections of value domains, full name (male, female), abbreviations (M, F), or numeric (1, 2). In these instances there would be a need to map the Male => M => 1, and Female =>F=>2. The purpose of value domain value structures is to support that mapping.

Figure 67 shows the values for the female gender and the relationship between them, that is, Female <is the same as> F. These three are in turn under the name, Gender Value Associations. To add an additional association among Female Value Associations, select the appropriate value domain value structure type name and press the Insert button. If a leaf value domain value structure type name is selected, the button works. Otherwise a message is displayed and the buttons are disabled.

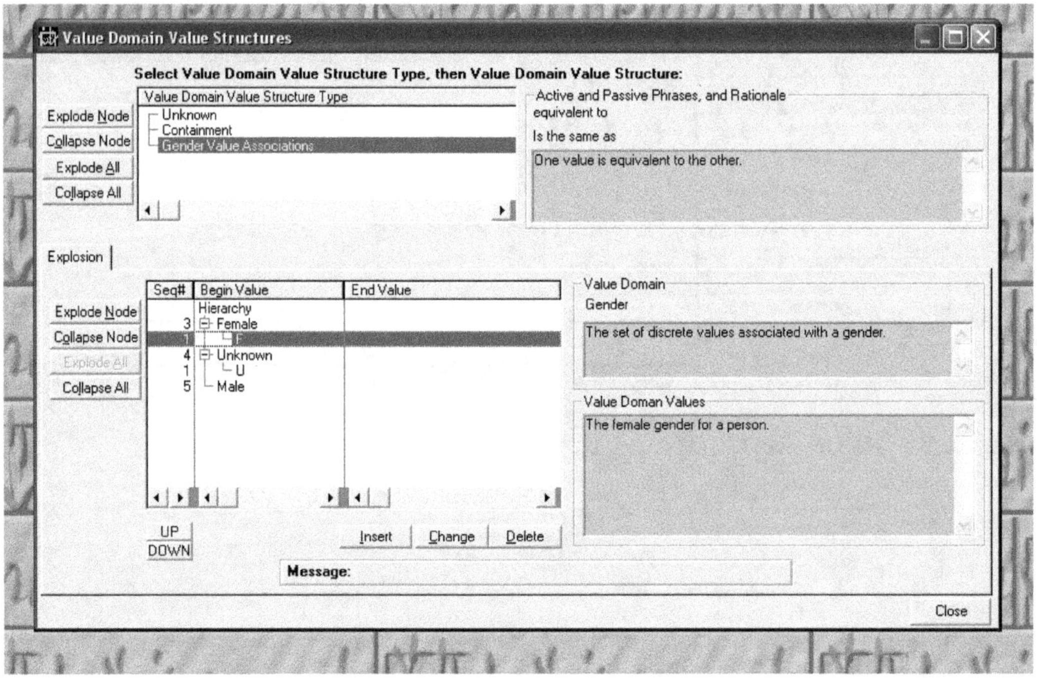

Figure 67. Value domain value structures.

If there is need to map contained values within a value domain, the Insert button can be pressed. That causes the newly inserted value to be shown within the root value. Under either scenario, Figure 68 is presented. The appropriate value domain value is selected and the select button is pressed. The select button is disabled if the value domain value that is to be inserted causes twins, recursion or an infinite recursion.

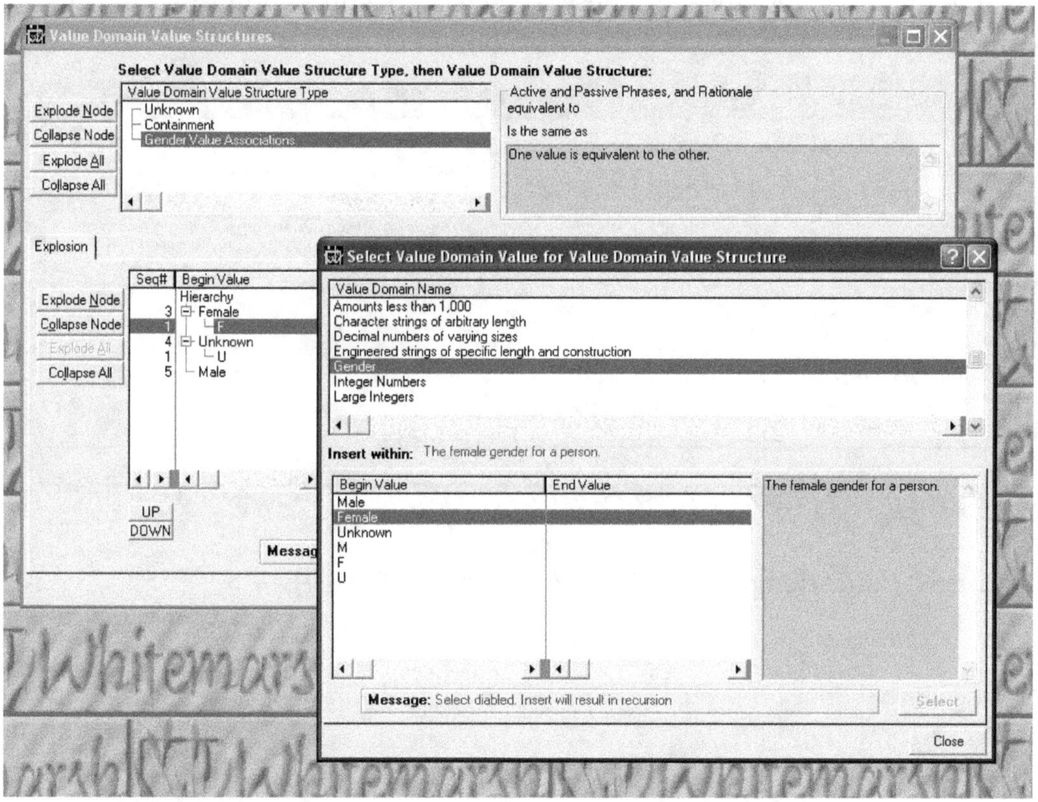

Figure 68. Value domain value structure update screen.

7.4.2.3 Value Domain Value Structure Types

The value domain value structure type provides the context for relating the structure of value domain values. Figure 69 shows a hierarchy of Gender value associations and within that Male value associations and Female value associations. In this example, to insert an entirely new structure, select "hierarchy" and press Insert. Otherwise, select the entry and press Insert to make a subordinate entry. In either case, Figure 70 is presented.

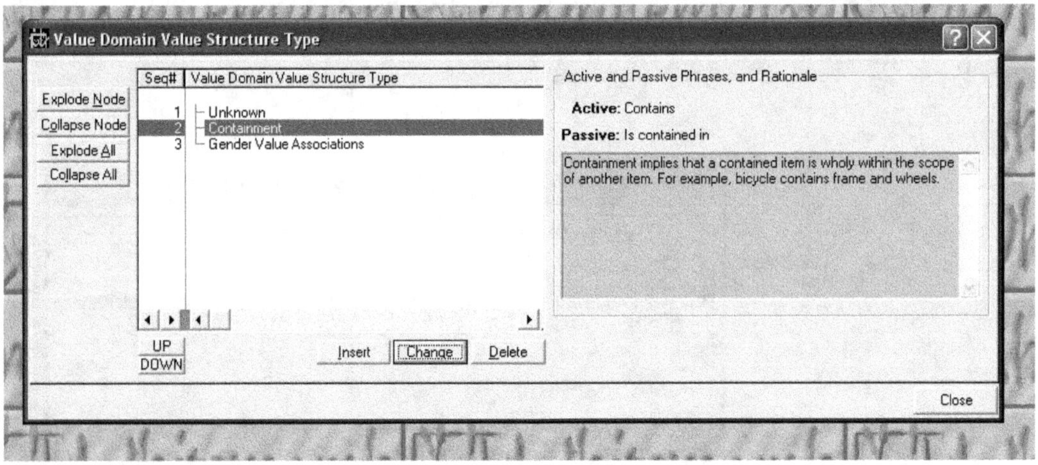

Figure 69. Value domain value structure types.

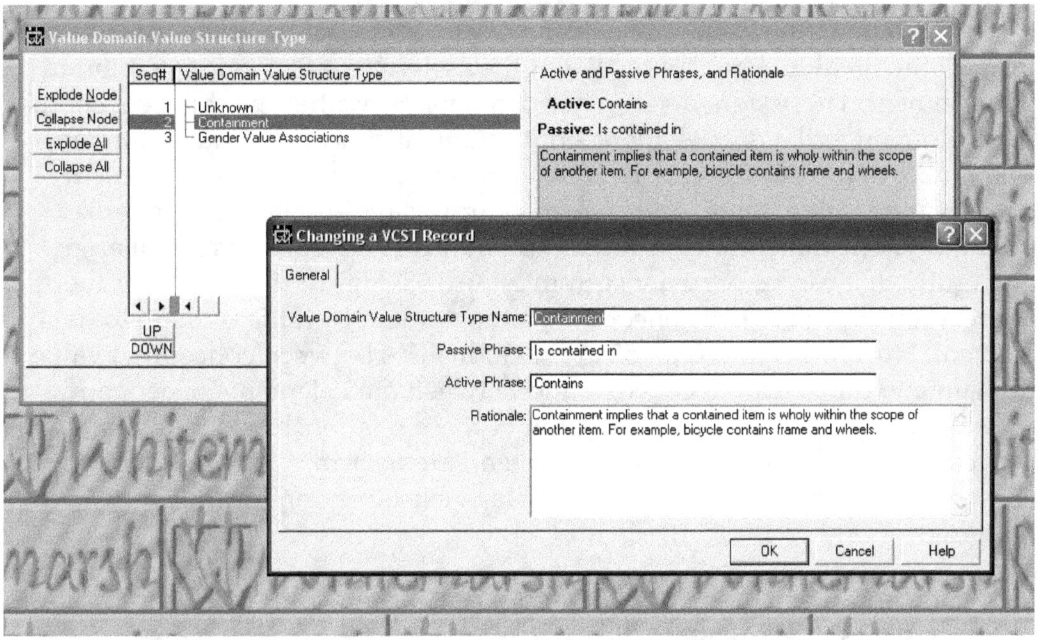

Figure 70. Value domain structure type update screen.

7.4.3 Value Domain Value Assignment

Value domains can be assigned to any of the following:

- Data Element.
- Attribute.
- Column.
- DBMS column.

What is actually being assigned, is not the value domain directly, but a reference to the value domain from within the chosen value domain structure. That is clearly shown in Figure 55. Before describing the approach and screens associated with value domain assignments, a critical data modeling ambiguity issue associated with the ISO 11179 standard must be addressed.

7.4.3.1 The ISO 11179 Data Model Ambiguity

Data element value domains represent the assignment of a value domain to a data element. The assigned value domain must be within the same conceptual value domain as the data element's data element conceptual value domain. The conceptual value domain browse window contains both data element concepts and value domains. The reason for this requirement is to avoid ambiguity. The data model of the ISO Standard 11179 contains an inherent ambiguity that can be partially seen in Figure 55. Figure 55 shows that there are relationships between conceptual value domain to value domain to data element. Not shown in Figure 55 are relationships between conceptual value domain and data element concept, and between data element concept and data element. Since the top and bottom of these relationship triples, that is, conceptual value domain and data element, are the same, Figure 71 clearly shows that a "diamond" set of relationships exists among the following meta entities:

- Conceptual value domain (a BOM structure).
- Value domain (a BOM structure).
- Data element Concept (a BOM structure).
- Data element.

Since the first three are BOM data structures, that is, conceptual value domain, value domain, and data element concept, the relationship proceeds from the "structure" records. While the data model is valid and does represent a correct solution to the modeling of these four meta entities, the only way to avoid the ambiguity is to create metadata processing software that detects potential ambiguities and prevents them. The problem is that there are two pathways between conceptual value domain and data element. The first is through data element concept, and the second is through value domain. There were three alternatives put forward to solve the potential ambiguity. These were:

- Ignore the ambiguity.
- Require strict inheritance.
- Permit family inheritance.

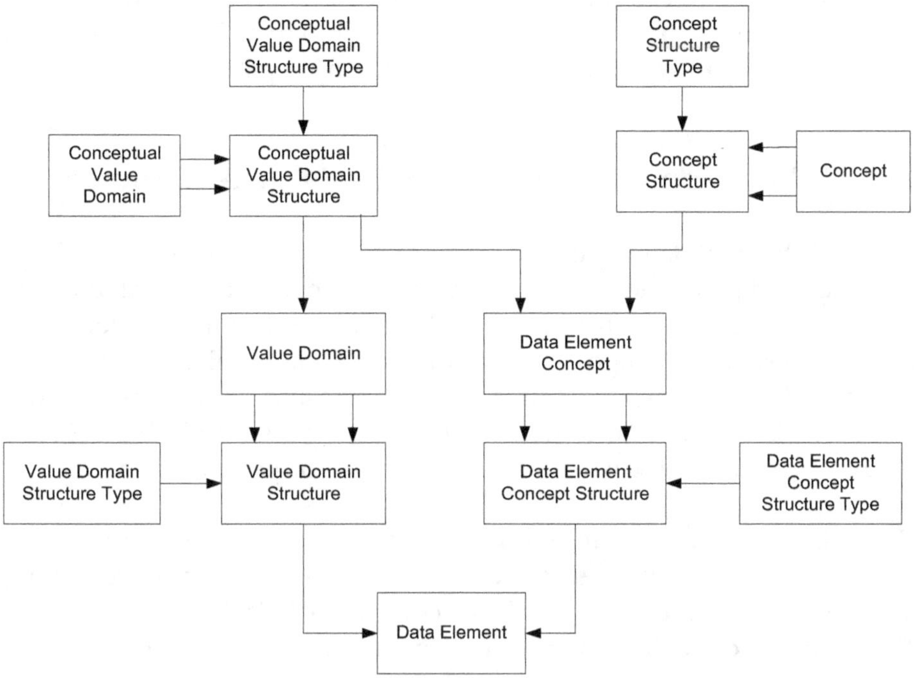

Figure 71. Value domain management diamond data structure.

The first, ignore the ambiguity, was immediately discarded because it is simply professionally irresponsible to place the full responsibility on the end-user to understand, detect, and arrive at exactly the right decision regarding the ambiguity.

The second, require strict inheritance, meant that when inheritance tracing is started at a specific data element and goes up to a particular conceptual value domain, the specifically selected conceptual value domain has to be exactly the same one as when inheritance tracing is started at the same data element and when up through the value domain. This alternative was intensively reviewed and it was decided that while it was the perfect solution, it was both not a practical solution, and it placed too strict a set of semantic restrictions.

The third alternative, permit family inheritance, required only that there be an ultimate common conceptual value domain ancestor when tracing up through data element concept and through value domain. This was accepted as having greater practicality. The determination of the chosen alternative was done by someone outside information technology whose PhD was in classical philosophy with a dissertation in ontologies.

7.4.3.2 Data Element Value Domain Value Assignment

Figure 72 presents the data element value domain assignment window. To arrive at the set of data element value domains, proceed left to right and top-down. Select the concept structure type, and concept. Select a data element concept structure type, and data element concept. That will list the data elements. Then, select a value domain structure type and value domain. At that point, the data elements and their value domains are within the context of concepts, data element concepts, and value domains appear at the bottom browse.

To assign a particular value domain to a particular data element, select the appropriate data element, the appropriate value domain, and press the Assign button (lower right). If an attribute, column or DBMS column is selected instead of a data element, the Assign and Set to Unknown buttons are disabled.

If the value domain is allowed to be assigned, a message to that effect is displayed in the message section. A value domain is allowed to be assigned under two conditions. First, if it is not already assigned to a contained attribute, column, or DBMS column, and second if it is a "superior" value domain to that already assigned to an attribute, column, or DBMS column.

Figure 73 shows a different data element value domain assignment screen. In this example, the value domain, Name Strings of Arbitrary Length, is attempted to be assigned to a data element. There is first a message at the bottom left that indicates that the Metabase System is evaluating the assignment. Then, there is the message that the assignment is rejected because the value domain had already been assigned to the data element.

Figure 74 shows another example, but this time it is an attribute value domain assignment screen. This assignment is again trying to assign the value domain, Name Strings of Arbitrary Length to the attribute, Movie Name. In this screen the assignment message at the bottom shows that the assignment failed. Also shown in this screen, above the assignment browse is the data element, Product Name, and the existing value domain assignment, Name Strings of Arbitrary Length.

Figure 75 is similar to that of Figure 74 except that the value domain that is begin assigned, Name Strings of Specific Lengths, is a value domain subset of Name Strings of Arbitrary Length. The nesting of these two value domains is shown in the browse, 5) Select Value Domain. In this case, because the assigned value domain is more restricted, the assignment was accepted.

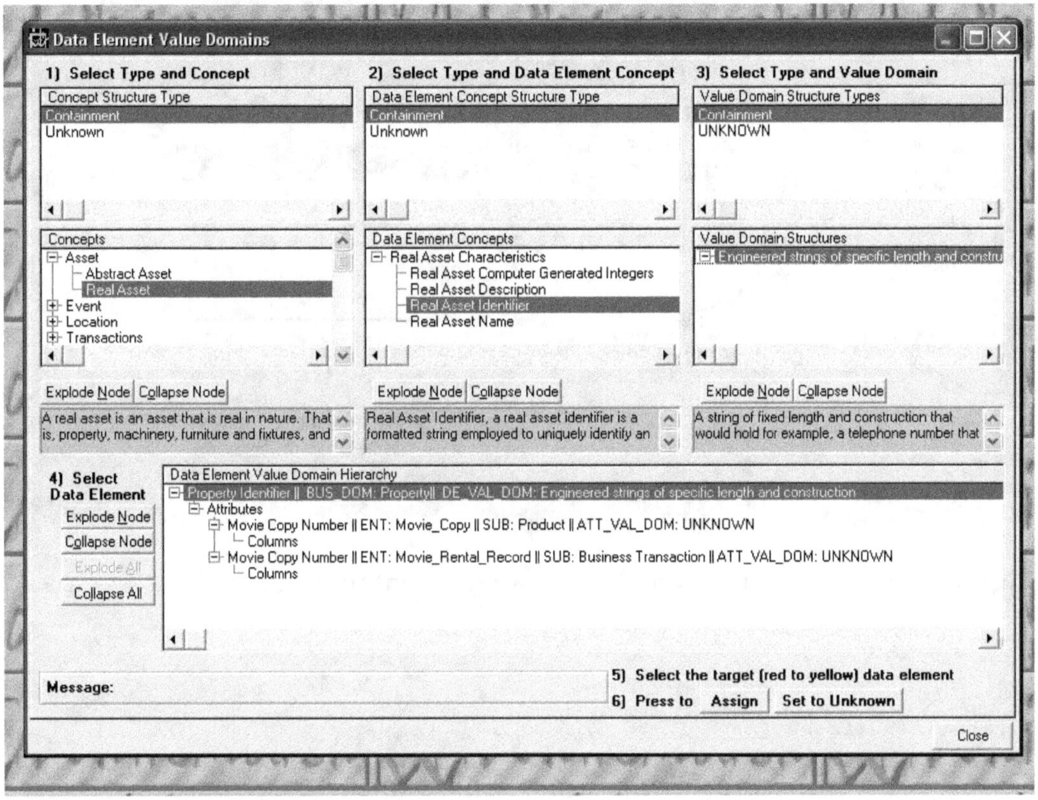

Figure 72. Data element value domain assignment.

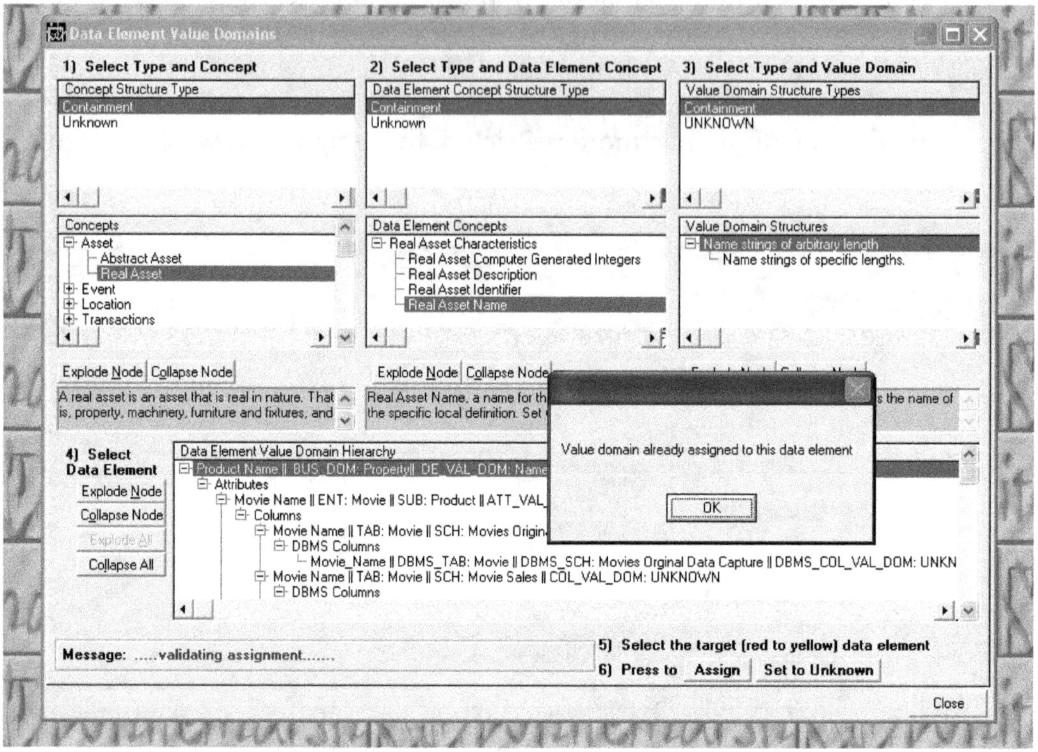

Figure 73. Data element value domain assignment messages.

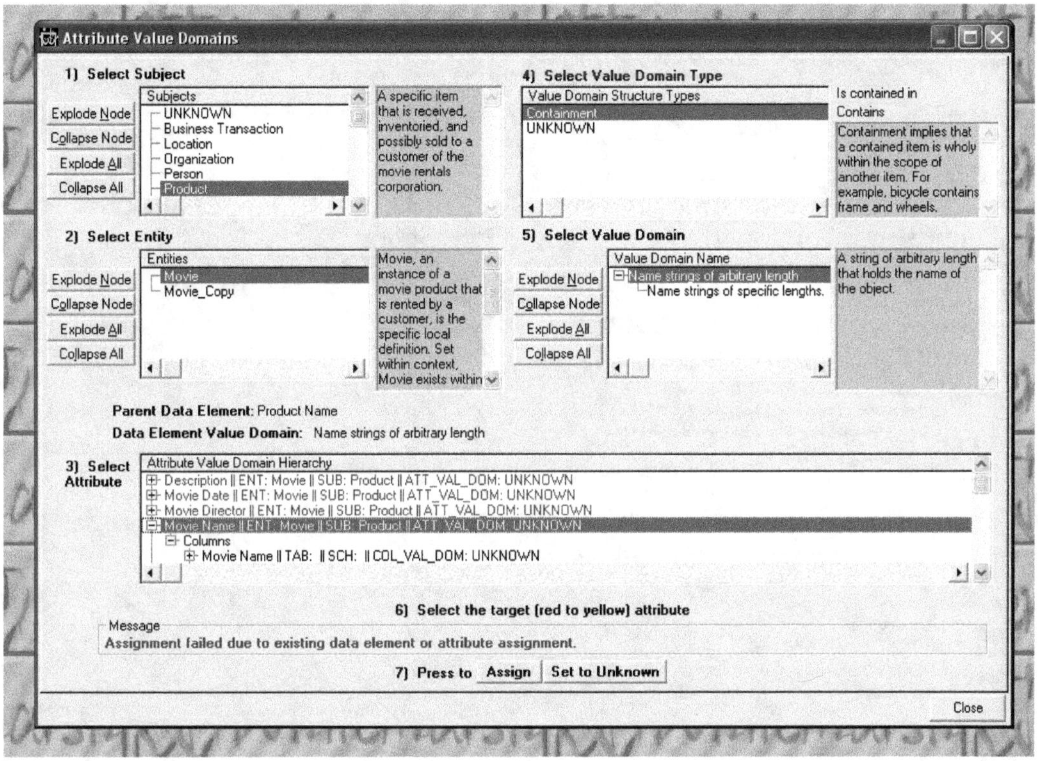

Figure 74. Attribute value domain assignment–duplicate assignment error.

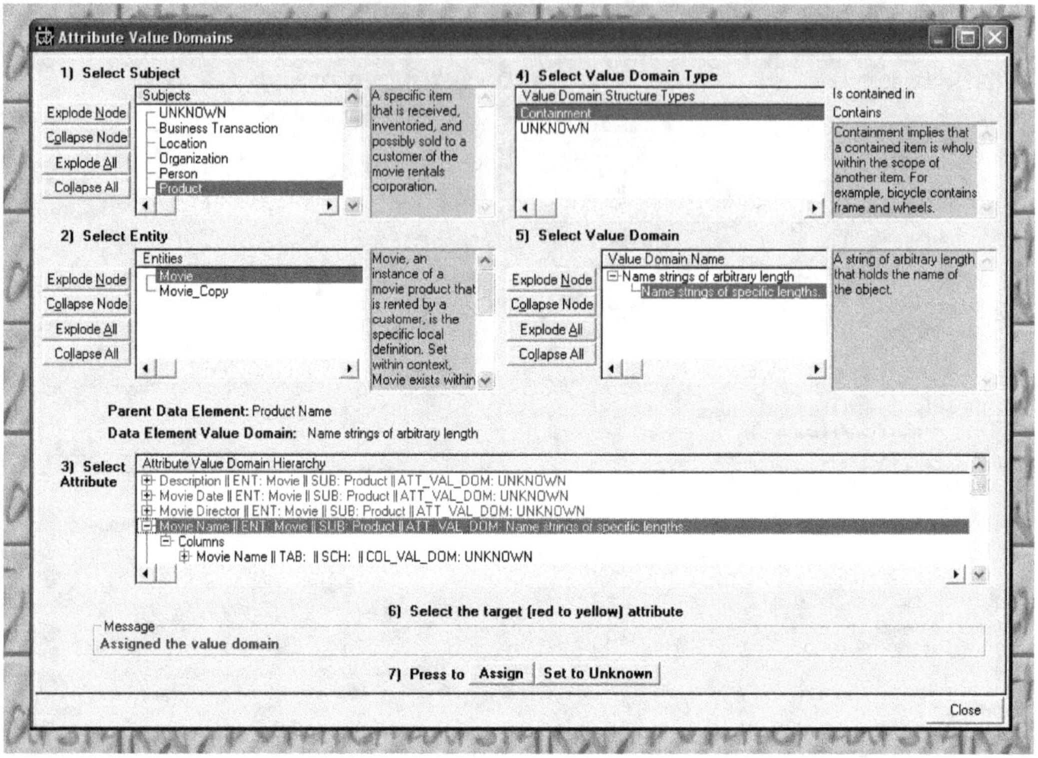

Figure 75. Attribute value domain assignment–assigning value domain subset.

7.5 Error Classes

Value domains are a very critical semantic component as they provide a very explicit first line of defense in stopping invalid values from entering any database. Value domains and their accompanying values can take on both type and values forms. For example, in a data warehouse database, the column, Region might take on the value, New England. While in another database, the column name might be New England Region. It all depends on the data modeling strategy. In either case, the value domain component of the Data Element Model has the values and the definitions from which either the names can be constructed, or the values that would become part of reference data can be produced. Thus, there should be a good effort to prevent the first class of errors and to prevent the second error class altogether.

If, in the first case, the column is Region, but for some records the value is New England and for other records it is New England Region, these non-standard values become an immediate source of Type I errors. Similarly, if in some domains, 1 means Male and in other domains 1 means Female, these become an obvious source of Type II errors.

The strategy to reduce value domain Type I and Type II errors involves paying attention to data modeling in which values become part of a name. Attention must also be paid to very minor value variants that cause whole collections of records to be missed from selection, sort, join, and other control-type SQL clauses. Great care must be taken to spot even the smallest of difference as might exist in for example, Althea Ln and Althea La. Since differences such as these can exist in the same organization, programmatic facilities must be created to prevent their creation and to constantly check and to catch them when they mysteriously occur.

7.6 Value Domain Management Summary

This chapter was a "can do" chapter. Increasingly, value domains are important to data semantics management. The Type II error cited above, that is, the ability to specify Gender = 1 as meaning Female in one environment while it means Male in another environment can be a critical source of errors that must be prevented. The first step in prevention is knowing about these possibilities, and being able to control and manage value domain value collections. The strategies that were set out in this Chapter enable this to take place.

Demanding that there be congruent value domain collections is just not possible, however. Nice, yes. Recommended, absolutely. But absolutely possible? No. This chapter set out strategies for setting down value domains, associated value collections, and the ability to map one value collection with another.

There is, however, no silver bullet here. If one collection has three values for a concept and another five values for a different instances of that concept, there can only be three mappings, and none of these might be exact. This is illustrated in the case of mapping "one-third's based values" to a set of "one-fifth's based values," or academic "letter" grades to precise "numeric" grades. Miracles are not possible.

Some domain value mappings are going to be unsatisfactory whenever square pegs are shoved into round holes. At least, this chapter provides a strategy to define the square to round mappings, and a Metabase System solution to know when and where the mappings are occurring.

This chapter showed the processes necessary to create value domains, value domain values, and to assign value domains to data elements. The process is the same for assigning value domains to attributes, columns, and DBMS columns. This chapter also showed how the Metabase System prevents both duplicative and conflicting value domain assignment.

The next chapter, Chapter 8, Fact Specification Cases, focuses on understanding all the different cases associated with fact specification. Some of the cases are simple, for example, single values for fields like Birth Date. Other cases require groups of fact collections such as a Full Name (First Name, Middle Name, and Last Name). Still others required facts like Life Span which would be from Birth to Death. Many of these fact cases result in virtual values such as a person's age ((Today - Birthdate)/365.25). Still others required sophisticated processes. Examples of all these data integrity rules that govern the fact cases are set out.

7.7 Questions and Exercises

1. This chapter asserts that values within restricted value domains are a form of semantics. Why would that be true?

2. How has your project, function, or organization attempted to perform value domain management? What has been your strategy? What are some of the problems?

3. Technology strategies for value domain management include: DBMS Schema clause specification, inclusion as database table row-based values, encoded in business information system programs. Compare and contrast the advantages and disadvantages of each.

4. How has your enterprise dealt with changes in value domain values in the following cases: 1 to 1 change in value, increased quantity of values over the same domain, reduced quantity of values over the same domain?

5. How has your organization dealt with changes in stored values in database backups over time? What has been the strategy? Cost? Resources?

6. What has been your organization's strategy when values need to be in different languages? How have you been able to ensure that there are acceptable and exact translations?

7. Examine Table 23. Has this happened to your and your organization? What examples can you discover? How did you discover them? How are valid values bound within your databases and business information systems. How difficult would it be to make changes to the way that you have valid values bound into your databases and business information systems? What strategy have you taken to resolve this problem.

8. What strategy have your and your organization taken to discover a complete set of values over a value domain? How have you resolved conflicting value sets from different databases within different projects, functions, and organizations? What strategy should be undertaken to merge value domain sets?

9. Compare and contrast direct value domains from Section 7.3.1 and indirect value domains from Section 7.3.2. What are the advantages and disadvantages of each approach. Which ones have you tried? What were the results?

10. Review the reference data model in Figure 54. Which kind of strategy is this? Direct or indirect? What problems does this type of strategy resolve?

11. What type of bureaucracy has your organization established to deal with value domain management similar to that in Figure 54?

12. Review the issues presented in Table 24. Is it a complete set? Are the results from factoring correct? Have you attempted these approaches? What has been the results?

13. Review the meta-model for value domains in Figure 55. Does this seem to address the complete set of value domain problems? How should this meta-model be modified to address the data structures contained in Figure 54?

14. How is master data, also called strategic data and/or golden source data similar to value domains? How is it different? Is it just a matter of data structure complexity?

15. Compare and contrast an approach for master data management versus traditional value domain management. Which is more complex? What should the overall set of steps be?

16. How should value domains and/or master data be managed? What is the effect on existing databases and business information systems that employ value domains and/or master data?

17. How should errors be handled for value domains and/or master data? That is, when data is entered and then rejected because of a value domain and/or master data different value, how should the error be handled? Simply rejected? What happens if the value that attempted to be entered is in fact correct? How do you handle this?

18. What should the effect of value domain and master data management across a functional area and across the enterprise have on business information system architecture? How should updates be handled to these types of data? Should all updates be driven to the authoritative data source for these? How? What should be the strategy for enterprise-wide updates and propagation? Must these types of data be non-redundant or can they be distributed? What are the advantages and disadvantages of each approach?

19. How should you approach the existence of data that becomes incorrect with valid values data? Should there be a set of systems that are constantly running that discovers data that is not correct? What would be the cost and effect of such a set of systems? How would you justify this to management?

20. What kind of valid value data management strategies do you have in effect to enterprise data database backups? Are database backups retrieved and updated on a regular basis to keep their valid value data uses current?

21. Under what situations should valid value data not be updated in current databases and database backups?

22 Under what situations should master data be centralized or decentralized? Under what situations should master data be duplicated? How would you deploy such alternatives? How would you manage master data updates? Real time? On a fixed schedule?

23. How should you approach the existence of data that becomes incorrect with master data? Should there be a set of systems that are constantly running that discovers data that is not correct? What would be the cost and effect of such a set of systems? How would you justify this to management?

24. What kind of master data management strategies do you have in effect to manage database backups? Are database backups retrieved and updated on a regular basis to keep their master data uses current?

25. Under what situations should master data not be updated in current databases and database backups?

26. Section 7.4.1 includes screen shots and supporting text. This section asserts that value domains need to be in a network data structure. Why is this true? Why won't a hierarchy work just as well? Compare and contrast a network data structure with a hierarchical one?

27. Figure 60 along with Tables 25 through 28 illustrate a simplified example of a network structure. What would have had to happen to make this a hierarchy? Which of the three tables (part, part structure, part structure type) contains the relationships? Could the relationships be contained exclusively in the part table? Why not? What would have happened to the network if that had happened?

28. What is the function of the Part Structure Type table? What would the effect be on a network structure if it is omitted? How could you distinguish among collections of structures without this table?

29. What are the reasons to justify the existence of the value domain values network structure that is described and shown via screen shots in Section 7.4.2?

30. Section 7.4.2.2 cites three types of problems that can arise unless there is sophisticated processing when networks are constructed. How would the Metabase System software have to be constructed to prevent these problems?

31. Section 7.4.3.1 indicates that there are real ambiguity problems with ISO standard 11179. Do you agree? Does this section make the case? What problems would arise of an ISO 11179 implementation was chosen that did not conform to Figure 71 for its data model?

32. The Error Classes section outlines the types of errors addressed by this chapter's materials. Do you agree that the error classes are addressed? How? What could be different? Has this chapter's contents and strategy provided an overall justification for its implementation? If yes, why? If now, why not? How would you quantity the benefits from having these error classes resolved? How would you present the approach, costs, and benefits to management in order to justify funding?

8

Fact Specification Cases

8.1 Introduction and Scope

Chapter 7, value domain Management, was a "can do" chapter. Value domains are important to data semantics management. The Type II error, that is, the ability to specify Gender = 1 as meaning Female in one environment while it means Male in another environment can be a critical source of errors that must be prevented. The first step in prevention is knowing about these possibilities, and being able to control and manage value domain value collections. The strategies set out in Chapter 7 enable this to take place.

This chapter, focuses on understanding all the different cases associated with fact specification. Some cases are simple, for example, single values for fields like Birth Date. Other cases require groups of fact collections such as a Full Name (First Name, Middle Name, and Last Name). Still others required facts like Life Span that would be from Birth to Death. Many of these fact cases result in virtual values such as a person's age ((Today - Birthdate)/365.25). Still, others required sophisticated processes. Examples of all these data integrity rules that govern the fact cases are set out.

This chapter not only identifies the different business fact cases but also shows how these facts are defined atomically within data elements, attributes, columns, and DBMS columns. Also shown is how to display traceability across all these contextual definitions. This chapter shows how fact cases can be bound into SQL Views where a given SQL column maps to multiple DBMS columns, and vice versa. Detailed also is how view columns can be mapped to both compound data elements, and to derived data elements.

This chapter sets out the different classes of data-based data integrity rules, sometimes also known as business rules. Since the rules in this book are data-centric, and since business rules, in contrast, are often contained in business information systems, the name used in this book and in this chapter remains Data Integrity Rules.

Shown in this chapter also are two data integrity rule meta-models, one for specifying data integrity rules, and another for binding data integrity rules in a Metabase System environment. The reader should be note, however, that as of 8/1/2008 neither of these meta models exist in the Metabase System. Both models can be implemented in just a few staff-months and are scheduled to be done during 2008.

This chapter concludes with a look into the near future where business rules can be subject to semantic compiling such that a rule specification that multiply distance by time would be prevented.

8.2 Concepts and Engineering

Six general fact cases need to be addressed:

- Simple: single fact with a restricted value set. Directly maps to the data integrity rule: Single table, Single column.
- Compound: single fact but with multiple contained facts.
- Group: Single concept with multiple contained facts such as Address. Directly maps to the data integrity rule: Single table, Multiple columns.
- Pair: Two facts that exist as matched pairs of codes and values. Directly maps to the data integrity rule: Single table, Multiple columns.
- Related: Two or more facts that require one or more data integrity rules to ensure consistency.
- Complex: A fact that contain more than one of the preceding five fact cases.

A more precise set of specifications of these general scenarios is provided in Section 8.3.1 through 8.3.7. Fact cases, 1, 3, and 4 map directly to the data integrity rules. Fact cases 2 and 5 map to multiple and different data integrity rules. Fact case 6 maps to multiple data integrity rules. Fact cases 3 and 4 can be addressed by table-based integrity clauses and/or stored procedures. Fact case 1 is a column constraint or a stored procedure tied to the column. Fact case 2 can be handled by a stored procedure tied to a complex (SQL:1999 or more recent SQL standard) column. Fact case 5 and 6 can be handled by stored procedures that are either tied to a single table, or to a table-based procedure that involves other tables.

As each fact scenario is discovered, the entire fact collection must be treated as a single semantic unit so that it can be counted, reported, and maintained as a unit. Engineering data-centric integrity starts with data integrity rule specifications. These data integrity rules are stored in the Metabase System and are subsequently bound into DBMS schemas as stored procedures, assertions, and triggers, or finally, process modules in business information systems.

8.2.1 Simple Facts

A simple fact is one that represents a single concept portrayed through a single set of values. For example, Employee City Name is portrayed through the values Mexico City, Dallas, New York, and Sao Paulo.

Rules governing values can also be established such as maximums, and minimums. For example, the value of Remaining Balance might be disallowed from being negative because that would represent a credit balance.

8.2.2 Compound Facts

A compound fact also represents a single concept but consists of multiple contained facts. For example, product number contains:

- Business Unit Code.
- UPC[8] Product Number.
- Variation Code.

---

8 UPC means a universal product code. It is a standard by which companies and their products can be known both by name, number, and by a specific computer scanable bar code.

Each of the contained facts (business unit number, UPC product number, and variation code) must be named, defined, and described. Further, the value set for each contained fact must be standardized. Finally, the rule for combining the contained facts (that is, business unit identifier and UP Product number and variation code) into the compound fact (product number) must be stated.

8.2.3 Group Facts

A group fact is one that represents a single concept like an address but is not perceived as a single value like a compound fact. Address, as a group fact, has to be standardized both as a collection of facts and as individual facts. The collection standardization means that addresses can contain fields such as:

- Honorific.
- Individual Name (First, Middle, and Last).
- Individual Title.
- Company Name.
- Street Address Number.
- Street Address Subunit Number.
- Street Address Name.
- Postal Code.
- City.
- State or Province.
- Country.

Sometimes, group facts can only be handed with complex data structures as in the standardization of Address. Address is inherently variable as different countries and country-subdivisions can have differently formatted addresses. In actuality, an address data structure is a bill-of-materials, where the type table of the bill of materials is hierarchical with Country at the apex, and country-subdivision as contained levels. Figure 76 illustrates a bill-of-materials structure for Address. The tables support this Address group fact specification are Tables 30, 31, and 32

In this group fact example, the address parts, for example, honorific (1), or postal code (8), or zip (9) are all stored in the Address Part table (see Table 30). From the example above, there would be 11 distinct parts. In the Address Part Structure table (see Table 31), a country, for example, the United States would be entered. All the Address Part Structure records "pointed to"

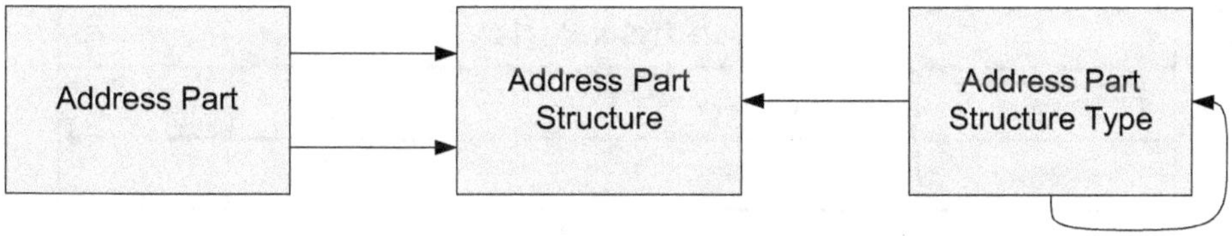

Figure 76. Address bill of materials data model.

by the United States Address Part Structure Type record (see Table 32) would be a United States Address. As a subordinate type of address, there could also be a Residential address and a Business address. To represent a complete bill-of-materials for Address, three extra parts are needed:

- Address.
- Residential Address.
- Business Address.

Table 31 contains the structure records and is read like this: <passive-id-record> is contained in <active-id-record>. For example, the structure record number 11 is read, "State or Province is contained in Residential Address." The top of any bill-of-materials structure (as implemented in the Metabase System) is presented by a "0" in the Active-Id column. Thus, Address (Passive Id 12 contained in structure record 1 of Table 31) is contained in "nothing." That is, it is at the top of the tree.

Address Parts	
ID	**Name**
1	Honorific
2	Individual Name (First, Middle, and Last)
3	Individual Title
4	Company Name
5	Street Address Number
6	Street Address Subunit Number
7	Street Address Name

Address Parts	
8	Postal Code
9	City
10	State or Province
11	Country
12	Address
13	Residential Address
14	Commercial Address

Table 30. Address parts table rows.

Address Part Structure Id	Active Id	Passive Id	Sequence Number	Address Part Structure Type Id
1	0	12	1	1
2	12	13	1	1
3	12	14	2	1
4	13	1	1	2
5	13	2	2	2
6	13	5	3	2
7	13	6	4	2
8	13	7	5	2
9	13	8	6	2
10	13	9	7	2
11	13	10	8	2

Table 31. Address parts structure table rows

Address Parts Structure Types		
ID	**Parent Id**	**Name**
1	0	United States
2	1	Residential
3	2	Commercial

Table 32. Address parts structure type table.

Each discrete address collection type, e.g., Residential Address, would be an address template that would be employed in the business information system process that collects the actual address part values. These values would be stored in the database and the template would be used to construct the address display and/or print label. While this is clearly more complicated than just AddressLine1, . . . , Zip, it enables a universal solution across all addresses regardless of type.

Beyond fact collections being standardized, the value set of these discrete fact collections must also be standardized. Thus, discrete fact values may exist independently, or in combination. For example, a birthdate is a single discrete fact, while another fact, postal code, automatically implies a specific city and state/province.

8.2.4 Pair Facts

A pair fact is one that represents a single concept but has multiple value representations. Marital status is the concept, and the two facts are *marital status code* (M or S) and *marital status name* (married or single). The concept must be named, defined, and described. The fact representation values must also be named, defined, and described along with the paired sets of data values:

- 1 and Married.
- 2 and Single.

A paired concept can be employed even when there are multiple languages. For example, while the agreed code value for married and single might be 1 and 2, the correspondent names in Spanish are: Casado and Soltero, and in English, Married and Single.

8.2.5 Related Facts

Two or more facts are related when their values--with respect to each other--exist as a semantically related pair. For example, every contract should have a *contract start date* and *contract end date*. The calendar difference between the two fields can be computed and interpreted as the contract's duration. While each fact exists alone with rules governing individual values such as "no new contract can have a start date in the past," the pair of contract date values is also subject to rules such as the valid *contract end date* cannot precede the *contract start date*. What would be the meaning of a negative contract duration?

Another example is Part Weight value and Part Weight units. Values from both columns are needed to know the meaning fully. In this case the values would be: 8, pounds.

8.2.6 Complex Facts

Some facts are more than one of the above five fact types. For example, the product number field contains the subfield, Business Unit, as a single alphabetic character, which, in turn, is a code for the business unit name and for the UPC code. In this example, the single letter, B (as taken from B0110101), really means both Blades Steel Company and represents the UPC code, 90000. In this type of situation, the Business Unit Identifier (it is really a code) must be discovered, defined, described, and has its value set standardized. Business Unit Identifier is also part of the fact pair Business Unit Identifier and UPC correlation, and the fact pair Business Unit Identifier and Business Unit Name. These two additional fact pairs must also be discovered, defined, described, and have their value sets standardized.

8.3 Data Integrity Rules

Simply put, a data integrity rule is a rule that, when executed, restores either integrity or tests whether integrity remains. A data integrity rule, sometimes also know as business rules, contains two parts: Data and process. Each must be carefully specified as business facts and the processes involved in the determination that the processed result is True or False. Figure 77 shows a general meta model for data-centric Data Integrity Rules. Figure 78 illustrates where Data Integrity Rules can be bound.

The Figure 77 meta-entities and a brief description of each are listed in Table 33. An example of a Data Integrity Rule might be to determine the status of a financial account. If it is 60 or more days over due, an overdue notice is to be sent out. In this situation, the data that comprises the Data Integrity Rule is illustrated in Table 33. In this example, all four data classes are needed.

In general, a Data Integrity Rule execution proceeds through its stages, i.e., Input, Select, Validate, and Action. If the result is unacceptable, a message would be presented and the transaction terminated. All the data integrity rules commonly proceed through these stages:

- **Input check**. That is, it would be validated for completeness. Are all the fields valued and present and do they meet surface validity rules such as a valid gender, etc.?
- **Selector Check**. That is, a subset of the fields would be used to select the data from the database. If a record is found, the test is passed, otherwise it fails.
- **Validator Check**. That the data in the database is acceptable with respect to the Input check data. For example, there is a valid Employee Id, but the name of the employee on the Input does not match that of the Database.
- **Action Check**. That a particular action regarding the data is carried out. That is, the database data is updated, the row deleted, a row is added, etc.

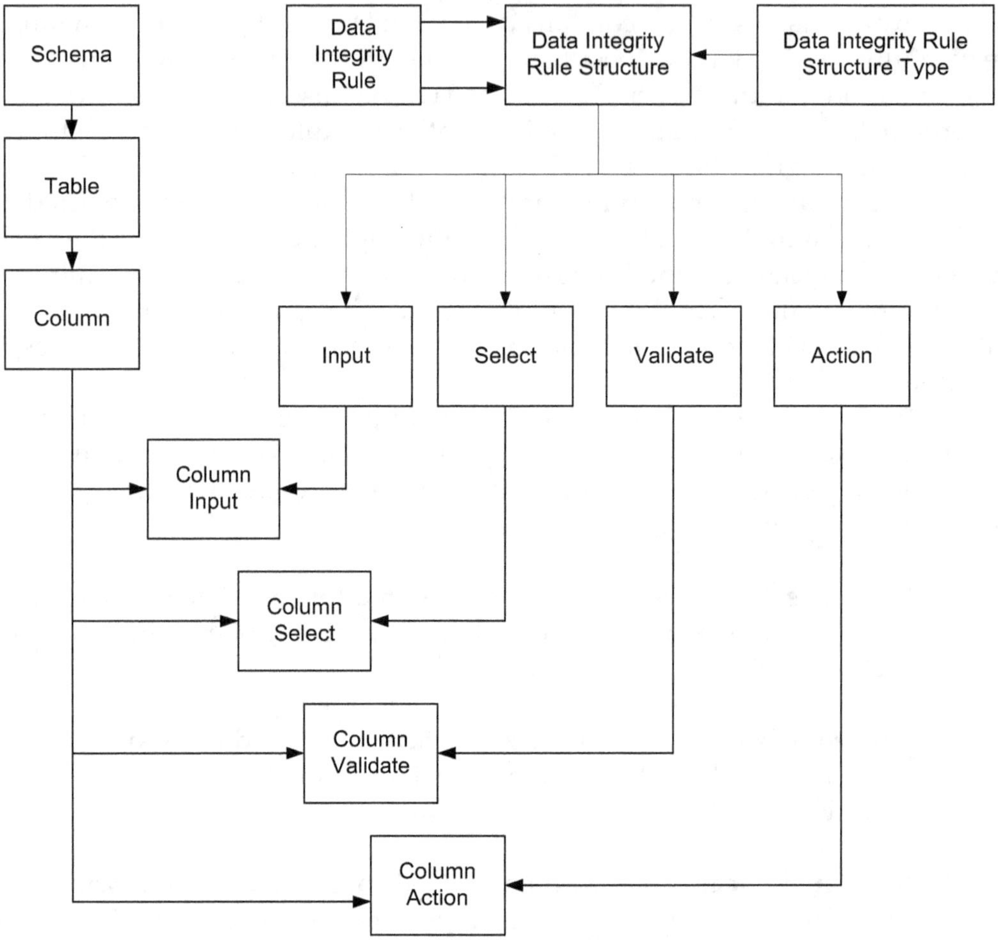

Figure 77. Data integrity rule specification meta model.

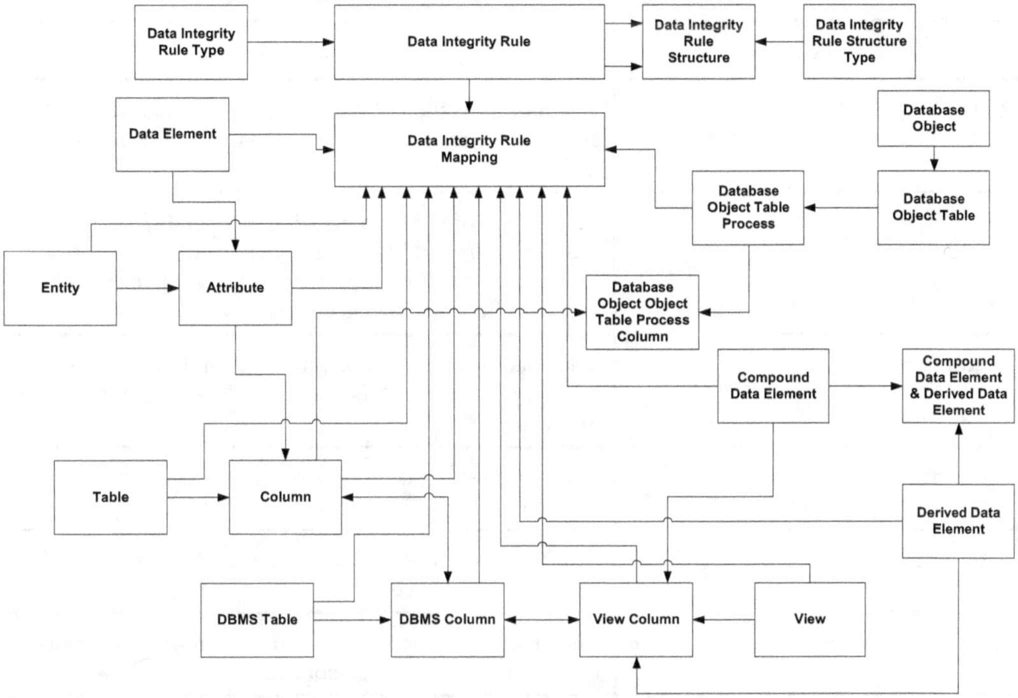

Figure 78. Data Integrity Rule binding meta model.

Data Integrity Rule Specification Meta Model	
Meta entity	**Description**
Data Integrity Rule	Name and description of the data integrity rule.
Data Integrity Rule Structure	Network structure among collections of data integrity rules
Data Integrity Rule Structure Type	Typed collections of data integrity rules such as table-based, column-based.
Input	The names of the database columns and/or variables that must be valued before the data integrity rule can be processed.

Data Integrity Rule Specification Meta Model	
Meta entity	**Description**
Select	The names of the database columns involved in data selection logic. Includes the actual SQL selection logic as well.
Validate	The names of the database columns that must be validated. Included are all the validation clauses, values that must be or not be met, etc. Includes the actual validation logic as well.
Action	The specifications of the actual action that is to be taken including insert, update, delete, calculations, and the like.
Column-input	The cross-reference between the input columns and the actual database columns.
Column-select	The cross-reference between the select columns and the actual database columns.
Column-validate	The cross-reference between the validation columns and the actual database columns.
Column-action	The cross-reference between the action columns and the actual database columns.

Table 33. Data integrity rule specification meta model.

Traditionally, data integrity rules are an integral component of the computer programs associated with data processing applications. With the arrival of database, computer programs have become less and less the central focus of the data processing. Computer programs have been replaced with query statements, automatically generated screen programs, and entire programs that are only ten to twenty statements long and are commonly known as stored procedures.

In addition to a complete change in program form, the quantity of opportunities for updating the rows of data within a DBMS table within a database environment has greatly increased. Because of these needs, data integrity rules must be centrally designed and made available in the repository.

If the information about the rules, as specified in Table 32 are stored in the Metabase System, a DBMS schema designer can know the integrity requirements for complete specification of table-based triggers and assertions and Column-based validations.

The next step is to bind the specified Data Integrity Rules into the appropriate places within the various meta-entities of the Metabase System. The meta model for binding data integrity rules into the Metabase System is shown in Figure 78. The meta-entity types supporting data integrity rules are:

- Column
- Column & data integrity rule
- Column & database view element
- Data element
- Data element & data integrity rule
- Data element domain
- Data element domain & data integrity rule
- Data integrity rule
- Data integrity rule & database view element
- Table

The specifications of a data integrity rules are illustrated in Figure 77. These specifications would be set as meta-entities related to the Data Integrity Rule meta entity. Figure 78 shows that a Data Integrity Rule is related to the Data Integrity Rule Mapping meta entity. This supports assignment dates, and the identification of which other meta entities that are subject to the specific Data Integrity Rule. Figure 78 shows, for example, that a given Data Integrity Rule can abe assigned to Entities, Tables, DBMS Tables, and Views. Data Integrity Rules can also be assigned to Data Elements, Compound Data Elements, Derived Data Elements, Attributes, Columns, and DBMS Columns.

Because of inheritance, a Data Integrity Rule wold not be assigned simultaneously to a Data Element, Attribute, Column, or DBMS Column. It would only need to be assigned to the highest level meta entity. If a Data Integrity Rule has a subordinate Data Integrity Rule, which, as shown in Figure 78, is clearly permitted, the subordinate Data Integrity Rule could be assigned to a Column while the parent Data Integrity Rule could be assigned to the Attribute. The sections that follow are a more precise specification of Data Integrity rules, which are:

- Single table, single Column
- Single table, multiple Column
- Single table, single Column, multiple row
- Single table, multiple Column, single row derived data
- Single table, single Column, multiple row derived data
- Multiple table derived data
- Multiple table, multiple Column (referential integrity)

Each Column might have one or more data integrity rules defined within the context of a table. A rule might be that the Column cannot be blank (or NULL) whenever the row is stored or subsequently updated. There might be other rules that are related to valid values, invalid values, ranges of valid and invalid values, and the like.

Tables themselves also have data integrity rules, in that all the rules associated with all the Columns are identified as belonging to the table.

A modified Backus Naur form of notation is used to illustrate data integrity rules. To indicate a particular column of a particular table that was retrieved from the database according to a specific select clause, the following notation is used:

C ::= column id
T ::= table type
SCn ::= select clause

Thus C.T.SC1(I) is the I-th column within the table (T) selected by the first select clause (SC). The number suffix on the SC denotes different select clauses. If different <table id number>s or different <column id>s are stored, either different tables and/or different columns are employed in the rule.

For convenience, the term <column> is used as a shorthand way to denote a specific <column> within a specific <table>. Each data integrity rule type is illustrated through a definition, example, and a grammar.

8.3.1 Single Table, Single Column

Definition: A data table that contains the valid and invalid values for columns.

Example: BIRTH_DATE column must contain only valid dates.

The requirements for defining column data integrity rules are the identification of the column, and the identification of whether the editing rule is for valid/invalid discrete values or for valid/invalid value ranges. These rules are already defined in the Metabase System under the category of edit tables. Included herein are the table structures for edit tables, illustrating the type of data needed to define column data integrity rules.

column:	<column (I)>
Valid Values:	<value> [,<value>] . . .
Range Values:	<FROM <value> TO <value>> . . .
Encode-Decode Values:	<ENCODE <value> DECODE <value>> . . .

8.3.2 Single Table, Multiple Column

Definition: Columns that are related to each other in a way not covered by any third normal rule.

Example: There are two different columns, CONTRACT_START_DATE and CONTRACT_END_DATE, and the dates are related; the CONTRACT_END_DATE must occur on or after the CONTRACT_START_DATE.

The requirements for defining this rule are the names of the two columns and the relational operator that exists between instances of the columns.

column:	<column (I)>
Binary Operator:	<binary operator>
column Name:	<column (j)>
<binary operator>s::=	| Less than (LT) | Less than or equal to (LE)

|Equal to (EQ)
|Greater than or equal to (GE)
|Greater than (GT)

Example: DEATH DATE GE BIRTH DATE

8.3.3 Single Table, Single Column, Multiple Row

Definition: An intra-table, single column, multiple row rule is one in which the rows from the same table are related to each other through the value instances of one column.

Example: Each CUSTOMER can have multiple CONTRACTS, but there is the rule that each CONTRACT_NUMBER's value must increase, thus installing a new contract with a CONTRACT_NUMBER that has a duplicate number or that is less than an installed number would be a violation of this rule.

The requirements for defining this rule are the name of the column to which the rule applies, the algebraic formula that governs the value of one instance of the column to the next, and the table selection clause that selects the records to which the rule applies.

column: <column (I)>

Algebraic Formula: <formula>

<formula>s::= <C.T.SC1(I)> <=> <C.T.SC2(I)> <formula operation>

<formula operation>::= programmable operation on the value of the column obtained from the second select clause before it is stored as the value of the column obtained from the first select clause.

<select clause>::=	selection criteria for selecting the row upon which the <formula operation> is to be executed before the new row is stored.

Example: SELECT LAST CONTRACT,
CONTRACT_NUMBER (CONTRACT (CURRENT)) =
CONTRACT_NUMBER (CONTRACT (LAST)) + 1

8.3.4 Single Table, Multiple Column, Single Row Derived Data

Definition: An intra-table derived data rule is one in which the value of a column in one row of one table is the result of a calculation of one or more columns within the same row of the same table.

Example: The term of the contract in days is equal to the number of days between the end date of the contract and its start date.

Requirements for defining this rule are the names of the two columns and the relational operator that exists between instances of the columns.

column:	<column (I)> . . .
Algebraic Formula:	<formula>
<formula>s::=	<C.T.SC1(I)> <=> <C.T.SC1(j)> [<formula operation> <C.T.SC1(k)>] . . .
<formula operation>::=	programmable operation on the value of <column (k)> obtained from the record selected before it is combined with the data value of <column (j)> from the same record before it is stored as the value of <column (I)> in that same row. This operation may iterate.

<select clause>::= selection criteria for selecting the row upon which the <formula operation> is to be executed before the new row is stored.

Example: TERM_OF_CONTRACT = CONTRACT_END_DATE - CONTRACT_START_DATE

8.3.5 Single Table, Single Column, Multiple Row Derived Data

Definition: A single table, single column, multiple row derived data rule is one in which the value of a column within the next-to-be-stored row has been computed from a data value stored in the most-recently stored row. Relationally, a serial number concept must be operating to select the most immediately stored row.

Example: A table contains a column REMAINING_OBLIGATORY_AUTHORITY. The value stored in this column is to be the value stored in the same column of the preceding row minus the amount of the current CONTRACT's value.

The requirements for defining this rule are the name of the column to which the rule applies, and the algebraic formula that governs the value of the column of the most recent row.

column: <column (I)> . . .

<formula>s::= <C.T.SC1(I)> <=> <C.T.SC2(j)> [<formula operation> <C.T.SC3(k)>] . . .

<formula operation>::= programmable operation on the value of <column (k)> obtained from the record selected by the second select clause before it is combined with the data value of <column (j)> from the record obtained from the third select clause, before it is

	stored as the value of <column (I)> in the row selected by the first select clause. This operation may iterate.
<select clause> ::=	selection criteria for selecting the row upon which the <formula operation> is to be executed before the new row is stored.

Example: REMAINING_BALANCE (CONTRACT (current)) = REMAINING_BALANCE (CONTRACT (last)) - PURCHASE_ORDER_VALUE

8.3.6 Multiple Table Derived Data

Definition: A multiple table derived data rule is one in which the value of a column in one table is the result of a calculation on one or more instances of the same or different columns in other tables.

Example: DEPARTMENT table has a TOTAL CURRENT SALARYs column. The value of that column would be affected by the EMPLOYEEs associated with the DEPARTMENT. As an employee is hired, transferred, or terminated, the TOTAL CURRENT SALARY column would have to be modified accordingly.

Requirements for defining this rule are the names of the two columns and the relational operator that exists between instances of the columns.

column: <column (I)> . . .

<formula>s::=	<C.T.SC1(I)> <=> <C.T.SC2(j)> [<formula operation> <C.T.SC3(k)>] . . .

<formula operation> ::= programmable operation on the value of <column (k)> obtained from the table type instance selected by the second select clause before it is combined with the data value of <column (j)> from possibly a different table type instance obtained from the third select clause, before it is stored as the value of <column (I)> in a table type instance of a type that may be different from either of the two other types obtained by the first select clause. This operation may iterate.

<select clause> ::= selection criteria for selecting the owner row, whose primary key is used to select the member rows across which the <member record operation> is to be executed.

Example: SELECT DEPARTMENT WHERE DEPARTMENT_NAME EQ "ACCOUNTS_PAYABLE"

SELECT EMPLOYEE WHERE EMPLOYEE DEPARTMENT NUMBER EQUAL DEPARTMENT_DEPARTMENT_NUMBER

DEPARTMENT_SALARY = SUM (EMPLOYEE_SALARY)

8.3.7 Multiple Table, Multiple Column (Referential Integrity)

Definition: Referential integrity is a data integrity rule that is an extension of the single table, multiple columns rule. Referential integrity involves multiple tables. Furthermore, the rule includes the specification of the automatic actions the DBMS is to undertake whenever an update is attempted that violates the rule.

Example: There is a CURRENT DEPARTMENT NUMBER column in the EMPLOYEE table. If an EMPLOYEE is transferred this field has to be updated with a DEPARTMENT NUMBER that is valid. The employee could not be transferred to a nonexistent department. Additionally, if a DEPARTMENT is dissolved, either the EMPLOYEEs must be "transferred" to a special nonexistent department (e.g., UNASSIGNED) or the EMPLOYEEs must be dissolved (probably an undesirable effect), or the dissolution of the DEPARTMENT must be disallowed. A referential integrity rule thus must not only specify the column basis for the referential integrity but also specify the operations effect of the rule (disallowed, dissolve, or NULL).

The requirements for defining this rule are the names of the two columns, one that acts as the primary key of the owner, and the other that acts as the foreign key in the member. Additionally, there is the requirement to state the actions that are to be taken whenever a referential integrity rule is violated. There are three actions possible:

- When the owner row is deleted, the member row's foreign key is updated to a NULL value, or the member row is deleted, or the owner action is rejected.
- When the member's foreign key is updated to a nonexistent owner primary key value, either the update is rejected or the value is set to a NULL value
- When a new member is inserted, but the value of the member's foreign key is a nonexistent owner primary key value, either the insert is rejected, or the value is set to a NULL value.

```
OWNER TABLE IS <owner table name>
MEMBER TABLE IS <member table name>
CONSTRAINT IS <owner table primary key> EQUAL
    <member column name>
ON DELETE <member action>
ON MODIFY <member action>
ON INSERT <member action>
```

```
EXAMPLE:    OWNER TABLE IS employee
        MEMBER TABLE IS dependent
        CONSTRAINT IS emp.ssn EQUAL dep.emp.ssn
        ON DELETE cascade
        ON MODIFY reject
        ON INSERT reject
```

8.4 Employing Fact Specification Cases

Facts can be employed in any of the following Metabase System components:

- Data elements.
- Derived data elements.
- Compound data elements.
- Attributes.
- Columns.
- DBMS columns.
- View columns.

8.4.1 Data Elements

Data elements are context independent business fact templates. Each data element represents a single value fact. Thus, a data element's definition is for simple single facts. The other fact cases, compound, group, pair, and related are accomplished within Specified, Implemented, and Operational Data Models. Table 34 enumerates the fact cases and how each is addressed by a data element.

Complete sets of data element semantics are explicitly assigned. Assigned are meta category values, data element concepts, data element classifications, and business domains. Its inherited semantics include those assigned to its containing data element concept and value domains. The value domains of the data element are specifically assigned.

Figure 79 shows the meta model for data elements. Shown in this model, data elements are the semantic joining of a data element concept via its data element concept structure, and a value domain via its value domain structure.

Fact Case applicability to a data element		
Case	**Case Name**	**Applicability Note**
Simple	Single fact and restricted value set. Directly maps to the data integrity rule: Single table, single column	Yes
Compound	Single fact but with multiple contained facts.	No. Data elements are defined singly and independent of any "container" such as a table.
Group	Single concept with multiple contained facts such as Address. Directly maps to the data integrity rule: Single table, multiple columns.	No. Data elements are defined singly and independent of any "container" such as a table.
Pair	Two facts that exist as matched pairs of codes and values. Directly maps to the data integrity rule: Single table, multiple columns.	No. Data elements are defined singly and independent of any "container" such as a table.
Related	Two or more facts that require one or more business rules to ensure consistency	No. Data elements are defined singly and independent of any "container" such as a table.
Complex	A fact that contain more than one of the preceding five fact situations	No. Data elements are defined singly and independent of any "container" such as a table.

Table 34. Fact case applicability to a data element.

Figure 80 shows an existing set of data elements. Subsets of data elements are selected by picking a new business domain. In Figure 80, the business domain, Finance, has a whole collection of related data elements.

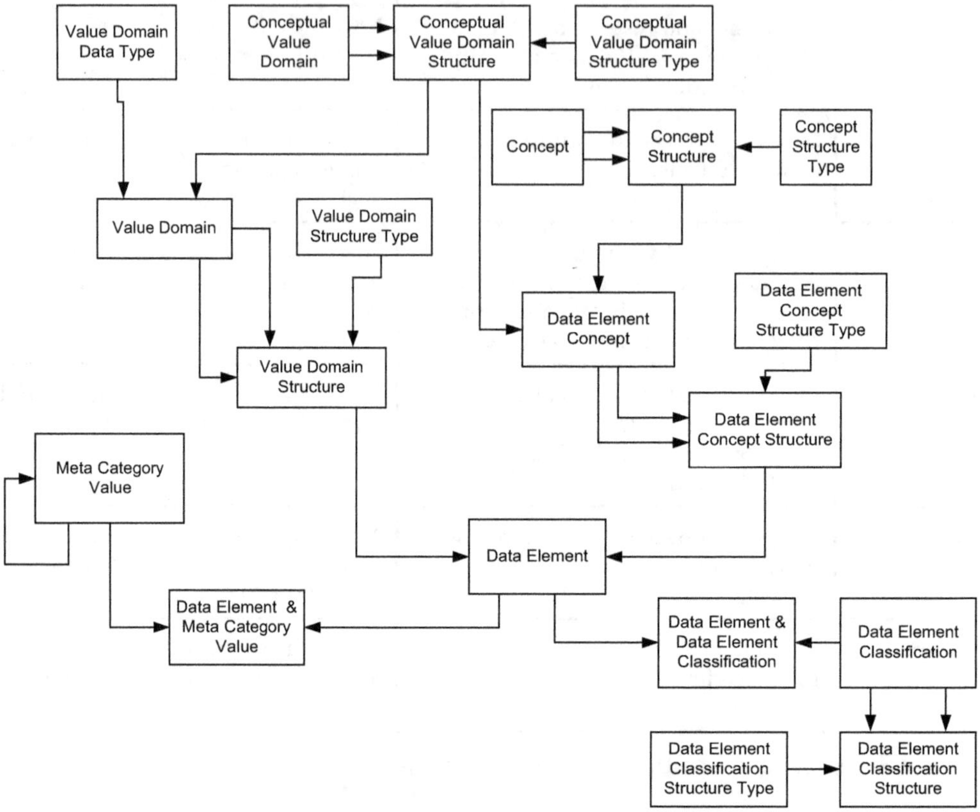

Figure 79. Data element meta model.

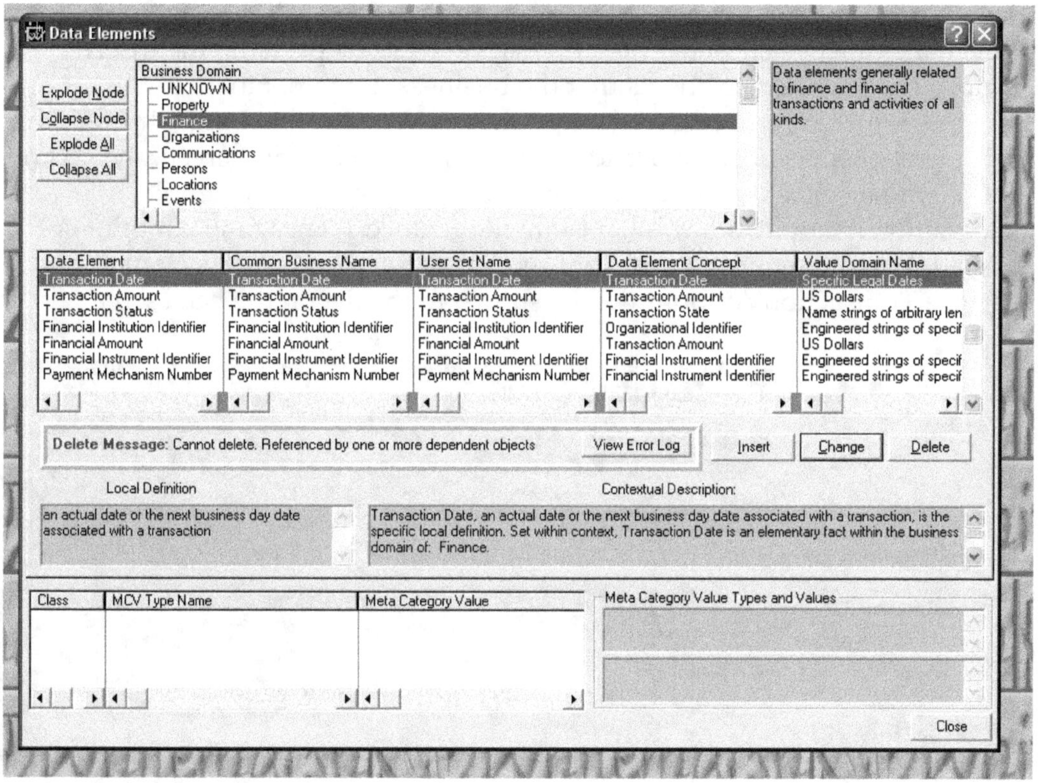

Figure 80. Data elements.

Figure 81 shows the screen for creating a new data element. The list of data elements is shown on the back screen. The data element update screen is the middle screen. The front screen shows a list of data element concepts from which one must be chosen. This process is started by pressing the Insert button. The back screen shows that the business domain, Finance, was selected, which now shows in the update screen. After "tabbing" through the data element concept structure Id entry field on the middle screen, the sets of data element concepts are shown. Once one is selected, for example, Transaction Amount, a new data element, Finance Transaction Amount is created. When the middle and front screens close, the new data element, Finance Transaction Amount will be shown in the list on the back screen.

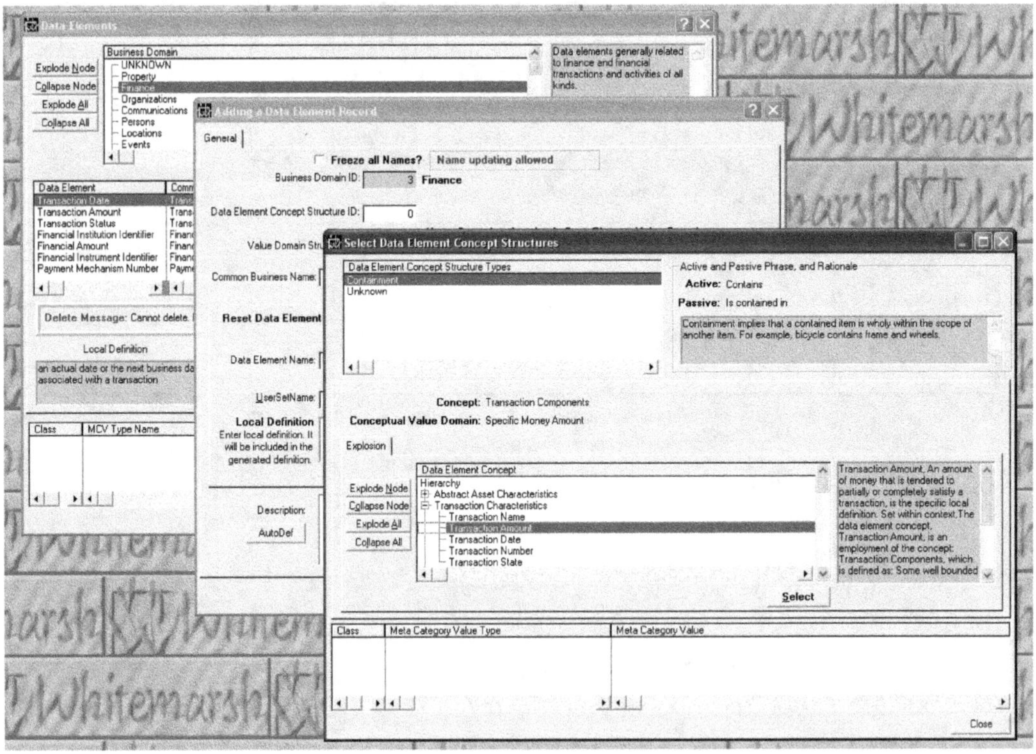

Figure 81. Data element insert screen.

8.4.2 Derived Data Elements

Derived data elements represent the result of some sort of calculation or transformation of either a collection of data elements or compound data elements or a combination of both. Derived data elements are a single fact case in which the value is derived through some calculation that involves other facts. A person's age or the annual sales value are obvious examples. Table 35 identifies the different fact cases handled by derived data elements.

Fact Case applicability to a derived data element		
Case	**Case Name**	**Applicability Note**
Simple	Single fact and restricted value set. Directly maps to the data integrity rule: Single table, single column	Yes. The result of the derivation is materialized in a single derived data element.
Compound	Single fact but with multiple contained facts.	No. This is handled by compound data elements
Group	Single concept with multiple contained facts such as Address. Directly maps to the data integrity rule: Single table, multiple columns.	No. This would be a compound data element.
Pair	Two facts that exist as matched pairs of codes and values. Directly maps to the data integrity rule: Single table, multiple columns.	No. There is no relationship between multiple derived data elements
Related	Two or more facts that require one or more business rules to ensure consistency	No. There is no relationship between multiple derived data elements
Complex	A fact that contain more than one of the preceding five fact situations	No. There is no relationship between multiple derived data elements

Table 35. Fact case applicability to a derived data element.

Figure 82 provides the meta model that supports the definition of derived data elements. Each is defined and related to the data elements that comprise it, or is related to one or more compound data elements. For example, if there is a derived data element, Total Department Salary, it would be the sum of all the employees assigned to that department. The data element involved would be Employee Salary. There would need to be selection logic and the Department table involved. That is all part of the data integrity rules.

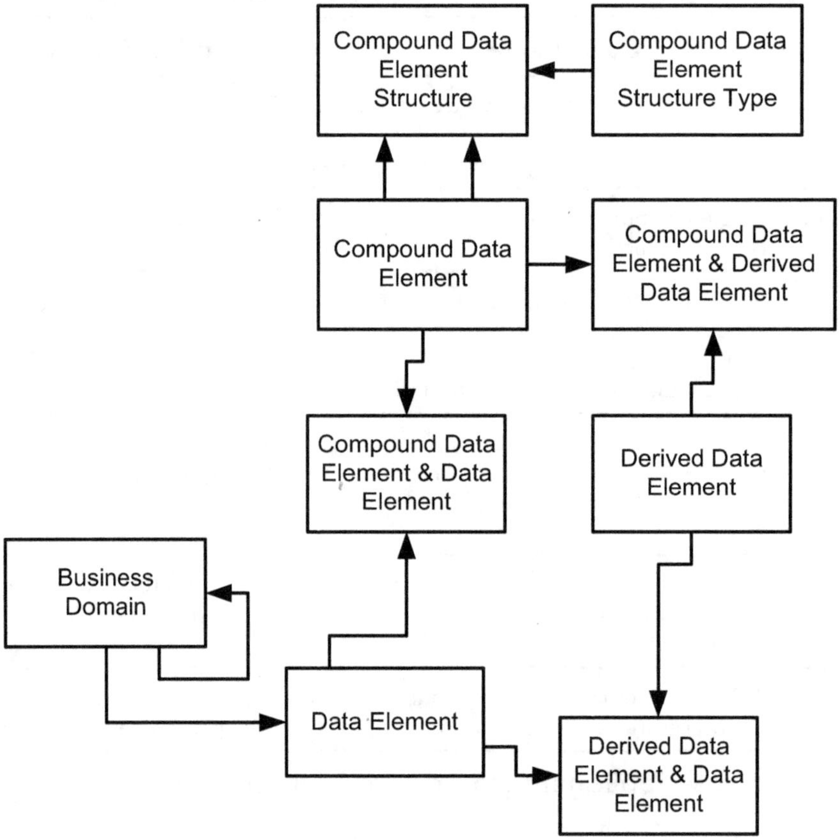

Figure 82. Derived and compound data elements model.

Figure 83 lists derived data elements. The one shown is Total Employee Payments. Shown in the bottom browse are the data elements involved in the creation of the value. Not all the data elements are involved in the actual calculation.

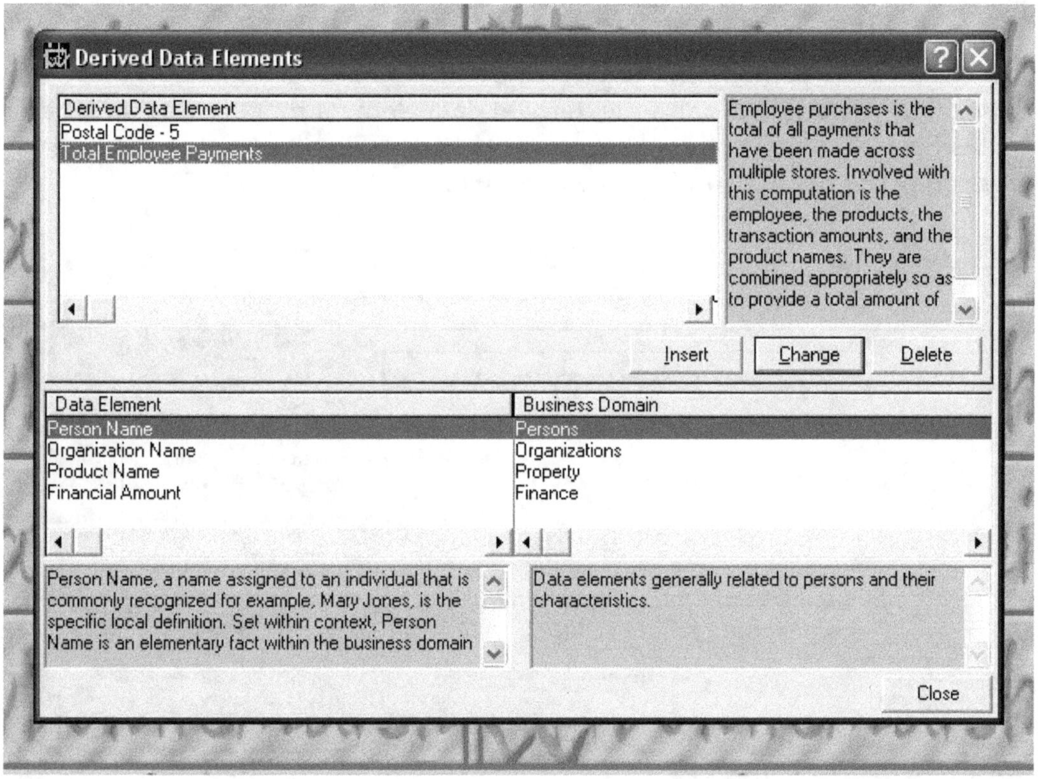

Figure 83. Derived data elements.

Figure 84 shows the update screen for a derived data element. Once a derived data element is defined, it has to be associated with one or more data elements. Figure 85 shows that screen. Tag the derived data element, and tag one or more data elements, and press the Assign button. Once the assignments are made, they show up in the bottom browse as shown in Figure 83.

Figure 86 shows the screen for assigning a derived data element to a compound data element. In this case, the compound data element is CityStateZipPostalCode, and the derived data element that is being assigned to it is Postal-5. Once both are tagged, press the Assign button to cause the creation of the relationship between the two.

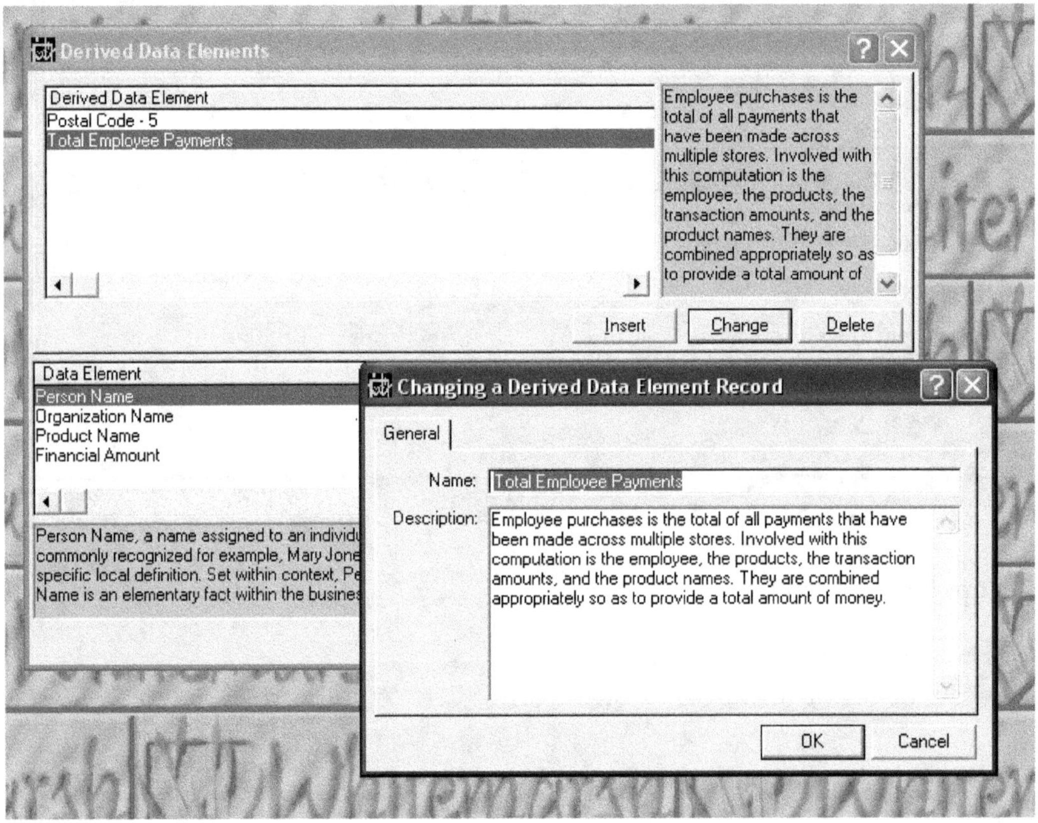

Figure 84. Derived data element update screen.

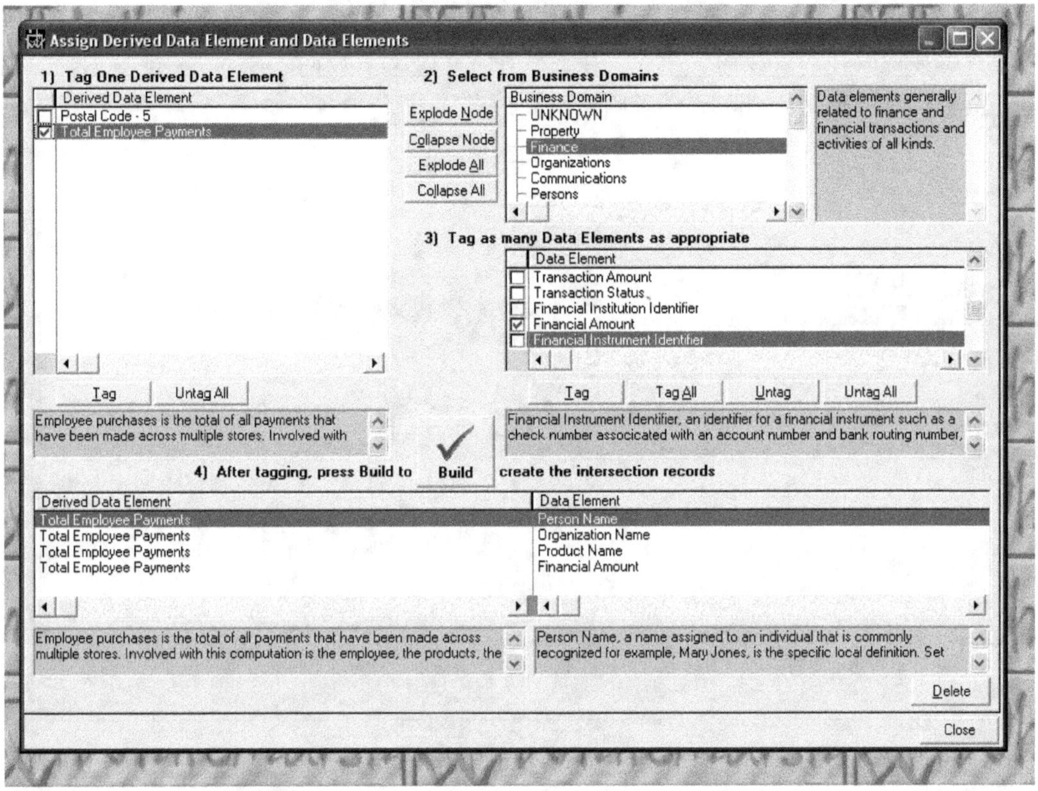

Figure 85. Derived data element data element assignment screen.

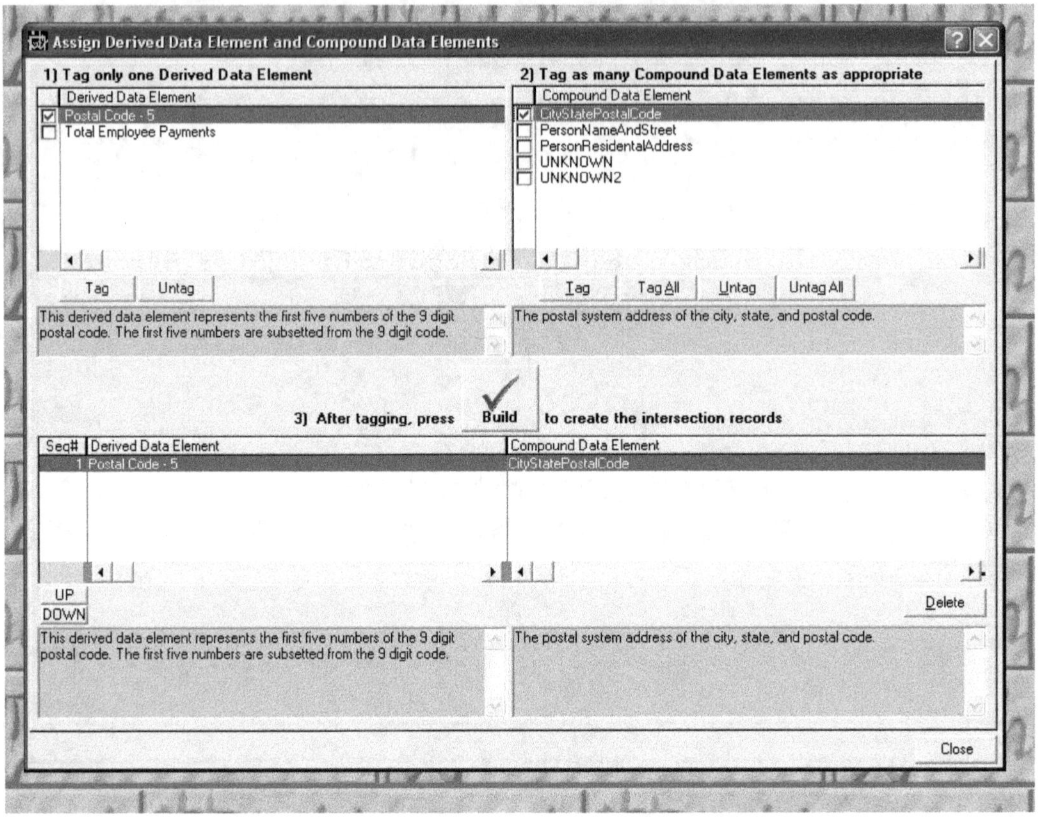

Figure 86. Compound data element derived data element assignment screen.

8.4.3 Compound Data Elements

Compound data elements consist of multiple data elements in a specific sequence. Compound data elements are a compound fact case because while its value appears to be a single value, it actually is composed of values from more than one other fact. Telephone Numbers, Addresses are obvious examples. The fact cases addressed by compound data elements are set out in Table 36.

Fact Case applicability to a compound data element		
Case	**Case Name**	**Applicability Note**
Simple	Single fact and restricted value set. Directly maps to the data integrity rule: Single table, single column	Generally no. This however may be required to establish compound data element structures. That is a compound data element that exists only to contain other compound data elements. See the Address example.
Compound	Single fact but with multiple contained facts.	Yes. That is the purpose of the compound data element.
Group	Single concept with multiple contained facts such as Address. Directly maps to the data integrity rule: Single table, multiple columns.	Yes. That is the purpose of the compound data element.
Pair	Two facts that exist as matched pairs of codes and values. Directly maps to the data integrity rule: Single table, multiple columns.	No. These are not two independent but "joined" facts as is required in a compound data element.
Related	Two or more facts that require one or more business rules to ensure consistency	No. These are not two independent but "joined" facts as is required in a compound data element.
Complex	A fact that contain more than one of the preceding five fact situations	No. These are not two independent but "joined" facts as is required in a compound data element.

Table 36. Fact case applicability to a compound data element.

Compound data elements may also consist of other compound data elements which too appear in a specific sequence. Figure 82 shows the data model for compound data elements as well. Figure 87 shows compound data elements. The compound data element, PersonNameAndStreet enumerates the data elements that are part of it. That is, Person Name and Street Address. Figure 88 shows the update screen for this compound data element. A new compound data element is named and defined.

Figure 89 shows that compound data elements are a bill-of-materials data structure. This figure shows the network structure for Person Residential Address. This compound data element, Person Residential Address contains the five different compound data elements:

- Person Name and Street.
- City State Postal Code.
    - City Name
    - State Name
    - Postal Code

The compound data element, City State Postal Code contains the three compound data elements, City Name, State Name, and Postal Code. The reason City Name and State Name are compound data elements is because they are both derived from the same data element, Geopolitical Name.

Each of the compound data element structures identify the data elements that comprise it. In Figure 83, Postal Code contains just one Element, Postal Code. If Person Name and Street had been selected, its contained data elements would have been Person Name and Street Name.

Figure 89 shows the two types of addresses: Business and Residential. Residential is exploded. If business had been exploded, the address would have additionally shown Company Name, and shown as well is a "re-use" of the compound data elements, Person, Street, and CityStatePostalCode. This is a key capability of Bills of Materials: Define-once, use-many-times.

Figure 90 shows the update screen for compound data element structure. The parent structure, City State Postal Code was selected and the Insert button was pressed. A set of possible selections is shown. The only selection that does not result in both an error message and a disabled Select button is Postal Code. Note that the message indicates that selection is "OK."

Figure 91 shows the process of allocating data elements to a compound data element. Tag the compound data element, and tag one or more data elements. In this example, Person Name and Street is tagged and

Person Name is also tagged. When the Assign button is pressed, the association between the two is created.

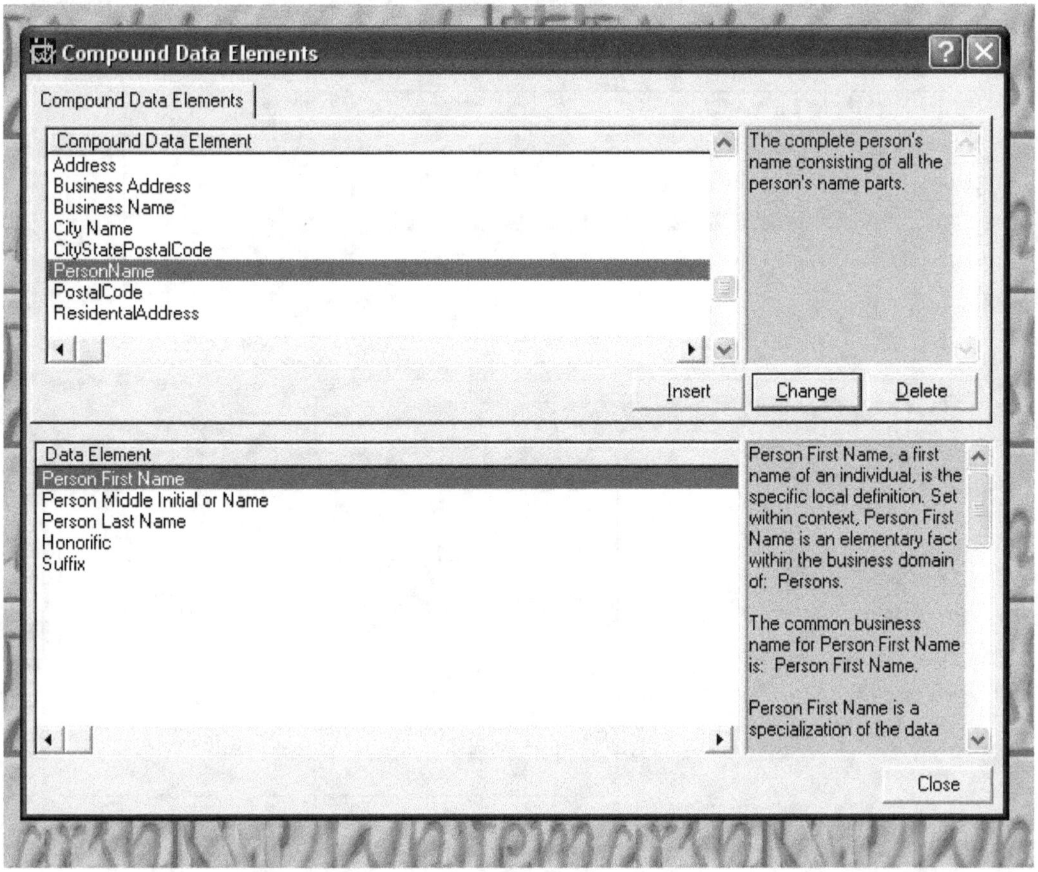

Figure 87. Compound data element screen.

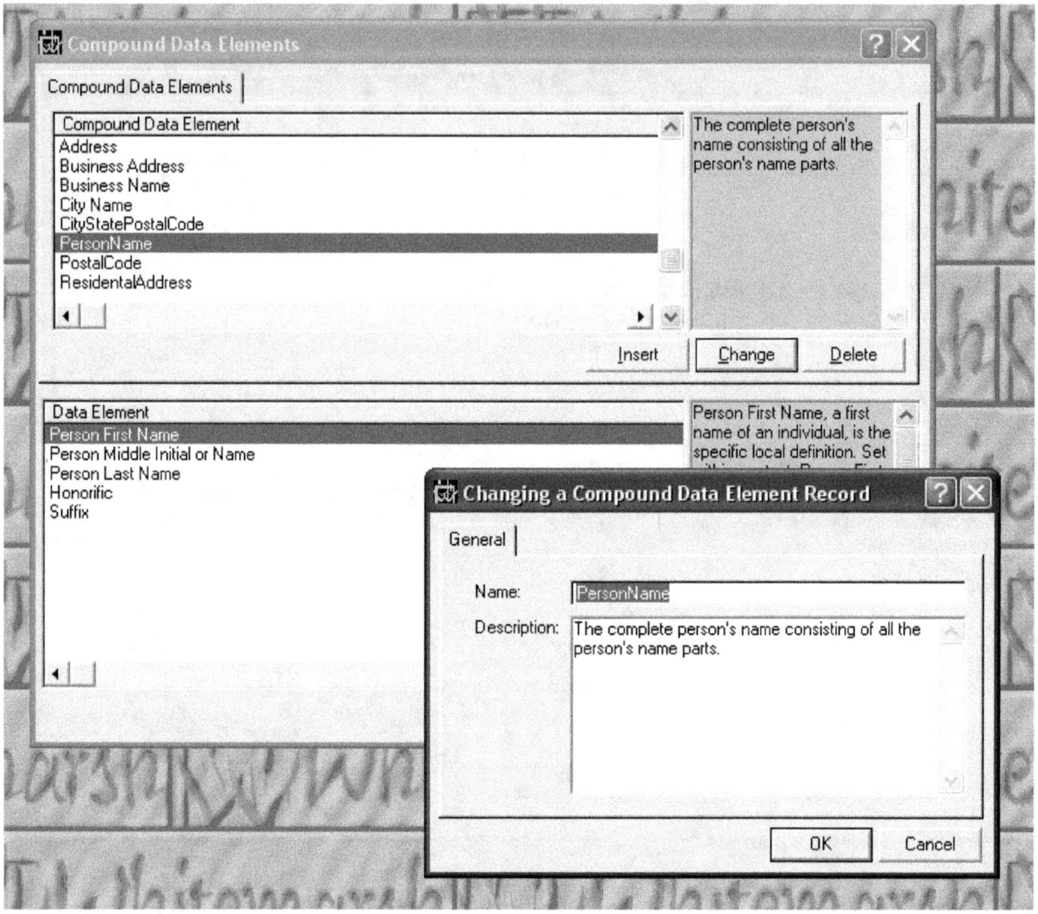

Figure 88. Compound data element update screen.

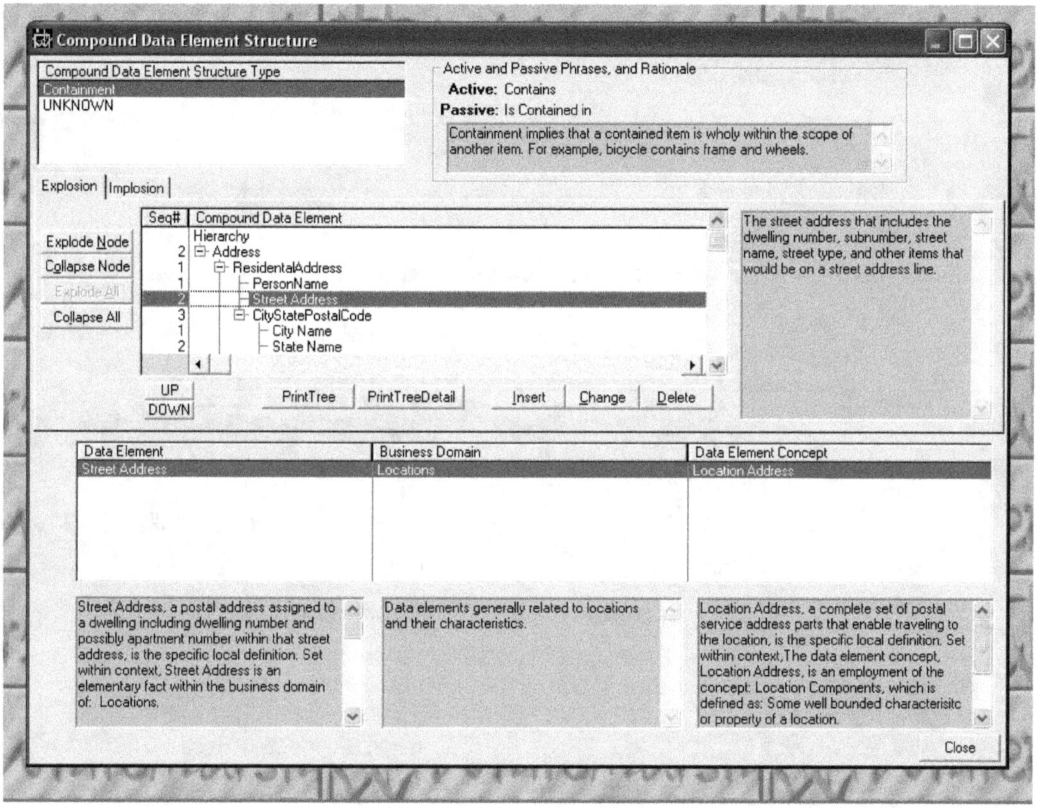

Figure 89. Compound data element structure screen.

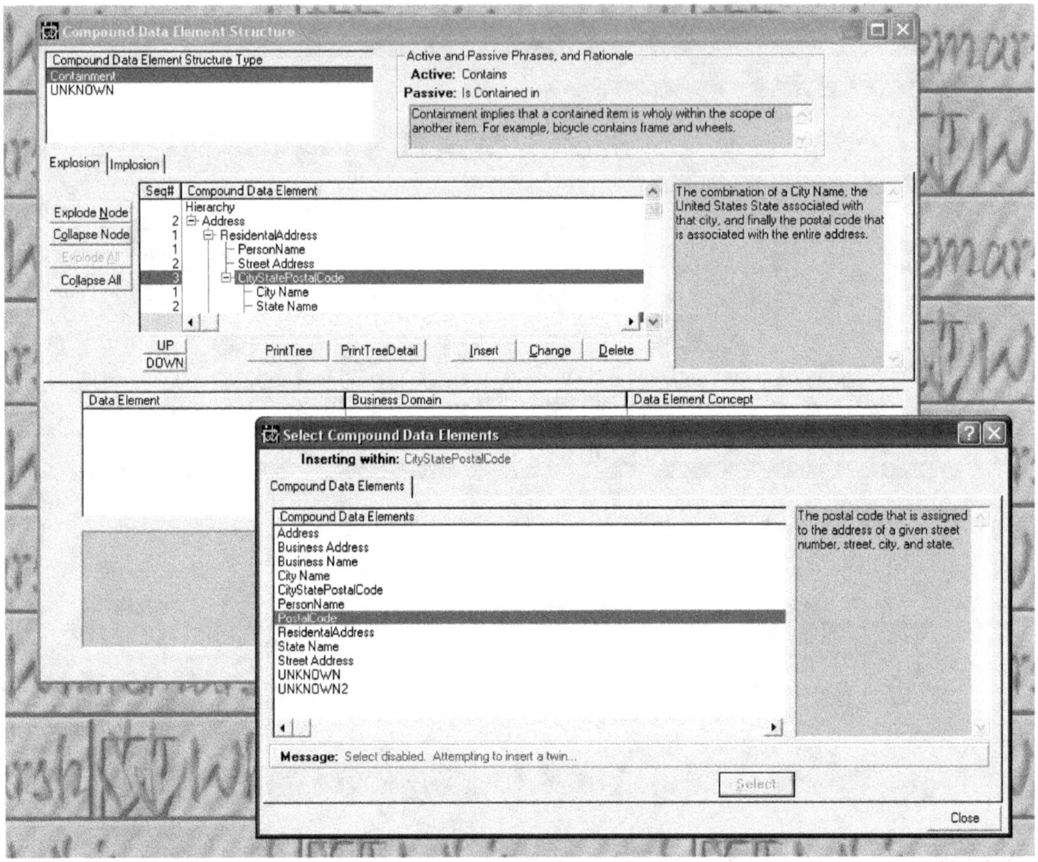

Figure 90. Compound data element structure update screen.

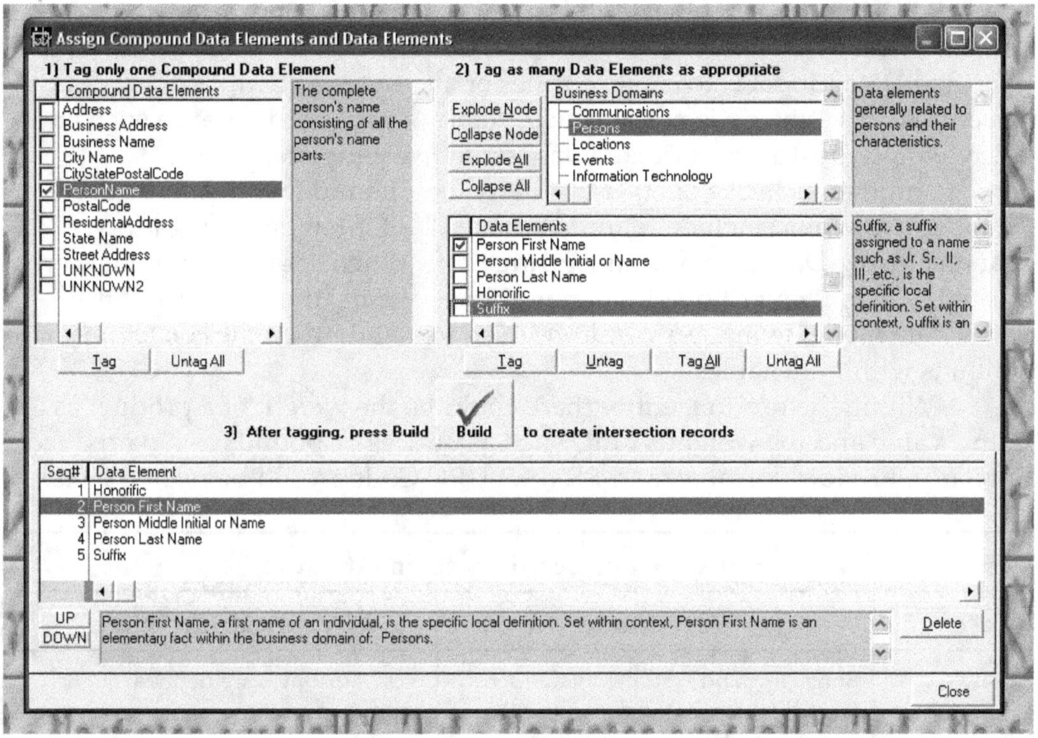

Figure 91. Compound data element to data element assignment screen.

8.4.4 Attributes

Attributes are the manifestation of the semantics of a data element within an entity of a subject. Attributes may have additional semantics that further refine it within the context of the data element. Attributes in the Metabase System are single valued simple facts.

The complete set of semantics for an attribute are those explicitly assigned, that is, meta category values. Its inherited semantics include those assigned to its containing data elements, entities, and value domains. The value domains of the attribute are specifically assigned. The proper assignment of the semantically nested meta category value and value domain is managed by the Metabase System.

The different fact cases that apply to attributes are set out in Table 37. A simple fact example would be Person First Name. Person Full Name would be a compound attribute that cannot be directly defined as a named collection of attributes. However, if the entity's name is Person Name, its attributes could be First Name, Middle Name, Last Name, etc. Since the objective of the Specified Data Model is to create entities of attributes as templates for data structures, this strategy is not only acceptable, it is needed because such an entity would form a standardized data structure template.

Another example of an entity-based compound fact would be Product Number that would include Manufacturing Plant, Manufacturing Line, Manufacturing Date, and Manufacturing Shift. When "real" data models are created, that is, a collection of tables within a schema, these subject-entity-based data model templates would be used to standardize the collections of columns within the tables.

Within the product entity there could be the weight of a product as both a value and the weight's units such as ounces or pounds as a paired fact. Similarly, Product Number, and Product UPC Code would be related facts.

Fact Case applicability to an attribute		
Case	**Case Name**	**Applicability Note**
Simple	Single fact and restricted value set. Directly maps to the data integrity rule: Single table, single column	Yes. Attributes are defined singly within an entity
Compound	Single fact but with multiple contained facts.	No. Attributes are defined singly.
Group	Single concept with multiple contained facts such as address. Directly maps to the data integrity rule: Single table, multiple columns.	Yes, given that the entity is the group's name such as Address.
Pair	Two facts that exist as matched pairs of codes and values. Directly maps to the data integrity rule: Single table, multiple columns.	Yes. There is however no semantic that explicitly ties the fields together. It would have to be done with a naming convention. E.g. Gender Code, Gender Name
Related	Two or more facts that require one or more business rules to ensure consistency	Yes. There is however no semantic that explicitly ties the fields together. It would have to be done with a "table constraint" that would be tied to both attribute implied values.

Fact Case applicability to an attribute		
Case	**Case Name**	**Applicability Note**
Complex	A fact that contain more than one of the preceding five fact situations	Yes. There is however no semantic that explicitly ties the fields together. It would have to be done with "column" or "table" constraints.

Table 37. Fact case applicability to an attribute.

Figure 92 shows the meta model for attributes. In this model, an attribute is the semantic intersection between an entity and a data element. The semantics of attributes can be refined through the allocation of meta category values (shown in Figure 92), and also by the assignment of value domains. These are both addressed in earlier chapters. Attributes can, of course participate in primary keys, Candidate/Alternate keys, and foreign keys. As foreign keys, the attributes form the relationships among entities within one or more Subjects.

What is critically different about attributes is that they exist within a container, the entity. The entity's nature imposes an intention of semantics. If the entity is Person, there is an intention that all fact-based attributes should be about a person. There are also two other classes of attributes about an entity: instance attributes, and audit attributes. Instance attributes relate to an entity's instance, such as create-date, and modify-date. There can also be audit-trail information such as the Project or Author of the Creation or Update. Data elements can be associated with entities to create in the following ways:

- The association of one data element and one entity creating one attribute.
- The association of one data element and multiple entities creating an attribute in multiple entities.
- The association of multiple data elements and one entity creating multiple attributes in one entity.

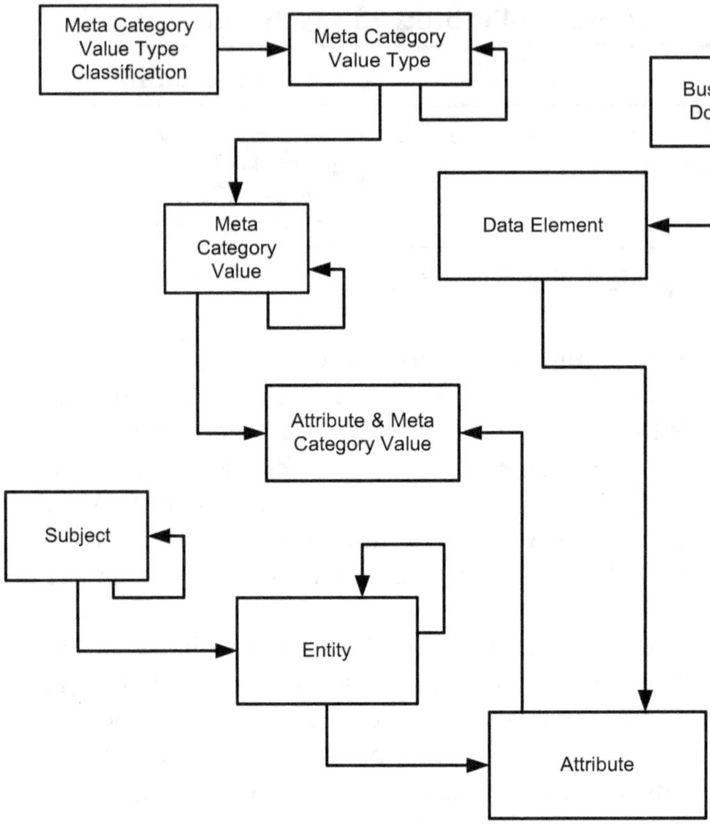

Figure 92. Attribute meta model.

Each of these alternatives is provided in the sections that follow. Usually, the manner in which attributes are created in the Metabase System is through the allocation of a data element to an entity. By allocation here, it is meant that the data element's semantics are employed as a template to "bootstrap" the semantics of the attribute.

Automatically inferred by the attribute is all the semantics of the data element. The attribute can be a further refinement of the data element's semantics. That is, it can have a more restricted set of assigned meta category values (See Chapter 6, Semantic Hierarchies) or it can have a more refined set of value domains (see Chapter 7, value domain Management).

8.4.4.1 Attribute Creation and Maintenance

Figure 93 shows the list of attributes for selected subject and entity. Below the selected attribute are all the assigned meta category values.

There are three buttons on this screen of special interest:

- Replace Underscores with Blanks.
- For Pkeys within Selected entity . . . Press.
- For all attributes within entity . . . AutoDef.

When the SQL DDL is imported from some outside source into the Specified Data Model, underscores are inappropriate. Blanks are much better because at

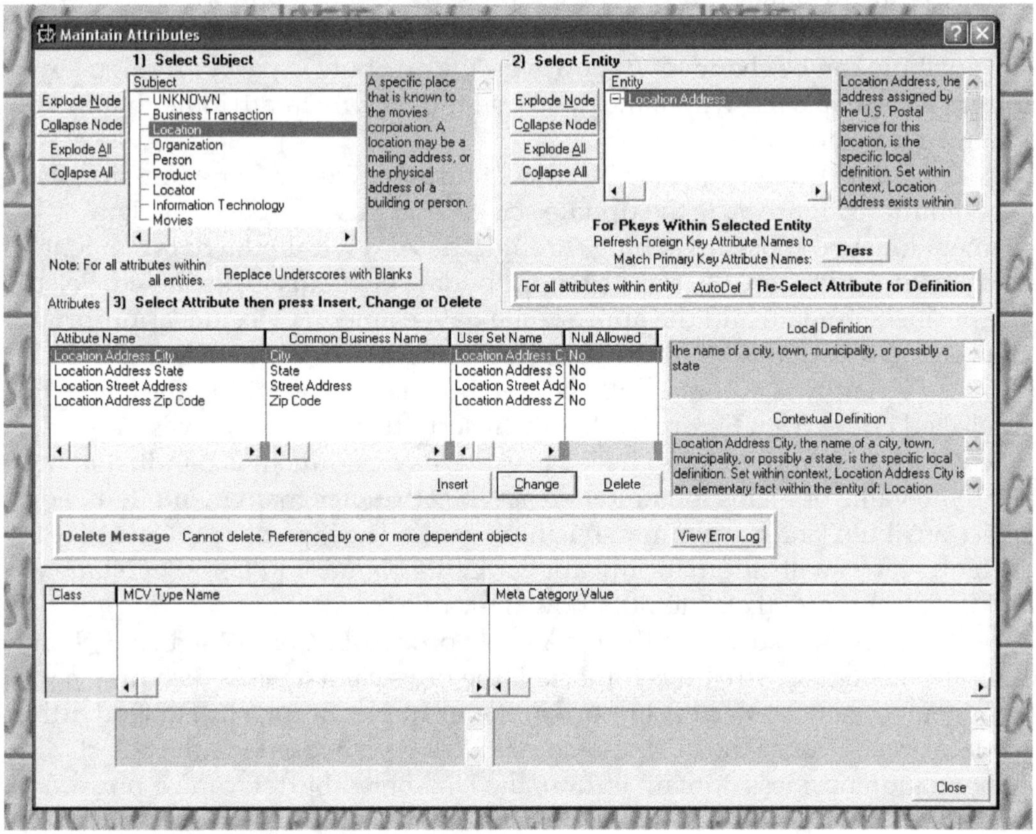

Figure 93. Attributes within an entity.

this level of specification, these models are just conceptual versus bound to a particular DBMS. Underscores are required for the DBMS-based models and the Metabase System accomplishes that automatically. Pressing Replace Underscores with Blanks does exactly what its name says for all attributes of all entities within the Selected subject.

The second button, For Pkeys within Selected entity . . . Press, this refreshes all the attributes that participate in a foreign key to match those in the source entity's primary key (Pkey).

The final button, For all attributes within entity . . . AutoDef, causes all the attributes in the Selected entity to have the AutoDef function execute. The AutoDef process takes the local definition of an entity and sets it within the context of all the meta objects that are the "parents" of the entity.

If an attribute is to be changed, press the Change button. Figure 94 is presented. If the attribute is not within a foreign key, all the meta-attributes of the attribute can be changed. If, however, the attribute is part of foreign key Figure 94 is displayed with a message that the only meta-attribute that can be changed is its description.

New attributes can also be created through Figure 93 by pressing the Insert button. Figure 94 is the update screen. Via this approach the data element must be chosen. Tabbing through the zero within the data element entry field, and Figure 95, causes the list of data elements is displayed. Select the business domain and the appropriate data element. Once the appropriate data element is identified, press the Select button.

The entity is automatically chosen because it was within the context of a selected entity from Figure 93 that the insert-attribute button was pressed. The value domain cannot be chosen because its assignment is handled in the attribute value domains menu item from the attributes main menu item. For every attribute, enter a common business name. Then, press the Reset button. Default names will automatically appear in the Name, and User Set Name fields. Select if null (i.e., "do not know the value").

Figure 94 also shows the UnAbbrev process button. When an SQL data definition language stream is imported, the physical names, which may be abbreviated, may be what is imported. For example, Empl_Hr_Dt. Under this update process, once the business domain has been chosen via the Select abbreviation business domain button, the UnAbbrev button can be pressed. If there has been a good quantity of abbreviations established on various business domains, the physical name will be automatically translated to a logical name. If there are multiple logical names for a given physical abbreviation, a screen presents itself and an appropriate logical name can be

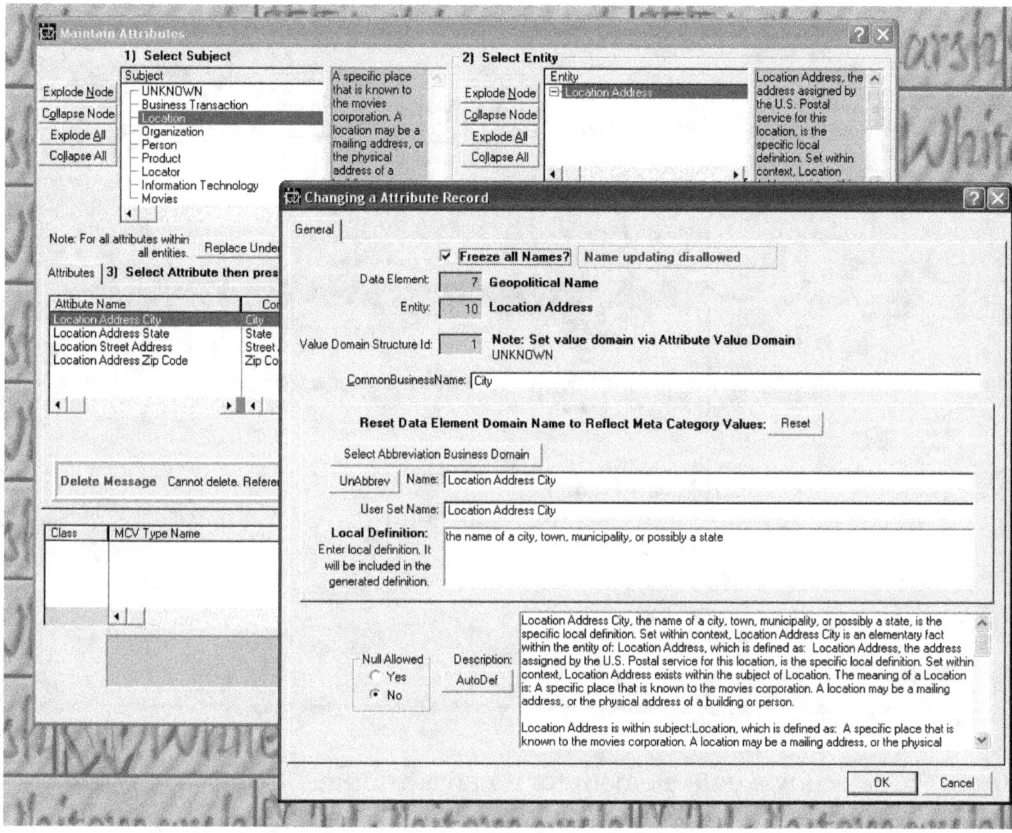

Figure 94. Entity attribute update screen.

chosen for a given word. In this example, the full name becomes, Employee Hire Date.

Finally, enter the Local Definition phrase. Keep it just as a phrase without a capital letter at the start and no period or comma at the end. For a complete contextual definition, press the AutoDef button. After all this, press the OK button and the attribute has been fully defined.

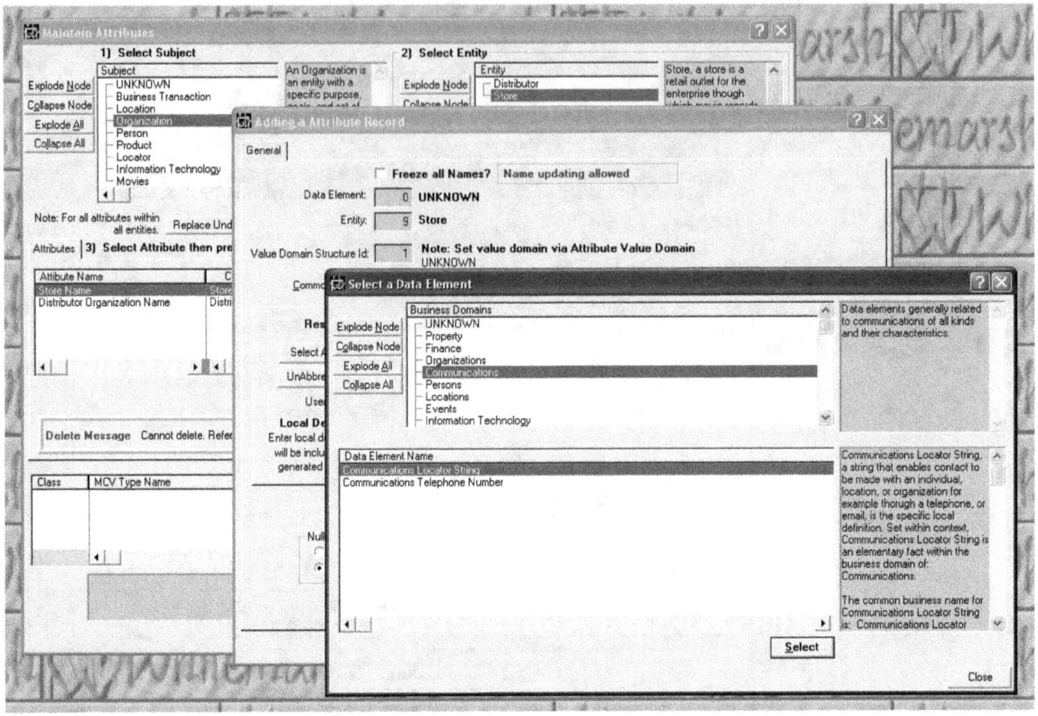

Figure 95. Selecting a data element for a new attribute.

8.4.4.2 One Attribute to Multiple Entities

There are many situations where a single attribute is to be created in multiple entities within one or multiple Subjects. An example is the addition of entity instance information within multiple entities such as create-date and modify-date, or the creation of audit-trail information such as the creation of the Project Name or Creator Name. In these examples, the target entities are tagged and one data element is tagged. Figure 96 illustrates this. In this figure, The business domain, organization is selected. Then, the data element, Organization Name, is tagged. On the right side of the window, the Subject, location, is Selected, and the data element, Location Address, is tagged. Other entities can be tagged as well. Once the Build button is pressed, the attributes are created and are shown in the bottom browse. Afterwards, the full sets of meta-attributes about the newly built attributes are created through attribute Maintenance described in Section 8.4.4.1.

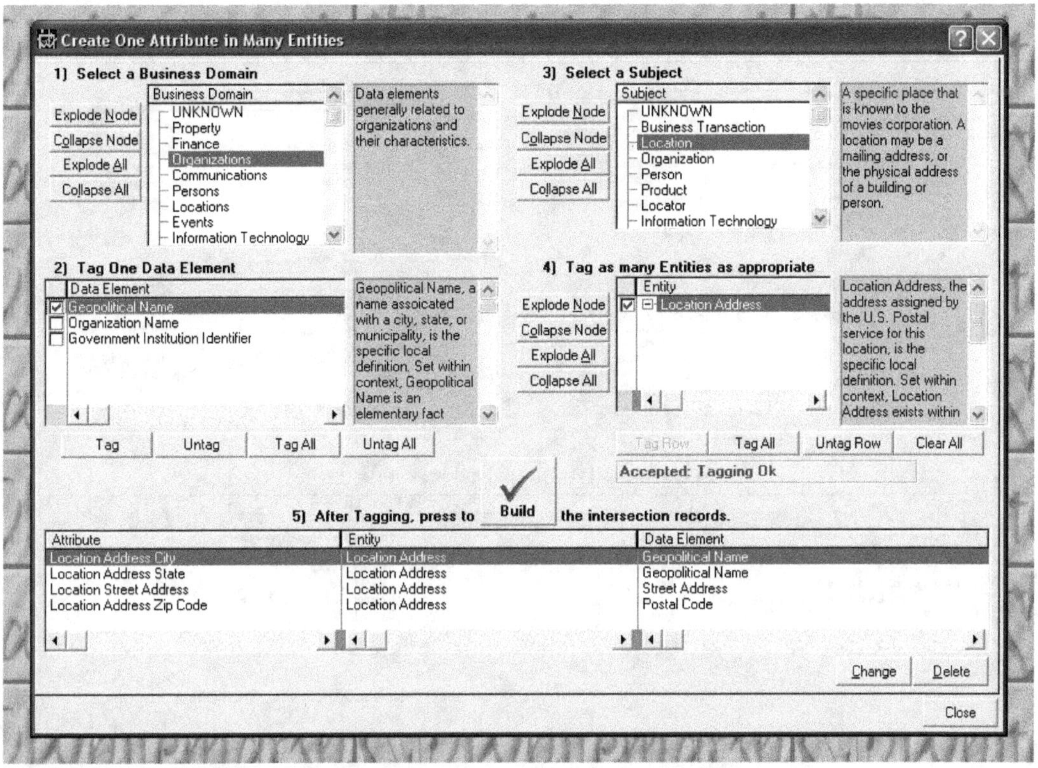

Figure 96. Create one attribute in many entities.

8.4.4.3 Multiple Attributes to One Entity.

The inverse to creating one attribute in many entities is creating many attributes in one entity. Here, it might be to install a complete set of Telephone Number related data elements including, for example, Country Code, Area Code, and Telephone Number. Figure 97 illustrates this. First the Subject, Business Transaction, is selected. Then, the entity, Payment, is selected and tagged. On the right side of the window, the business domain, Finance, is selected, and three different data elements, Financial Instrument Identifier, Payment Mechanism Number, and Payment Mechanism Name are selected and tagged. Once the Build button is pressed, three attributes are created. Once created the remaining meta-attributes of the newly created attributes are created through the processes set out in Section 8.4.4.1.

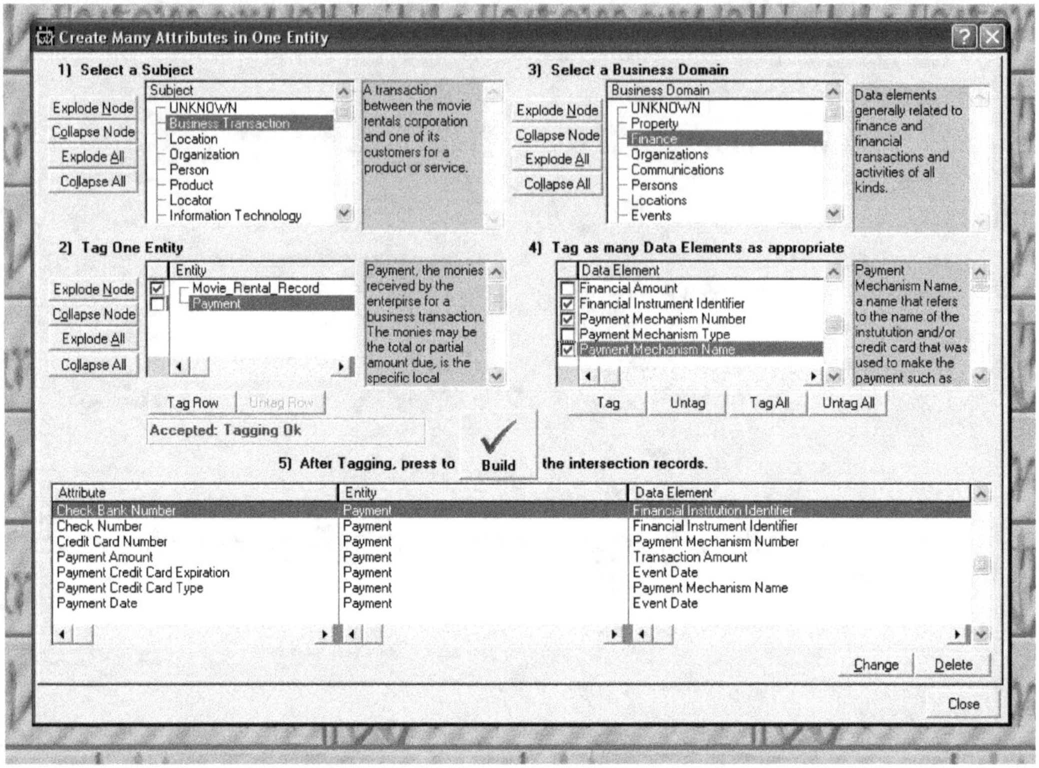

Figure 97. Create many attributes in one entity.

8.4.5 Columns

Columns are the manifestation of the semantics of a data element within a table of a schema. Additionally, a column is a deployment of the semantics of an attribute. The fact cases that apply to a column are listed in Table 38.

The complete sets of semantics for a column are those explicitly assigned, that is, meta category values. Its inherited semantics includes those assigned to its containing data elements, attributes, tables, and value domains. The value domains of the column are specifically assigned.

Columns in the Metabase System can be simple facts, compound facts, group, paired, related, or complex facts. A simple fact example would be Employee First Name. Employee Full Name would be a compound fact. The weight of a product as both a value and units would be a paired fact. Employee Birth and Death Date would be related facts. Finally, in SQL:1999 or later, the column, Address, could contains subordinate columns, and within that, the Postal zip code could contain 5 and 4 digit parts.

Fact Case applicability to a column		
Case	**Case Name**	**Applicability Note**
Simple	Single fact and restricted value set. Directly maps to the data integrity rule: Single table, single column	Yes. Columns can be defined singly within a table
Compound	Single fact but with multiple contained facts.	Yes, columns can be defined as having multiple contained columns to any depth. This is a capability of SQL:1999 and more recent SQL-based DBMSs
Group	Single concept with multiple contained facts such as Address. Directly maps to the data integrity rule: Single table, multiple columns.	Yes. A parent column would be defined as the group's name, and subordinate columns would represent the elements contained in the group. There could be Address, contained elements for Address Street, Address City, Address State, and Address Postal Code. In turn, Address Postal Code could be subdivided into Zip-5 and Zip-4.
Pair	Two facts that exist as matched pairs of codes and values. Directly maps to the data integrity rule: Single table, multiple columns.	Yes. A parent column would be the pair's name, and the subordinate DBMS columns would represent the pair of columns of the pair. The pair name would be Gender and the contained columns would be Gender Code, Gender Name.
Related	Two or more facts that require one or more business rules to ensure consistency	Yes. A parent column would be the pair's name, and the subordinate columns would represent the related columns of the pair. The related name would be Life Dates and the contained columns would be Birth Date, and Death Date.

Fact Case applicability to a column		
Case	**Case Name**	**Applicability Note**
Complex	A fact that contain more than one of the preceding five fact situations	Yes. There is however no semantic that explicitly ties the fields together. It would have to be done with "column" or "table" constraints.

Table 38. Fact Case applicability to a column.

Figure 98 depicts the column meta model. In this meta model, columns are the semantic intersections a data elements and a table. Columns may have additional semantics that further refine the column within the context of either the attribute or the data element. The order of processing these additional semantics is that the column must first be a subset of the attribute, which in turn must be a subset of the data element. The semantics of the column, like attributes, can be restricted by value domains. The proper assignment of the semantically nested meta category value and value domain is managed by the Metabase System. What is different from attributes is that a column can have subordinate structures. Finally, not all the columns of a table must map to attributes from a single entity.

Similar to attributes, columns can be created singly, or a single column can be created in multiple tables, or a multiple columns can be created in single table.

8.4.5.1 Creation and Maintenance of Columns

Creation or maintenance of columns starts with a list of columns. Figure 99 shows the list of columns for selected subject and table. Below the selected column are all the assigned meta category values. The Print Tree button displays the tree structure of a column if it is complex.

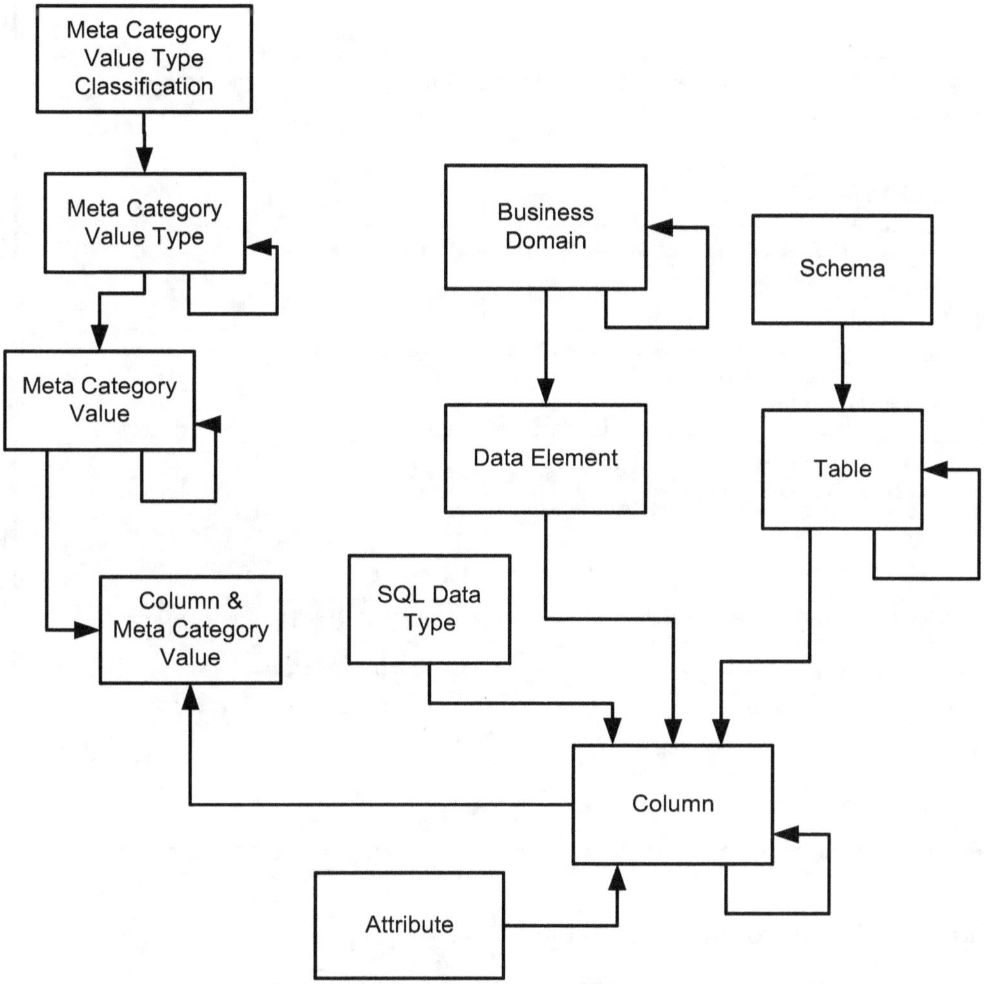

Figure 98. Column meta model.

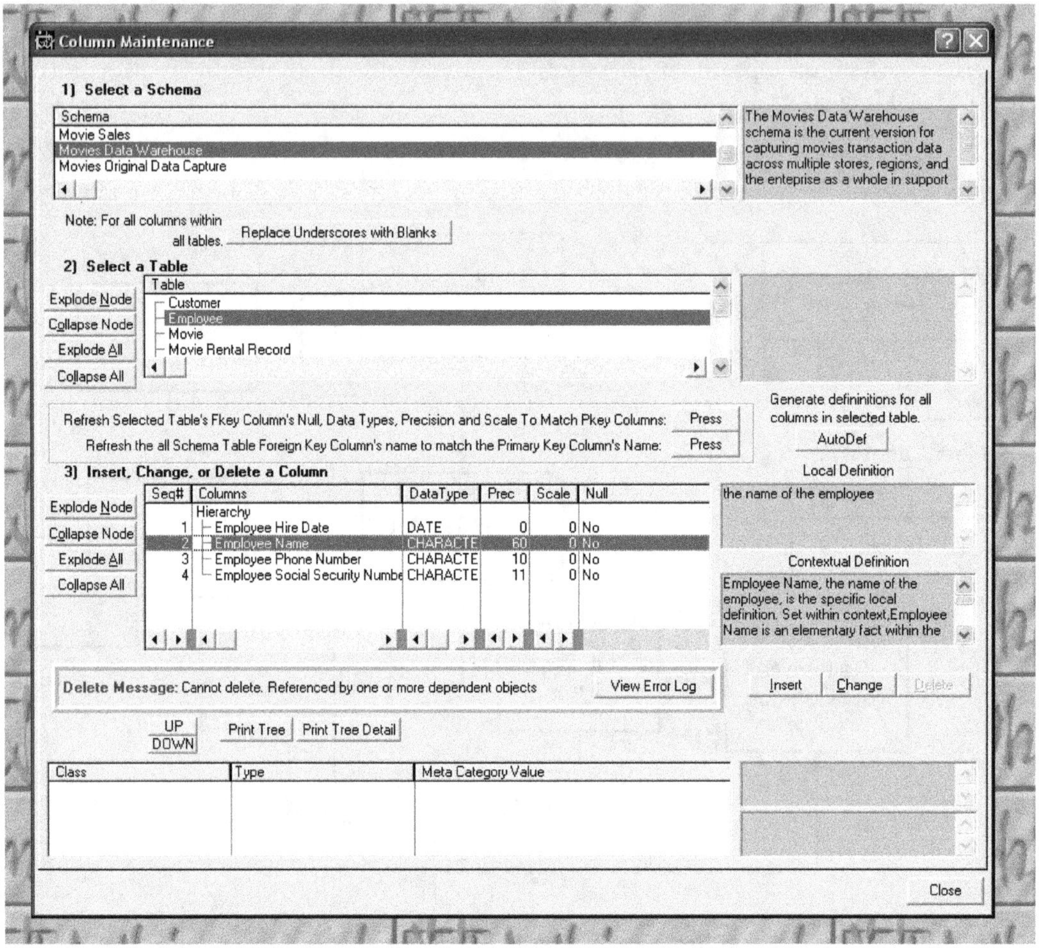

Figure 99. List of columns in a table.

There four buttons on this screen of special interest:

- Replace Underscores with Blanks.
- Refresh Selected table's Fkey . . . Press.
- Refresh the Foreign key . . . Press.
- Generate definitions . . . AutoDef.

In the event that SQL DDL is imported from some outside source into the Implemented Data Model, underscores are inappropriate. Blanks are much better because at this level of specification, there models are just logical versus bound to a particular DBMS. Underscores are required for the DBMS-based models and the Metabase System accomplishes that automatically. Pressing Replace Underscores with Blanks does exactly what its name says for all attributes of all entities within the Selected schema.

The second button, Refresh Selected table's Fkey . . . Press, this refreshes all the columns that participate in a foreign key to match those in the source table's primary key (Pkey).

The third button, Refresh the Foreign key . . . Press, changes the column names in a foreign key to match those in the corresponding primary key.

The final button, For all attributes within entity . . . AutoDef, causes all the columns in the Selected table to have the AutoDef function execute. The AutoDef process takes the local definition of a table and sets it within the context of all the meta objects that are the "parents" of the table.

If a column is to be changed, press the Change button. Figure 100 is presented. If the column is not within a foreign key, all the meta columns of the column can be changed. If, however, the column is part of foreign key, a message is displayed that indicates that the only meta-attribute of the column that can be changed is its description.

New columns can also be created through Figure 99 by pressing the Insert button. Via this approach the data element must be chosen. Figure 101 presents the list of data elements that is displayed after tabbing off the data element Id field. Select the business domain and the appropriate data element. In this case, the selected data element is: Communications Locator String. Once the appropriate data element is identified, press the Select button.

Columns, as shown in Figure 98 are also related to attributes. The only correct attribute that can be assigned is one that employs the same data element that has just been selected. Figure 102 shows this process. The select screen from Figure 101 shows first that the targeted data element is: Communications Locator String. Just any attribute will not do, of course. So, in Step 1, the right subject, Locator, is selected. Then, in Step 2, the Locator entity, Telephone Number, is selected. At that point, because the targeted data element is already known, the only attributes shown are those that map to the data element, Communications Locator String. Here, the selected attribute is Telephone Number Phone Number. In the attribute maintenance screen this obtuse user-set name could be changed to just Telephone Number.

Figure 100 also shows the UnAbbrev process. When an SQL data definition language stream is imported, the physical names, abbreviated name may be imported. For example, Empl_Hr_Dt. Under this update process, once the business domain has been chosen via the Select abbreviation business domain button, the UnAbbrev button can be pressed. If there has been a good quantity of abbreviations established on various business domains, the physical name can be automatically translated to a logical name. If there are multiple logical names for a given physical abbreviation, a screen presents itself and an appropriate logical name can be chosen for a given word. In this example, the full name becomes, Employee Hire Date.

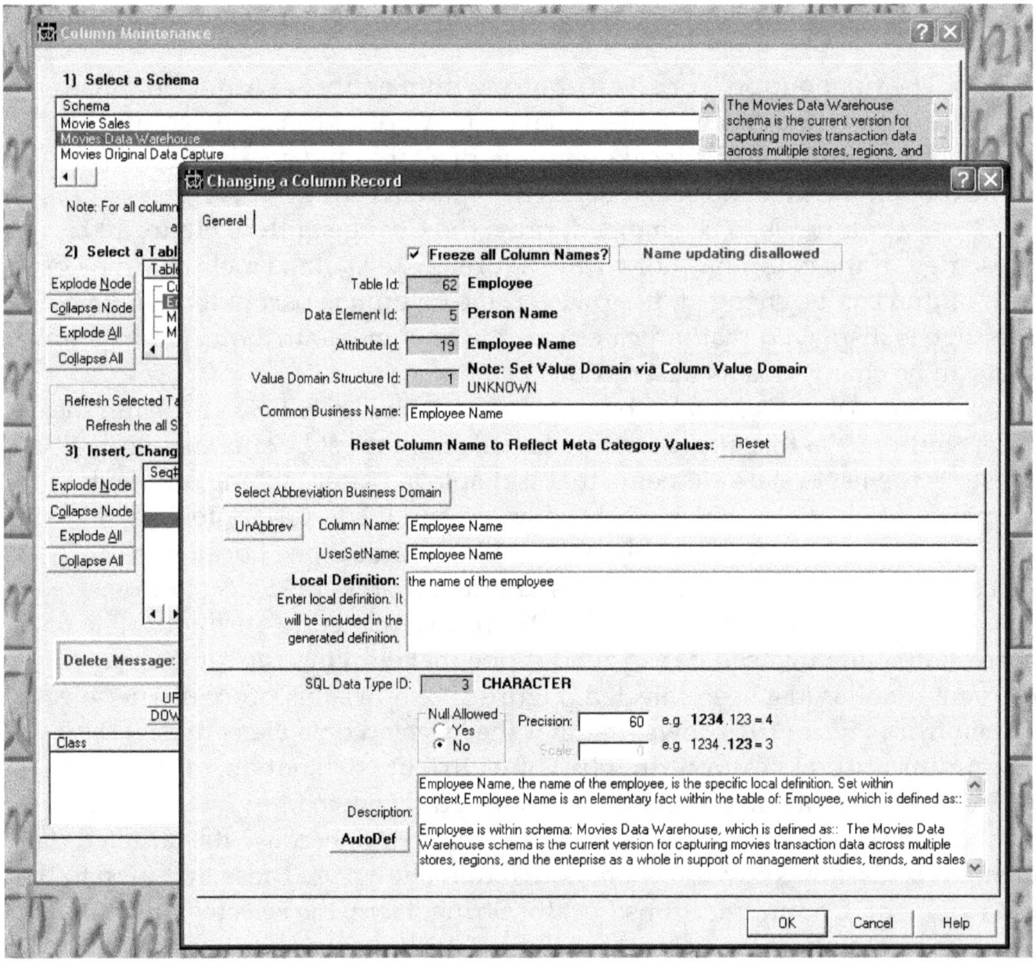

Figure 100. Screen for creating or updating a column.

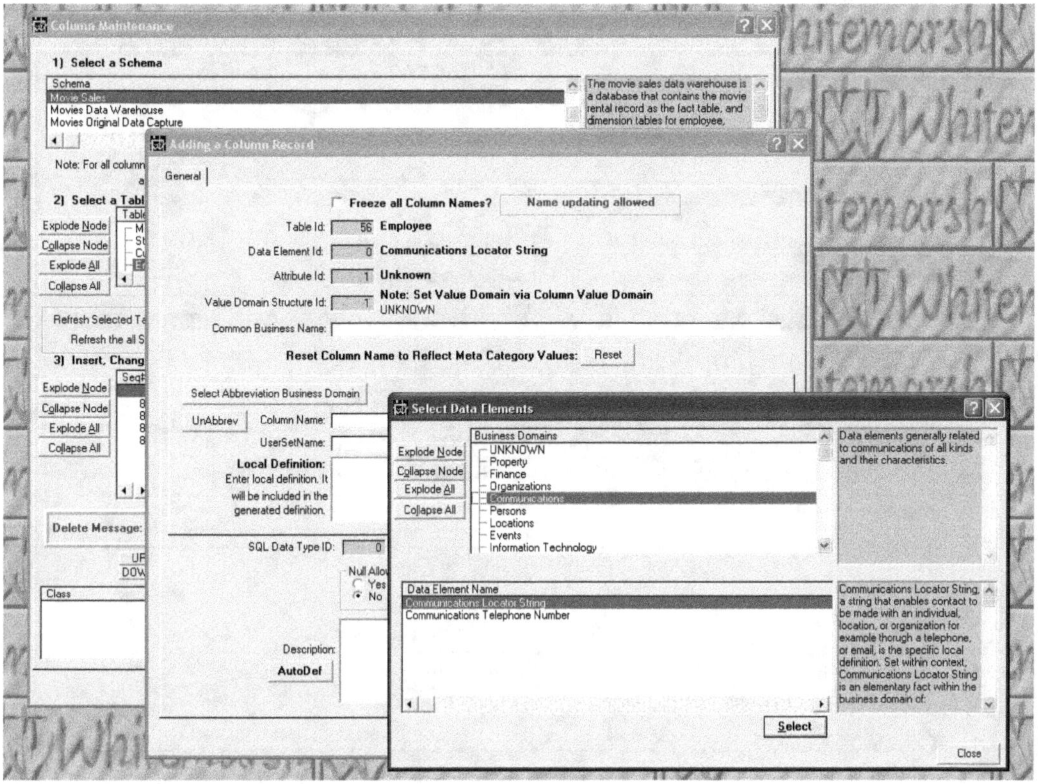

Figure 101. Selecting data element for column.

Finally, enter the Local Definition phrase. Just keep it a phrase without a capital letter at the start and no period or comma at the end. For a complete contextual definition, press the AutoDef button. The process employs the local definition phrase and blends it into the table's, schema's, meta category value's, and value domain's definition associated with the column to construct its complete definition.

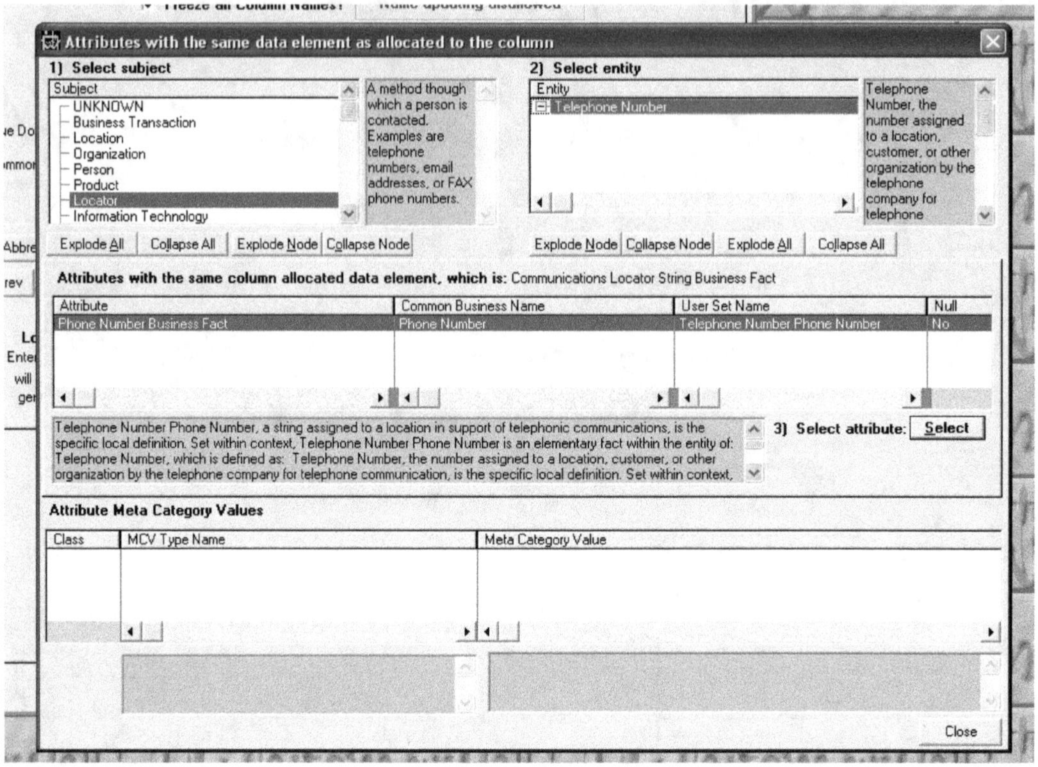

Figure 102. Selecting attribute reference for new column.

Figure 103 shows the creation of subordinate column parts to the Customer Phone Number. Added are:

- Customer Country Code.
- Customer Area Code.
- Customer Phone Number.
- Customer Extension.

These subordinate columns are created by first selecting the parent column, Customer Phone Number, and pressing Insert. The normal set of column screens, Figures 100, 101, and 102 are presented.

A natural question at this point is: What are the relationships between a compound data element, an entity of attributes, and a column with a subordinate set of columns as shown in Figure 103? The answer is as follows: First, the compound data element, which, in turn, points to atomic and non-

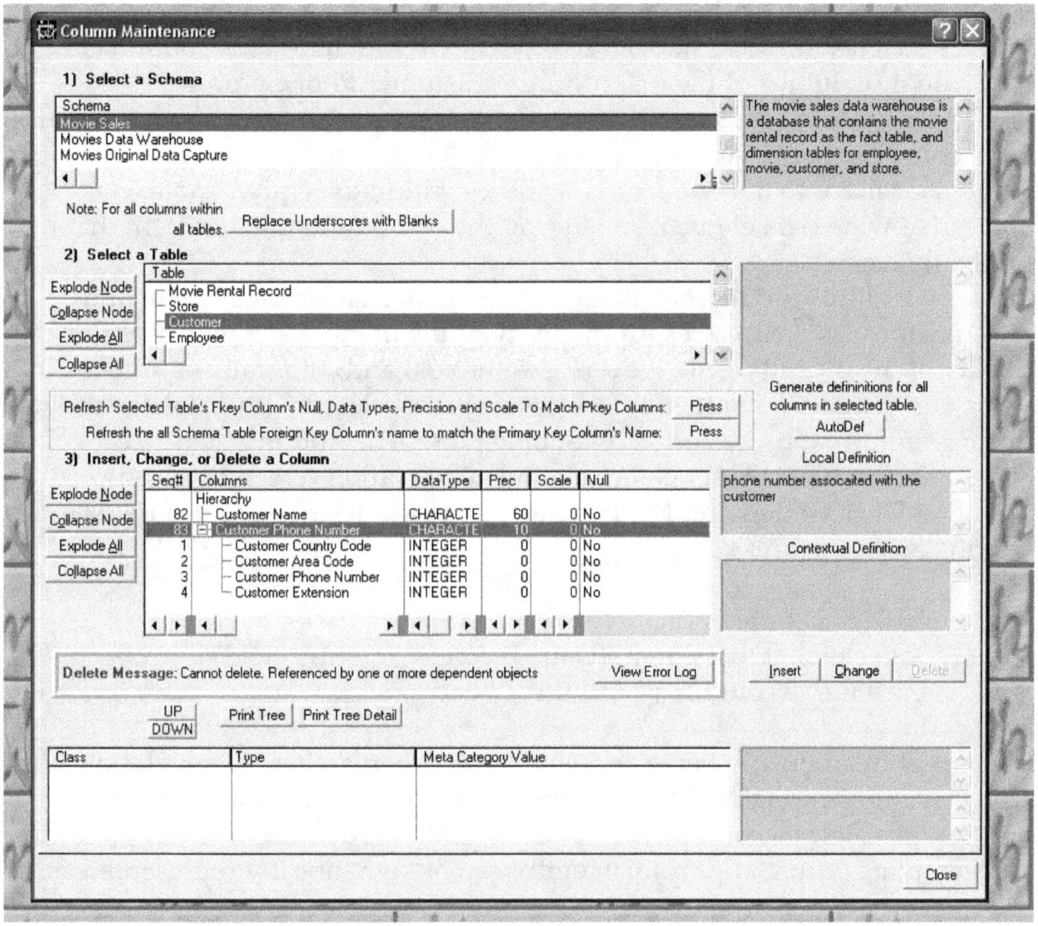

Figure 103. Subordinate column components to a defined column.

derived data elements is a template for a semantically valuable set of data elements. This template can later be used for constructing an entity, or a set of columns in a table. Thus, there could well be a compound data element for Telephone Number.

Second, there could be an entity, Phone Number within the Subject, Communications that would have the attributes that map to the atomic and non-derived data elements that comprise the compound data element, Phone Number.

Third, when a table is created in the Implemented Data Model that employs the attributes from the entity Phone Number, these would appear as standalone columns within the table. A Metabase System re-engineering process enables the data modeler to "move" these stand-alone columns to be subordinate columns of the real column, Customer Phone Number.

Once all this is done, there is a mapping from all the different schema-table-columns that have Phone Number under their many different local names and maps to the one set of attributes within the entity, and to the enterprise-wide data elements including the compound data element that was originally created.

All of this is data semantics management in its most efficient and effective form as it enhances productivity of the modelers, it increases the quality of the overall result, it decreases the cost and/or totally eliminates value and semantics conflict. It reduces risks associated with database and business information systems creation, deployment, and maintenance.

Figure 100 represents the relationship creation screen between Specified Data Models and Implemented Data Models. The screen fields for relating a column are:

- Table to column. (Table_Id entry field)
- Data element to column. (Data_Element_Id entry field)
- Attribute to column. (Attribute_Id entry field)

There is no relationship between Subject and schema, there is no relationship between entity and table. Rather, there are relationships between attribute and column. Because of that, the Specified Data Models are true templates. As true templates, attributes from an entity can be semantically represented as:

- Columns in different tables of the same or different schemas.
- Columns in the same table.
- Subordinate columns within a group column in a table.

8.4.5.2 One Column to Multiple Tables

There are a number of situations where a single column is created for multiple tables and even across multiple schemas. An example is the addition of table instance information such as create-date and modify-date, or the creation of audit-trail information such as the creation of the Project Name or Creator Name. In these examples, the schemas, and tables within a selected schema are tagged and one data element is tagged. Figure 104 illustrates this. In this Figure, The business domain, Finance, is selected. Then, the data element, Finance Amount, is tagged. On the right side of the window, the schema, Movie Sales, is Selected, and the table, Movie Rental Record, is tagged. Other tables across multiple schemas can be tagged as well. Once the Build button is pressed, the columns are created and are shown in the bottom browse. Thereafter, the full set of meta-attributes about the newly built columns are created through column Maintenance described in Section 8.4.5.1.

8.4.5.3 Multiple Columns to One Table

The inverse to creating a column in many tables within one or across multiple schemas is creating many columns in one table. Here, it might be to install a complete set of Telephone Number related data elements including, for example, Country Code, Area Code, and Telephone Number. Figure 105 illustrates a different example. First the schema, Movie Data Warehouse, is selected. Then, the table, Movie Rental Record, is selected and tagged. On the right side of the window, the business domain, Finance is selected, and two different data elements, Transaction Date, and Transaction Amount are selected and tagged. Once the Build button is pressed, three columns are created. Once created the remaining meta-attributes of the newly created columns are created through the processes set out in Section 8.4.5.1.

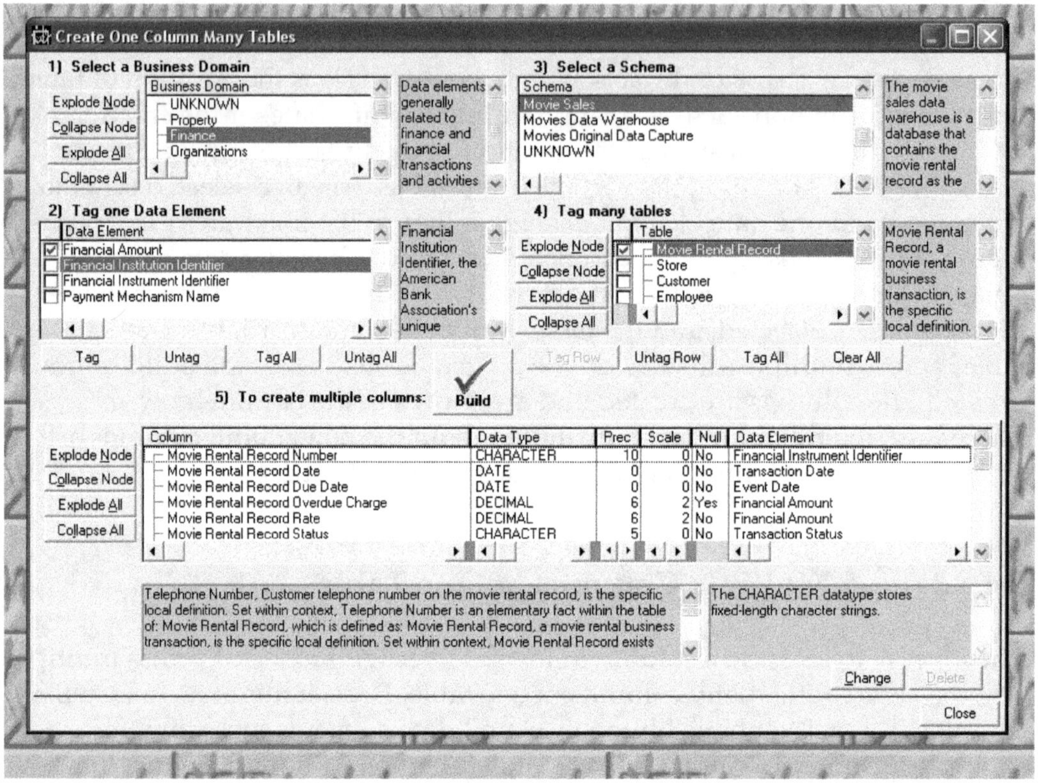

Figure 104. One column to multiple tables.

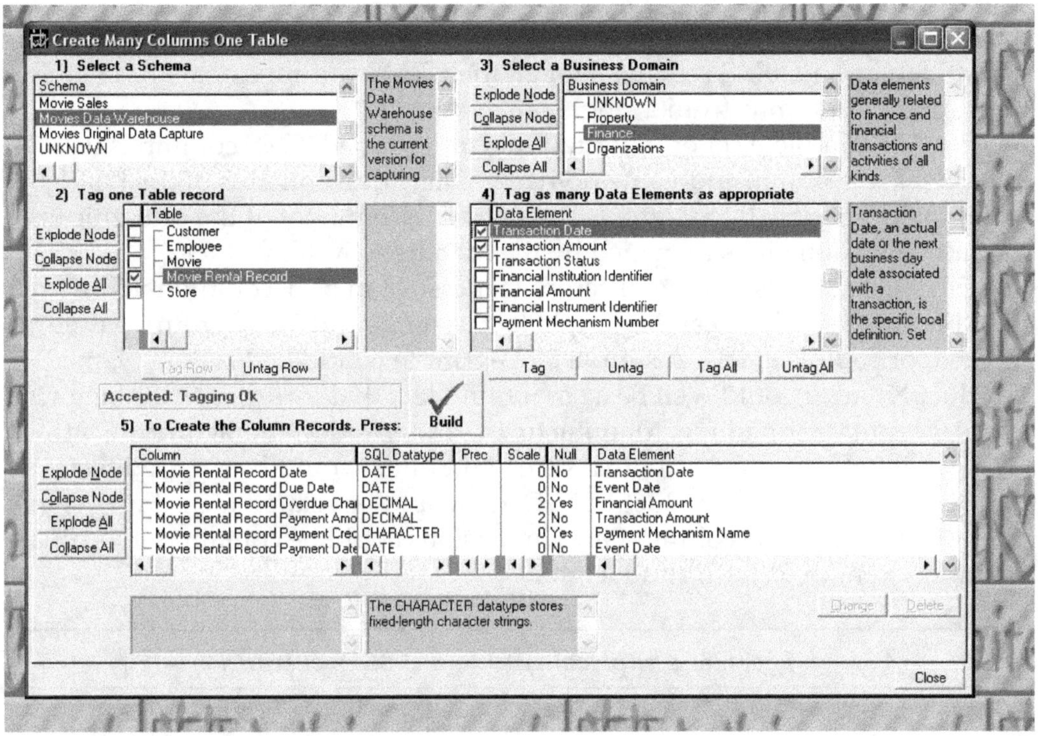

Figure 105. Multiple columns to one table.

8.4.6 DBMS Columns

DBMS columns are the manifestation of the semantics of a column within a DBMS table of a DBMS schema. Not all the DBMS columns of a DBMS table must map to columns from a single table.

The complete set of inherited semantics for a DBMS column is its DBMS table, column, and value domains. The value domains of the DBMS column are specifically assigned. The proper assignment of the semantically nested value domain is managed by the Metabase System.

Table 39 sets out the different fact cases can be accommodated with DBMS columns, which can be simple facts, compound facts, group, paired, related, or complex facts. A simple fact example would be Product Name. Product Number could well be a compound fact that includes Manufacturing Plant, Manufacturing Line, Manufacturing Date, and Manufacturing Shift. The weight of a product as both a value and units would be a paired fact. Product Number, and Product UPC Code would be related facts. Finally, in SQL:1999 or later, the column Part Dimensions could be contains subordinate columns, Length, Width, and Height.

Fact Case applicability to a DBMS column		
Case	**Case Name**	**Applicability Note**
Simple	Single fact and restricted value set. Directly maps to the data integrity rule: Single table, single column	Yes. DBMS columns can be defined singly within a DBMS table
Compound	single fact but with multiple contained facts.	Yes. DBMS columns can be defined as having multiple contained DBMS columns to any depth. This is a capability of SQL:1999 and more recent SQL-based DBMSs
Group	Single concept with multiple contained facts such as address. Directly maps to the data integrity rule: Single table, multiple columns.	Yes. A parent DBMS column would be defined as the group's name, and subordinate DBMS columns would represent the elements contained in the group. There could be Address, contained elements for Address Street, Address City, Address State, and Address Postal Code. In turn, Address Postal Code could be subdivided into Zip-5 and Zip

Fact Case applicability to a DBMS column		
Case	**Case Name**	**Applicability Note**
Pair	Two facts that exist as matched pairs of codes and values. Directly maps to the data integrity rule: Single table, multiple columns.	Yes. A parent DBMS column would be the pair's name, and the subordinate DBMS columns would represent the pair of DBMS columns of the pair. The pair name would be Gender and the contained DBMS columns would be Gender Code, Gender Name.
Related	Two or more facts that require one or more business rules to ensure consistency	Yes. A parent DBMS column would be the pair's name, and the subordinate DBMS columns would represent the related DBMS columns of the pair. The related name would be Life Dates and the contained DBMS columns would be Birth Date, and Death Date.
Complex	A fact that contain more than one of the preceding five fact situations	Yes. There is however no semantic that explicitly ties the fields together. It would have to be done with "column" or "table" constraints.

Table 39. Fact case applicability to a DBMS column.

Figure 106 depicts the DBMS column meta model. The semantics of the column, like attributes, can be restricted by value domains. In this model, DBMS columns may have subordinate columns.

The relationship between the Implemented Data Model to the Operational Data Model is the same as between Specified and Implemented, that is, through the relationship between columns and DBMS columns. Because of this form of relationship, the Implemented Data Models can be employed as templates. As templates, columns from a table can be semantically represented in the Operational Data Model as:

- DBMS columns in different DBMS tables of the same or different DBMS schemas.
- DBMS columns in the same DBMS table.

- Subordinate DBMS columns within a group DBMS column in a DBMS table.

The practical application of this is the creation of data warehouse data models. That is because data warehouse data models most often consist of data structures of column collections from multiple Implemented Data Models. Through this approach there is a many-to-one relationship between columns of an Implemented Data Model and DBMS columns of the Operational Data Model.

Another practical use of Operational Data Models is the creation of alternate versions of data models that operate under different DBMSs, operating systems, and hardware configurations.

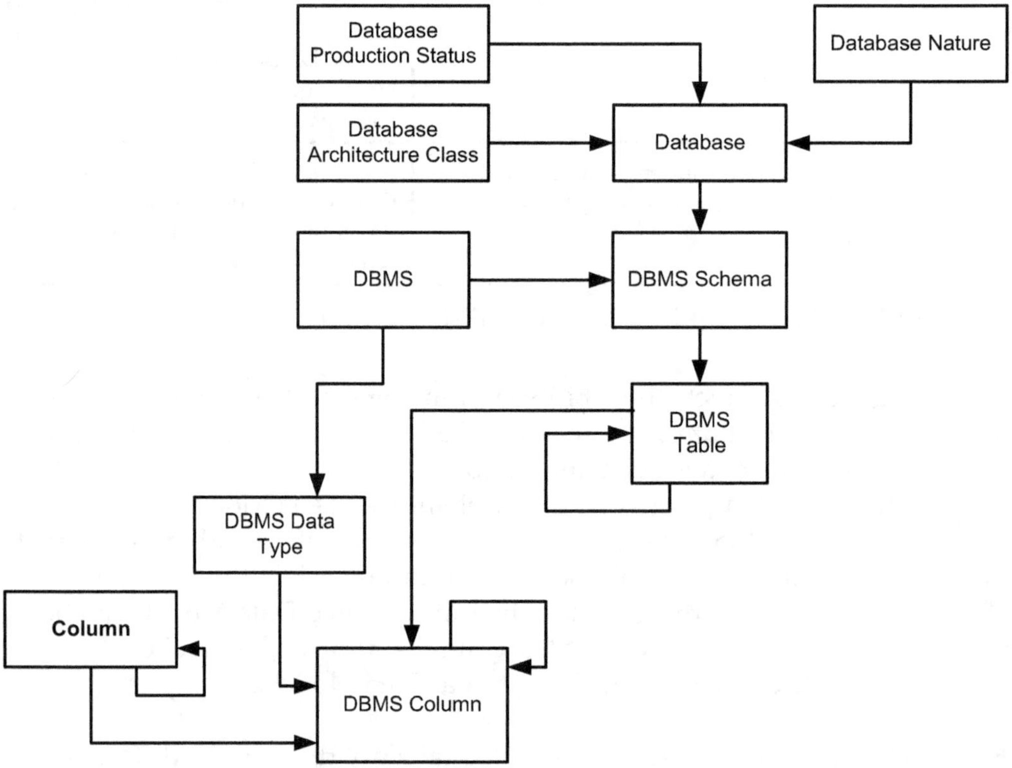

Figure 106. DBMS Column meta model.

8.4.6.1 Creation and Maintenance of DBMS Columns

Creation or maintenance of DBMS columns starts with a list of DBMS columns. Figure 107 shows the list of DBMS columns for selected DBMS Schema and DBMS table. If a DBMS column is to be changed, press the Change button. Figure 108 is presented. New DBMS columns can also be created through Figure 107 by pressing the Insert button. An Operational Data Model DBMS column is not the use of a data element within a DBMS table as it is within a column within a table of an Implemented Data Model. Rather, it is entirely new and independent.

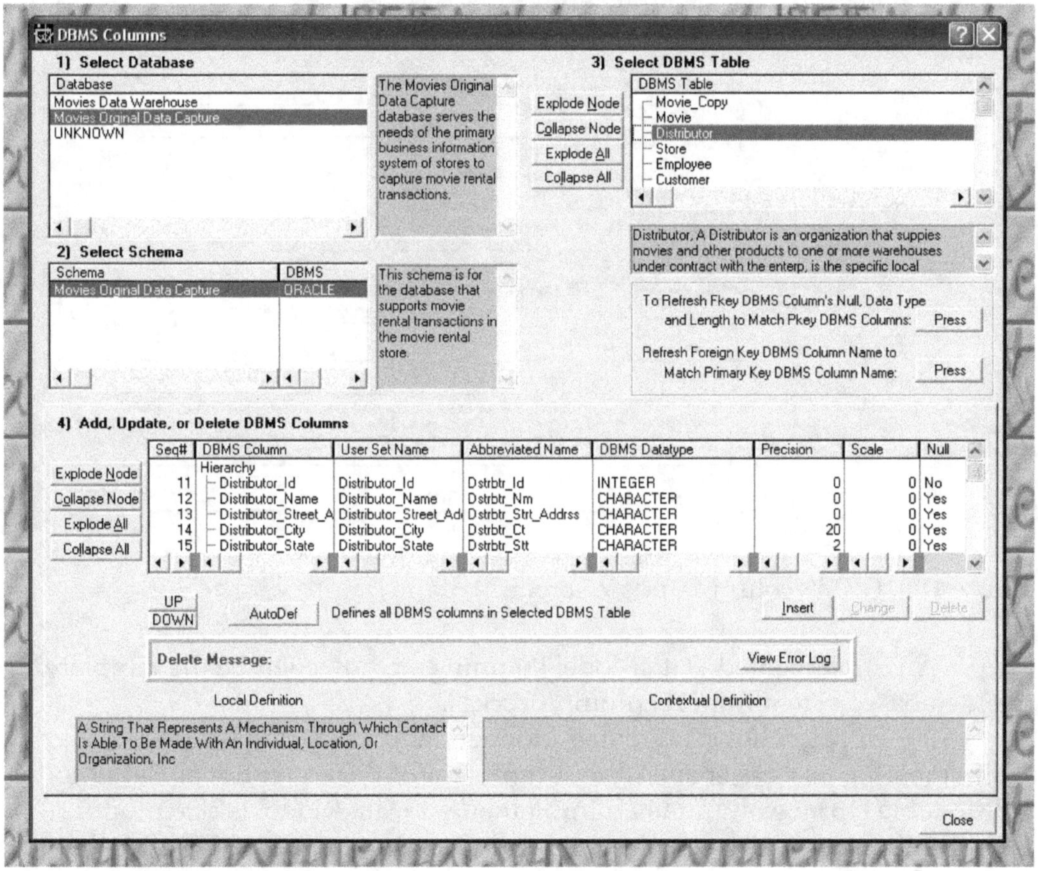

Figure 107. DBMS columns.

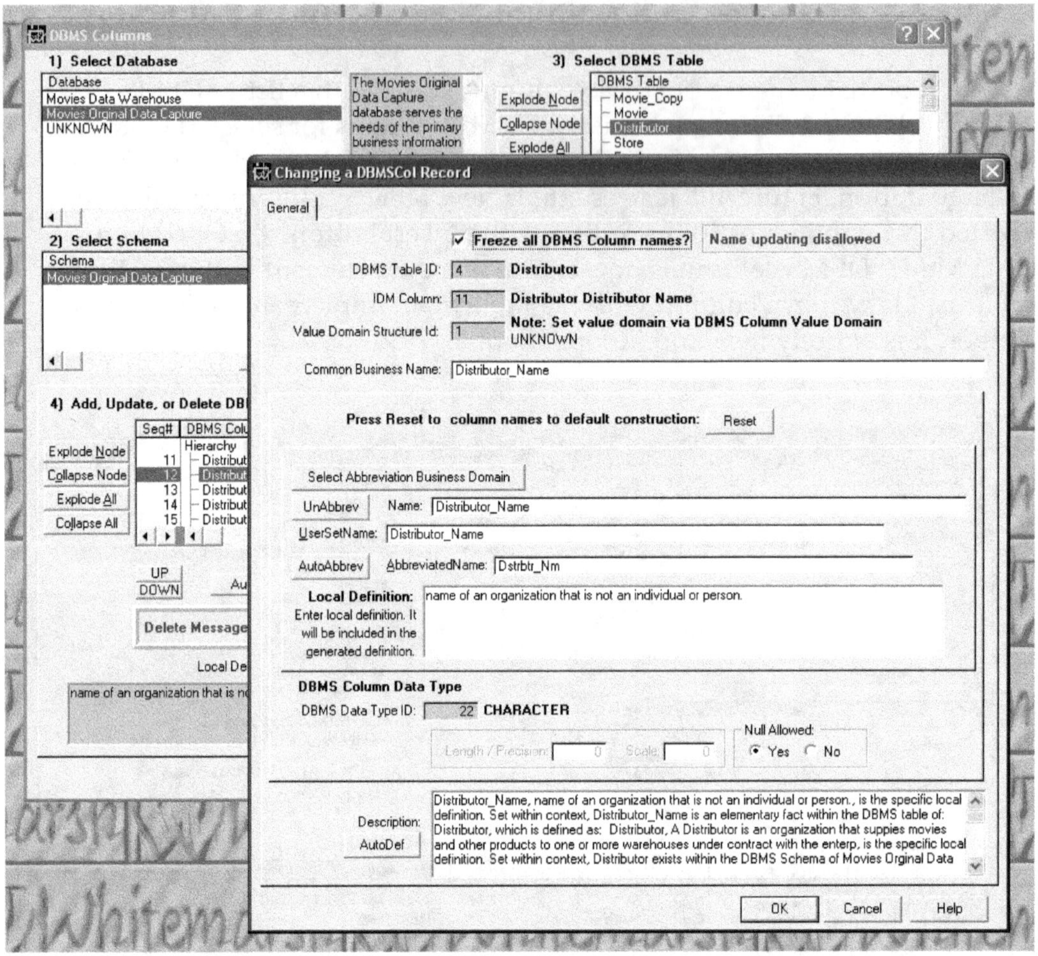

Figure 108. DBMS column update screen.

An Implemented Data Model column can, however, be the template, the Metabase System Import column process.

If a DBMS column is created independently, a Metabase System ReEngineering process enables the assignment of the Operational Data Model DBMS-table DBMS-column to an Implemented Data Model table-column. DBMS columns can also remain unrelated to an Implemented Data Model column. In this case, the column reference is shown as "unknown."

If the DBMS column is not within a foreign key, all the meta DBMS columns of the DBMS column can be changed. If, however, the DBMS column is part of foreign key a message is displayed that indicates that the only meta DBMS column that can be changed is its description.

There are three buttons on this screen of special interest:

- Refresh Selected DBMS table's Fkey . . . Press.
- Refresh the Foreign key . . . Press.
- Generate definitions . . . AutoDef.

The first button, Refresh Selected DBMS table's Fkey . . . Press, this refreshes all the DBMS columns that participate in a foreign key to match those in the source DBMS table's primary key (Pkey).

The second button, Refresh the Foreign key . . . Press, changes the DBMS column names in a foreign key to match those in the corresponding primary key.

The final button, For all DBMS columns within a DBMS table . . . AutoDef, causes all the DBMS columns in the Selected DBMS table to have the AutoDef function execute. The AutoDef process takes the local definition of a DBMS table and sets it within the context of all the meta objects that are the "parents" of the DBMS table.

Figure 107 shows the creation of subordinate DBMS column parts to the Distributor Street Address can be added, for example:

- Dwelling Street Number.
- Dwelling SubUnit Number.
- Street Name.
- Street Name Type.

These subordinate components are created by first selecting the parent DBMS column, Distributor Street Address, and pressing Insert. The normal DBMS column screen, Figure 108 is presented.

Figure 108 represents the relationship creation screen between Implemented Data Models and Operational Data Models. This relationship is not based on a schema-DBMS schema relationship, nor a table-DBMS table relationship. Rather, it is based on a column-DBMS column relationships. Because of that the Implemented Data Models can be templates for Operational Data Models. As templates, columns from a table can be semantically represented as:

- DBMS columns in different DBMS tables of the same or different DBMS schemas.
- DBMS columns in the same DBMS table.
- Subordinate DBMS columns within a group DBMS column in a DBMS table.

When a new DBMS column is created, the Implemented Data Model table column must be selected. This is accomplished by "tabbing through" the Implemented Data Model column entry that in a new DBMS column insert process is initially valued as zero. Figure 109, the Select column screen, is presented. Pick the appropriate schema, table, column and press the Select button.

Figure 108 also shows the UnAbbrev process. When an SQL data definition language stream is imported, the imported physical names may be

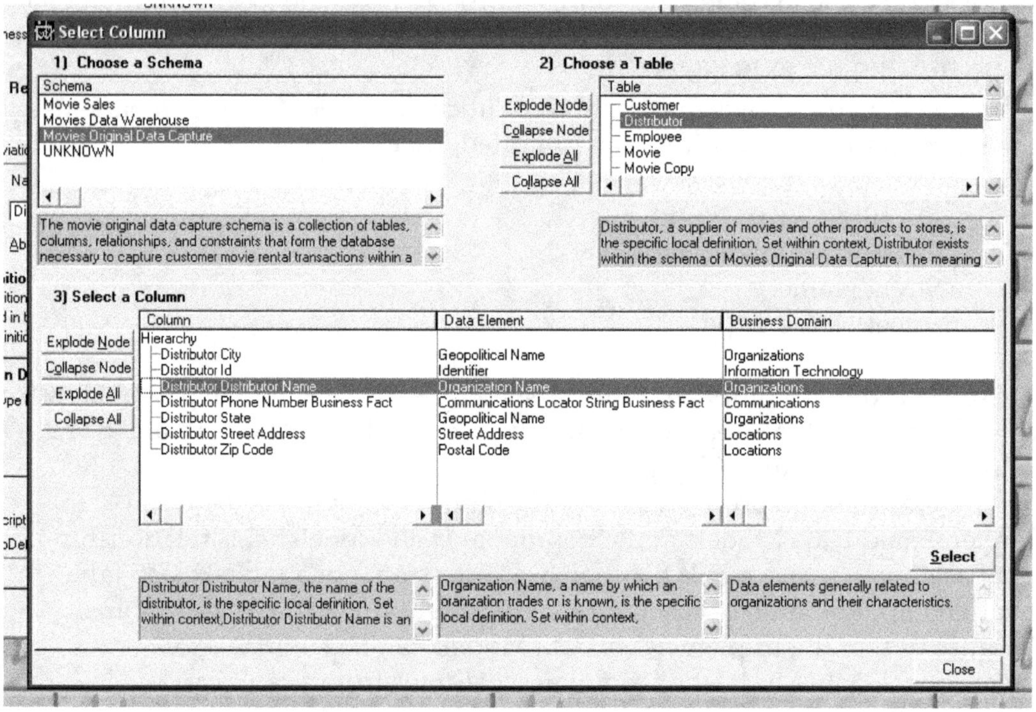

Figure 109. Select column screen.

abbreviated. For example, Empl_Hr_Dt. Under this update process, once the business domain has been chosen via the Select abbreviation business domain button, the UnAbbrev button can be pressed. If there has been a good quantity of abbreviations established on various business domains, the physical name can be automatically translated to a logical name. If there are multiple logical names for a given physical abbreviation, a screen presents itself and an appropriate logical name can be chosen for a given word. In this example, the full name becomes, Employee Hire Date.

Finally, enter the Local Definition phrase. Just keep it a phrase without a capital letter at the start and no period or comma at the end. For a complete contextual definition, press the AutoDef button.

8.4.7 View Columns

A view column is mapped to one or more compound data elements, or one or more derived data elements, and one or more DBMS columns. A view column also forms the basis for the interrelationship among views. Table 40 sets out the fact cases addressed by view columns including examples. The complete set of inherited semantics for a view column is its view. Inferred semantics include the various compound data elements, derived data elements, or DBMS columns to which a view column is related. View columns also are relatable to different view columns.

View column fact cases range through all the cases from simple to complex. Much depends on the specific SQL DBMS as to how complex a view column can be. Unique for view columns is that a value can be completely virtual. Thus, the view column, Age, could be materialized from the subtraction of a person's persistent BirthDate value from "Today" divided by 365.25 to produce an age in years. A view column can also be generated through functions, some programmed directly into the View and others provided through the SQL language functions such as Maximum, Minimum, and the like.

Fact Case applicability to a View		
Case	**Case Name**	**Applicability Note**
Simple	Single fact and restricted value set. Directly maps to the data integrity rule: Single table, single column	Yes. View columns can be defined singly within a View
Compound	single fact but with multiple contained facts.	Yes. View columns can be defined as having multiple contained view columns to any depth. This is a capability of SQL:1999 and more recent SQL-based DBMSs
Group	Single concept with multiple contained facts such as Address. Directly maps to the data integrity rule: Single table, Multiple columns.	Yes. A parent view column would be defined as the group's name, and subordinate DBMS columns would represent the elements contained in the group. There could be Address, contained elements for Address Street, Address City, Address State, and Address Postal Code. In turn, Address Postal Code could be subdivided into Zip-5 and Zip
Pair	Two facts that exist as matched pairs of codes and values. Directly maps to the data integrity rule: Single table, Multiple columns.	Yes. A parent view column would be the pair's name, and the subordinate view columns would represent the pair of view columns of the pair. The pair name would be Gender and the contained view columns would be Gender Code, Gender Name.
Related	Two or more facts that require one or more business rules to ensure consistency	Yes. A parent view column would be the pair's name, and the subordinate view columns would represent the related view columns of the pair. The related name would be Life Dates and the contained view columns would be Birth Date, and Death Date.
Complex	A fact that contain more than one of the preceding five fact situations	Yes. There is however no semantic that explicitly ties the fields together. It would have to be done with "column" or "table" constraints.

Table 40. Fact case applicability to a database object table process column.

Figure 110 presents the meta model for views. This model has views and view columns. View columns are related to DBMS columns in a many-to-many relationship. A DBMS column may be related to multiple view columns in different Views, and a view column may be mapped to different DBMS schemas.

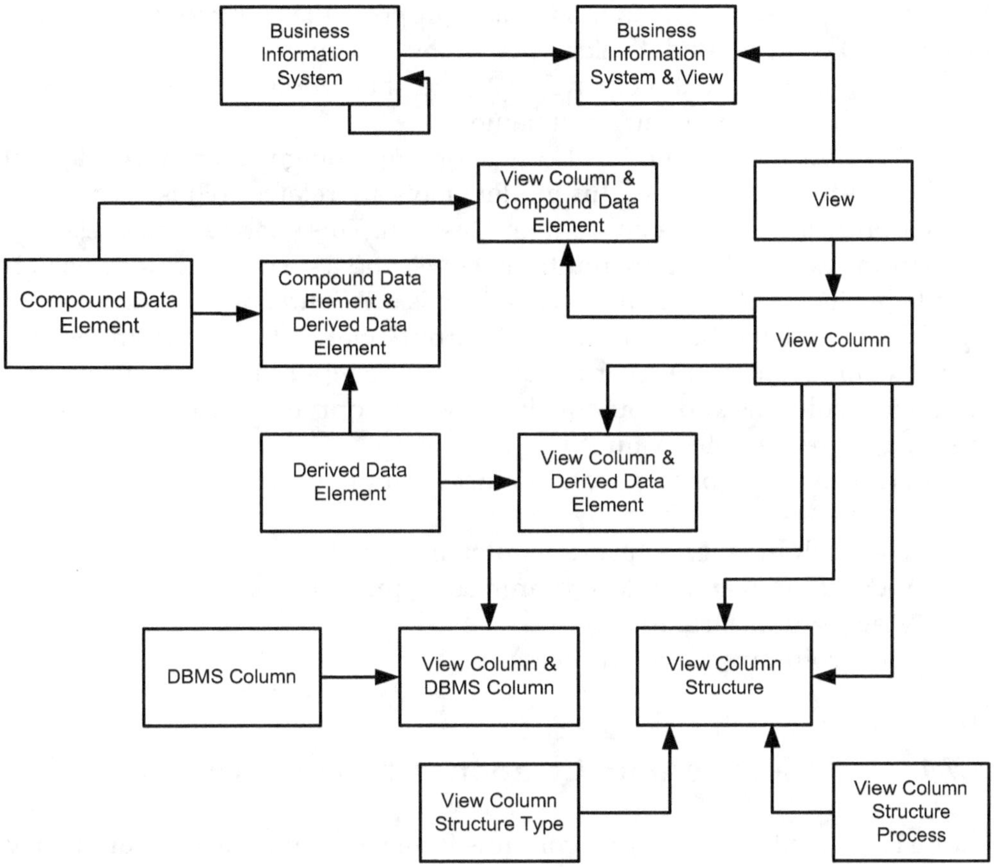

Figure 110. View column meta model.

View columns may also be related to each other through a bill-of-materials structure, view column structure. View column structures are additionally supported by view column structure processes. The purpose of view column structures is to support the mapping between and among view columns. For example, a view column might be Employee Full Name, and be related to the concatenation of the strings, Full Name, Middle Initial, and Last Name. Employee Full Name is virtual and its value is dynamically constructed by the process that concatenates the various Person Name Parts that comprise it. The view column structure process would set out the mapping to the view columns and the process necessary to concatenate the name part strings.

The view column meta model also supports mapping to derived data elements and compound data elements. In the example above, there would be a compound data element, Person Full Name, that would identify the person name parts that participate in a full name.

The view column model also supports mapping view columns to each other from different views. In this model views are related to business information systems in a many-to-many fashion. This enables a business information system to employ multiple views, and a given view to be used by multiple business information systems. In this configure, one view can be an "input" view and another can been an "output" view. The specifications of the view column structure records are to map the input view columns to the output view columns and would include, as appropriate, any processes to transform the represented values.

The processes for view column fact creation include:

- View column creation and maintenance.
- View column and DBMS column assignment.
- View column structures.
- View column structure processes.

8.4.7.1 View Column Creation and Maintenance

Figure 111 presents a list of view columns for a view and the associated DBMS columns for each view column. To enter a new view column, press Insert. To modify a view column, select it and press enter.

Once a view column is created, it may be updated. To update one, select it via Figure 111, and press enter. Figure 112 presents the screen for view column updating.

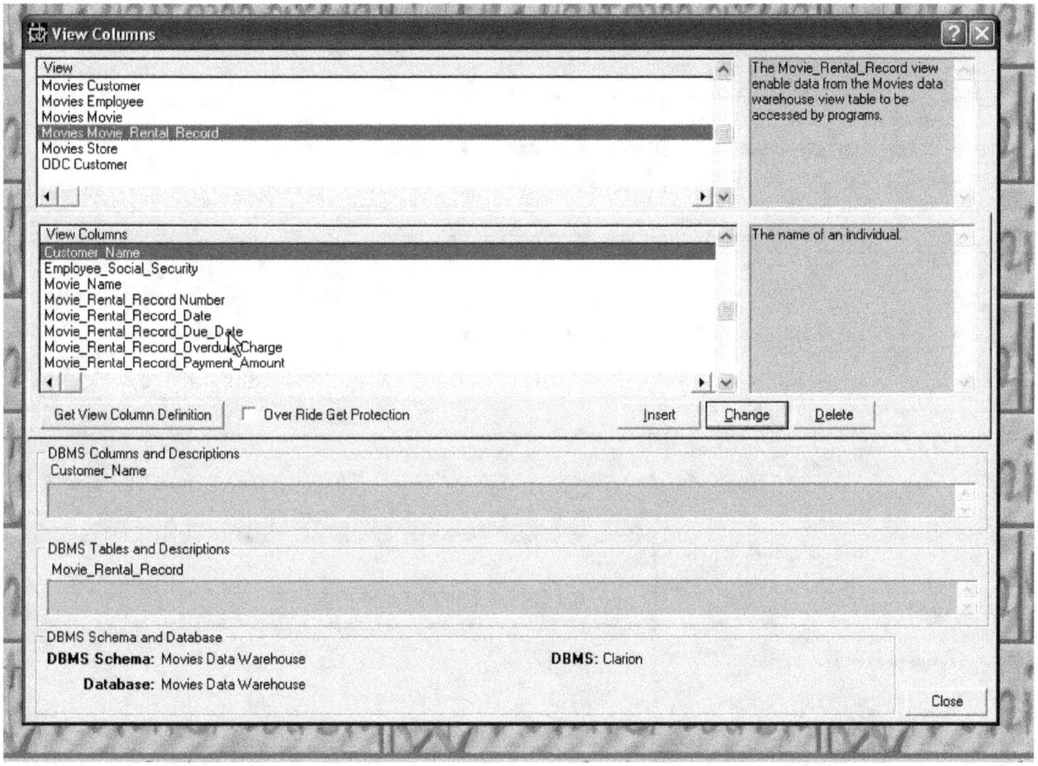

Figure 111. View columns.

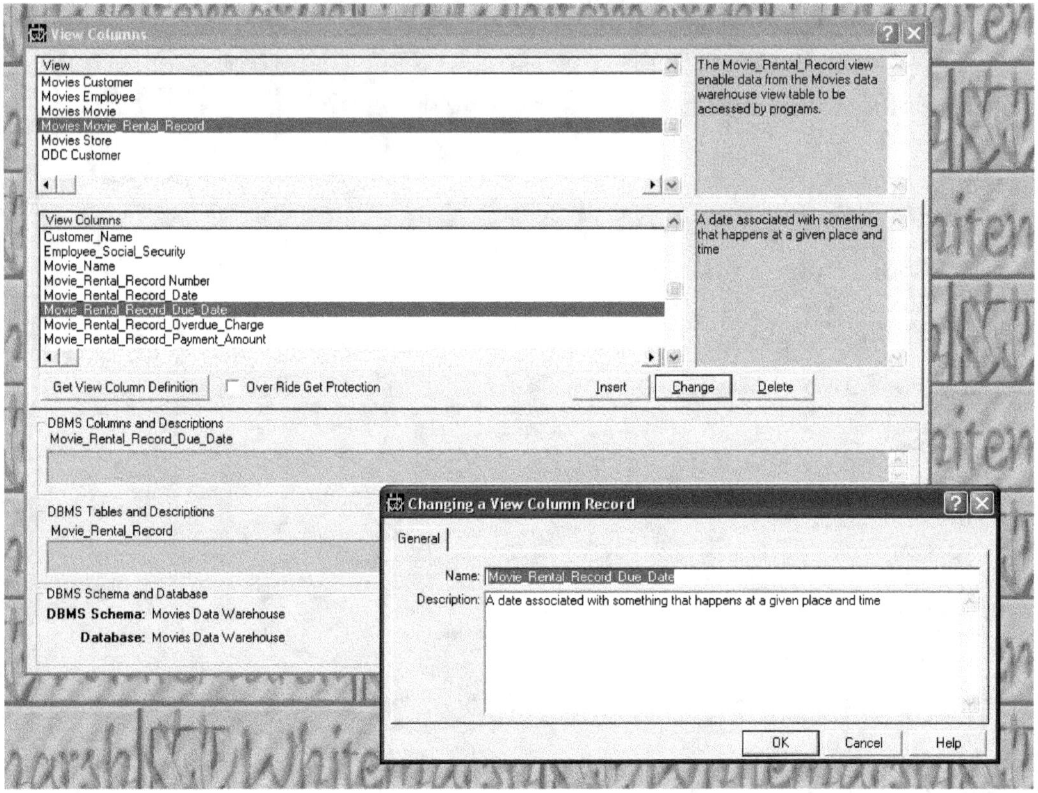

Figure 112. View column update screen.

8.4.7.2 View Column and DBMS Column Assignment

Figure 113 presents the screen that supports the process of assigning DBMS columns to a view. Select and tag a view column. Tag one or more DBMS columns from one or more DBMS tables. Then, press the build button. Mapped DBMS columns appear in the bottom window.

A view column may be complex. That is, it may either be a compound data element, for example, like Person Full Name. In this case, the full set of semantics of the view column should be mapped to the appropriate compound data element. The assignment process is illustrated in Figure 113.

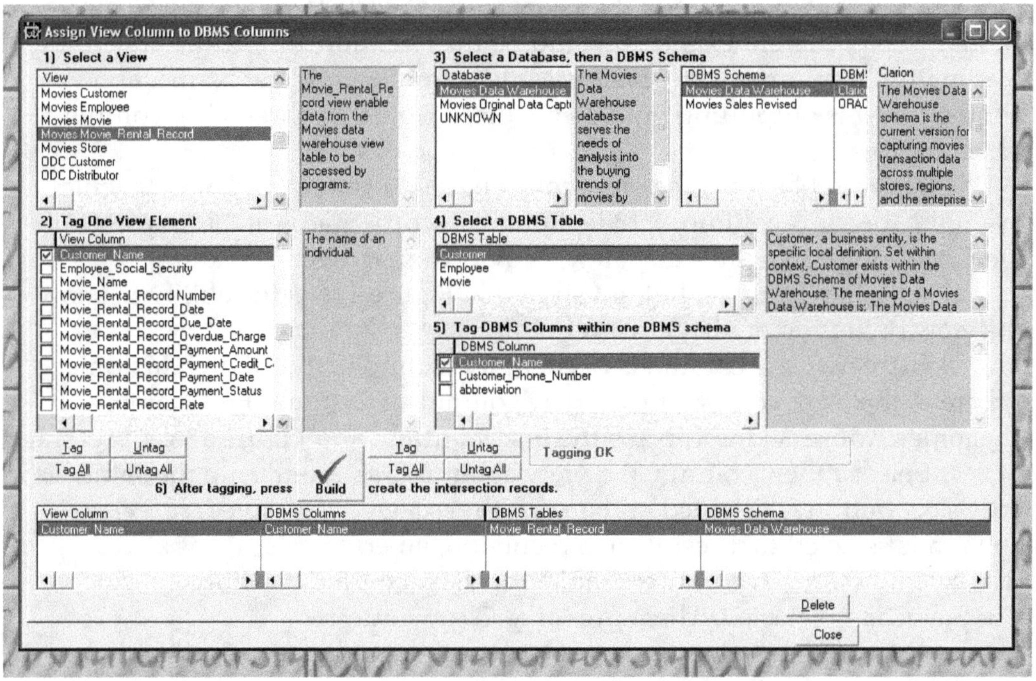

Figure 113. View column DBMS column assignment.

A view column may be complex. That is, it may either be a derived data element, for example, like Person Age. Here, the full set of semantics of the view column should be mapped to the appropriate derived data element. The assignment process is illustrated in Figure 113.

8.4.7.3 View Column Structures

View column structure is the relationship between two view columns of different views governed by a specific view column structure process. A view column structure can also represent the relationship between view columns in the same view as is required when constructing virtual view columns. A similar but a different example would be the situation where one view is reading three columns, First Name, Middle Initial, and Last Name, and another view is outputting Full Name, there would be three view column structure rows.

Figure 114 shows the view column structure screen. This particular example, the view column, Movie_Rental_Record_Status is the "parent" of a view column from a different View, Movie_Rental_Record_Status.

Another example, but not shown in Figure 114 is the ability to show the view definition of virtual view columns.

If a new view column is needed, for example, the ODBC Customer Address, that is to consist of two view columns, Customer Full Name, and Customer Address Line 1, press the Insert button, and Figure 115 is presented.

Select a view column. If a view column is selected from the same view, the Select button is disabled and an error message is displayed. If a view column is selected that results in a recursion, an error message is also displayed. In this particular example, the view column, Customer CityStateZip is selected. Then, press the Close button.

Once an inter-view column structure is completed, select the top of the structure and press the Select view column structure process button. A set of processes, like those in Figure 110 are displayed. Select the appropriate one and press the Select button.

The final step in this process is to reselect the view. At that point, all the appropriate changes are displayed.

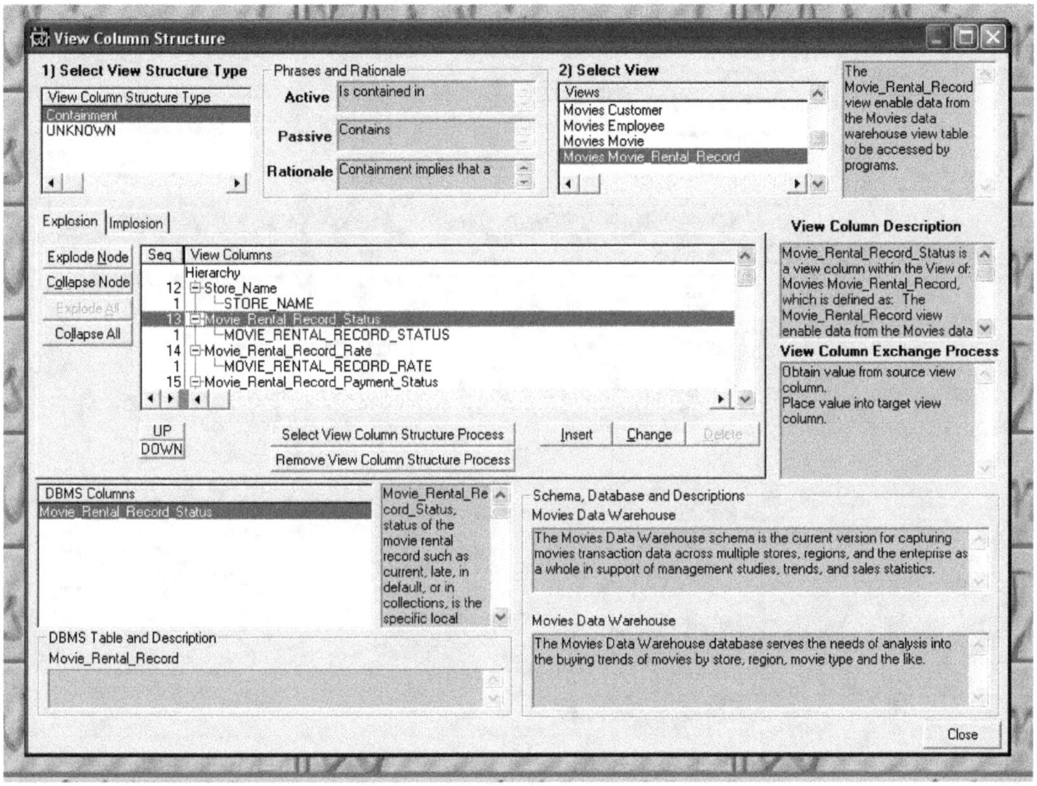

Figure 114. View column structures.

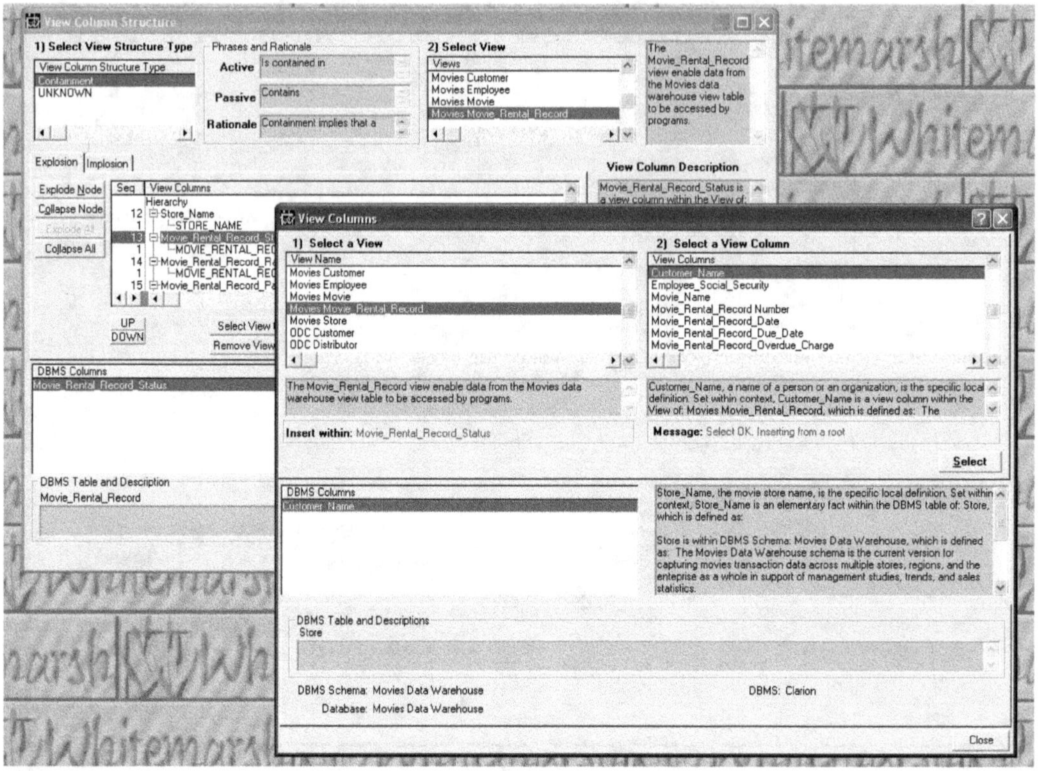

Figure 115. View column structures update screen.

8.4.7.4 View Column Structure Processes

A view column structure process is the pseudo code that defines the relationship among all the view columns referenced in a set of view column structure records. Figure 116 shows a list of the view column structure processes. To add, delete or change one, press the appropriate button. On insert or change, Figure 117 is presented. Make the appropriate entries and close the screen.

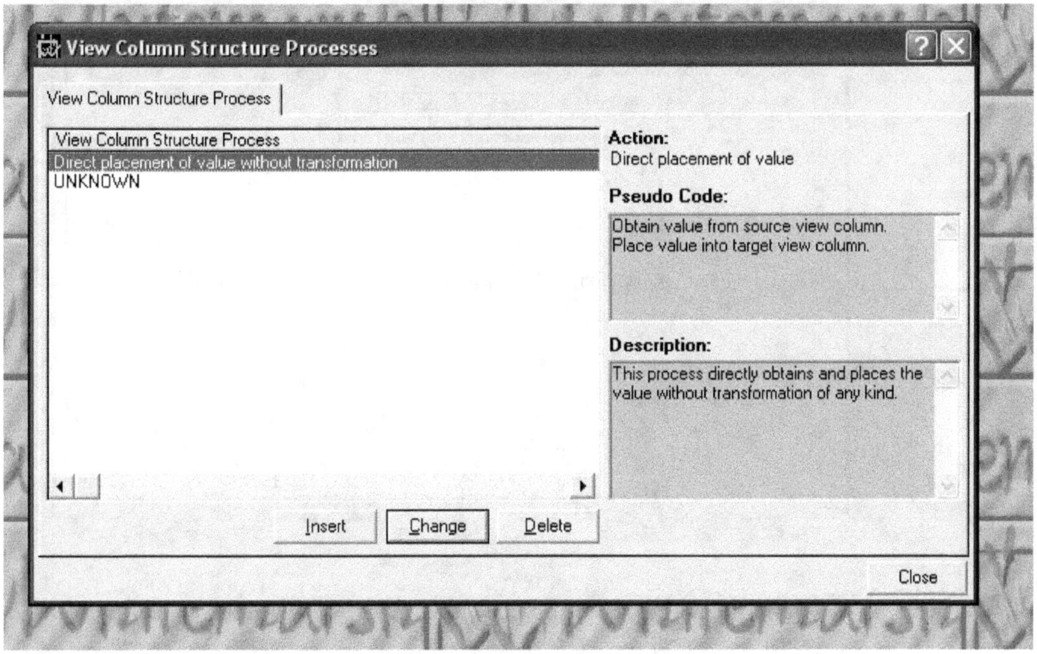

Figure 116. View column structure processes.

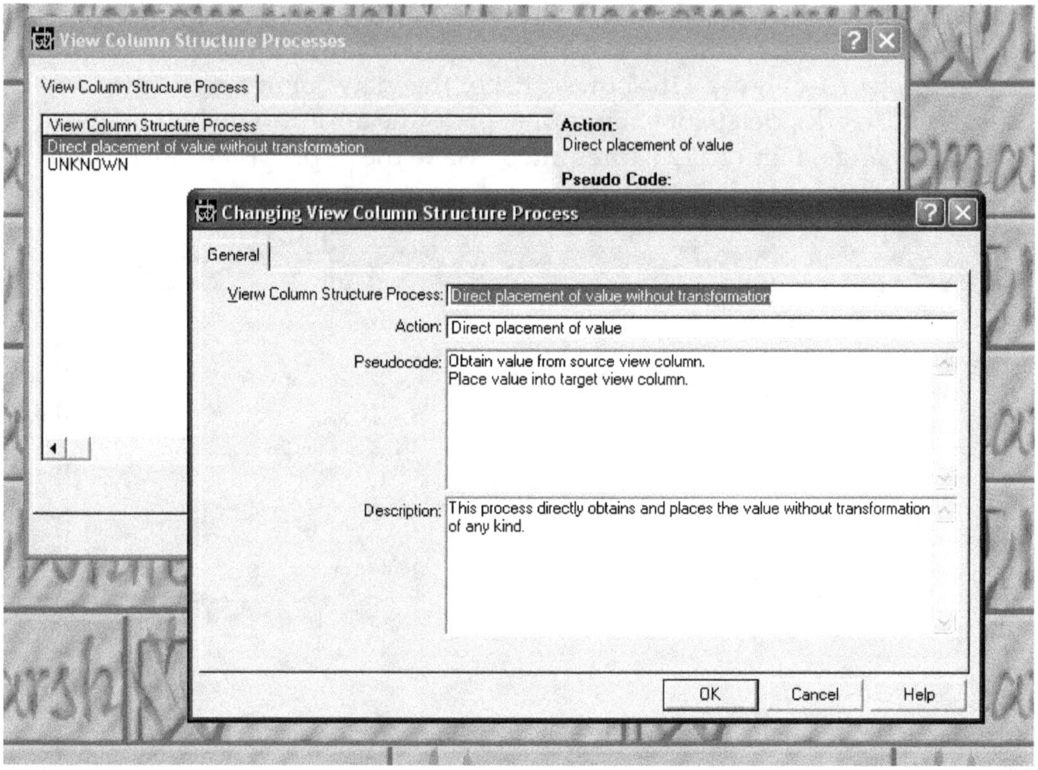

Figure 117. View column structure process update screen.

8.4.8 Database Object Table Process Columns

A database object table process column is an association of a specific database object table process and a specific column of a table. The meta model for database object table process columns is presented in Figure 118. This model clearly shows that database object table process columns are clearly set within the context of Implemented Data Model columns and database object table

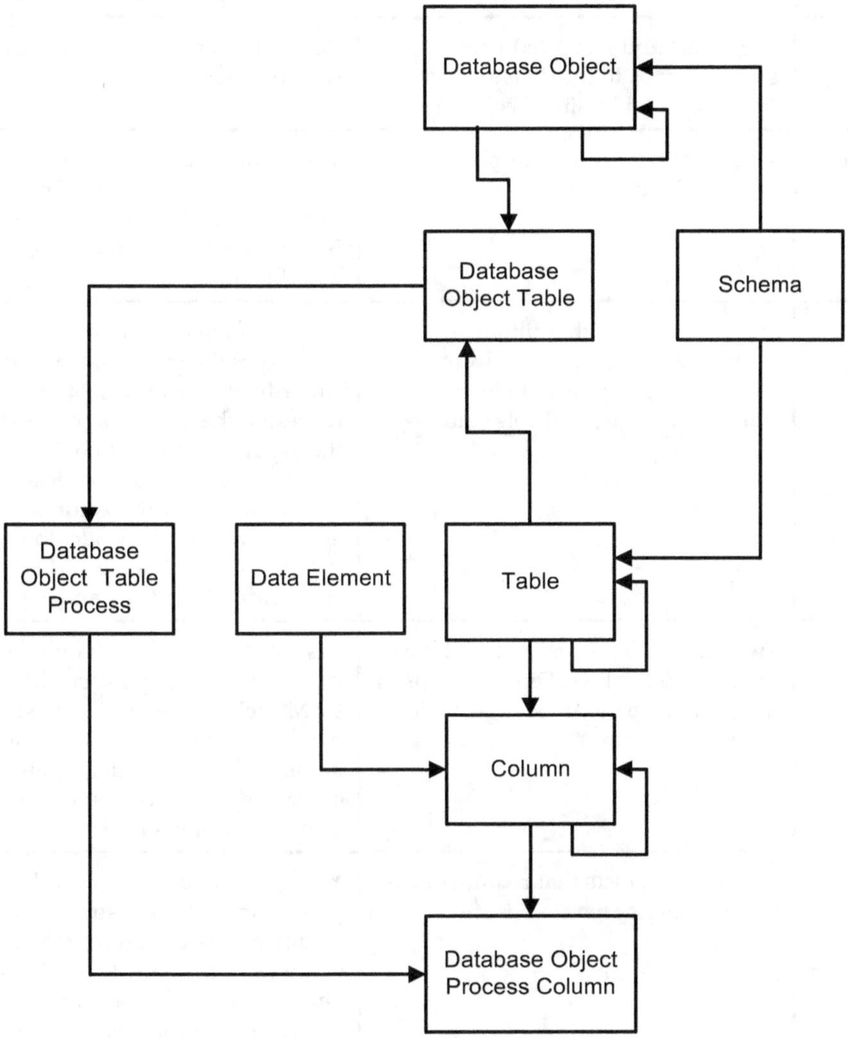

Figure 118. Database object meta model.

processes. In actuality, database object table process columns are not really fact case definitions, but are references to Implemented Data Model columns upon which a database object table process executes. The fact cases applicable to a database object table process column are set out in Table 41.

Fact Case applicability to a database object table process column		
Case	**Case Name**	**Applicability Note**
Simple	Single fact and restricted value set. Directly maps to the data integrity rule: Single table, single column	Yes. Columns can be defined singly within a table
Compound	single fact but with multiple contained facts.	Yes, columns can be defined as having multiple contained columns to any depth. This is a capability of SQL:1999 and more recent SQL-based DBMSs
Group	Single concept with multiple contained facts such as address. Directly maps to the data integrity rule: Single table, multiple columns.	Yes. A parent column would be defined as the group's name, and subordinate columns would represent the elements contained in the group. There could be Address, contained elements for Address Street, Address City, Address State, and Address Postal Code. In turn, Address Postal Code could be subdivided into Zip-5 and Zip-4.
Pair	Two facts that exist as matched pairs of codes and values. Directly maps to the data integrity rule: Single table, multiple columns.	Yes. A parent column would be the pair's name, and the subordinate DBMS columns would represent the pair of columns of the pair. The pair name would be Gender and the contained columns would be Gender Code, Gender Name.
Related	Two or more facts that require one or more business rules to ensure consistency	Yes. A parent column would be the pair's name, and the subordinate columns would represent the related columns of the pair. The related name would be Life Dates and the contained columns would be Birth Date, and Death Date.

Fact Case applicability to a database object table process column		
Case	**Case Name**	**Applicability Note**
Complex	A fact that contain more than one of the preceding five fact situations	Yes. There is however no semantic that explicitly ties the fields together. It would have to be done with "column" or "table" constraints.

Table 41. Fact case applicability to a database object table process column.

The database object table process column assignments' window is shown in Figure 119. In this window, database object table processes are assigned to database object tables.

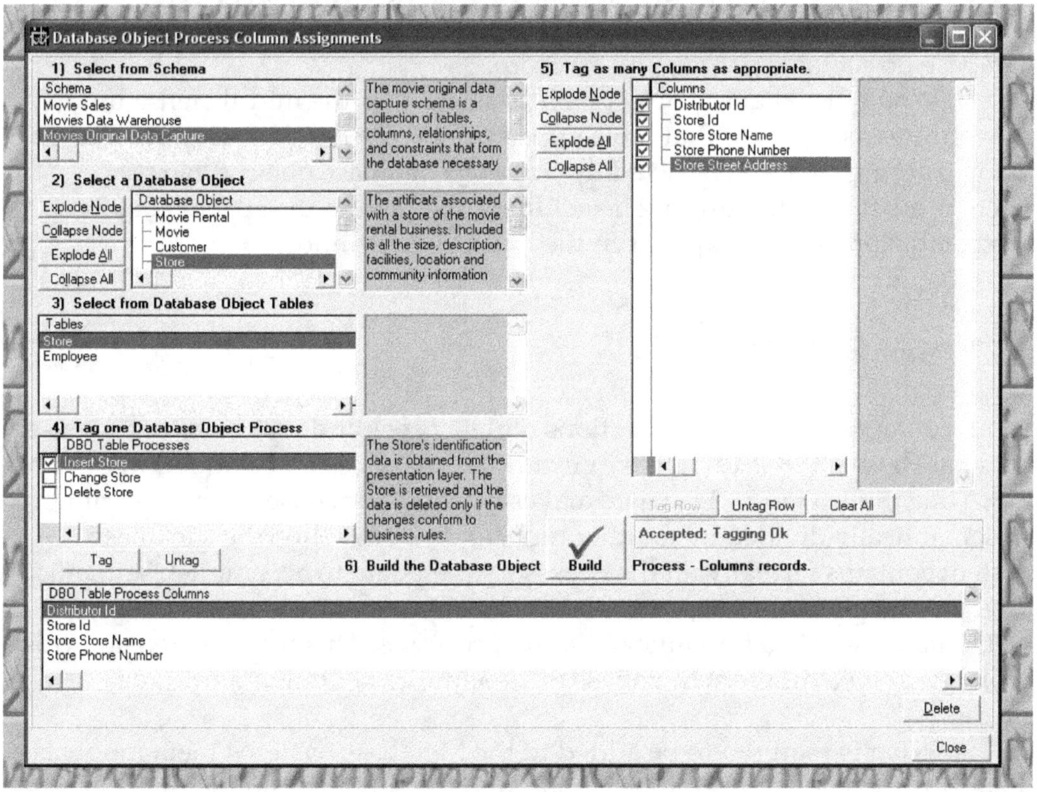

Figure 119. Database object table process column assignment.

The database object table process column assignments are created by identifying the appropriate database object table process, and tagging the appropriate database object columns. The database object table process is selected by selecting the database object in the upper left window. At that point, the assigned database object tables appear in the middle left window. Select the appropriate one, and the assigned database object table processes appear.

Tag only one database object table process. In this example, on the left side of the window, a schema, Movies Original data Capture, is selected. Selected is the database object, Store. Finally tagged is the database object table process, Insert Store. On the right side of the window, the columns for the database object, Store, are shown. In this example they are all tagged. When the Build button is pressed, the four column to database object table process associations are created and appear in the bottom browse. Accomplished is the identification of the four columns involved in the database object table process, Insert Store.

When the database object table is selected in the middle left window, the columns assigned to that table appear in the upper right window. Tag as many columns as are to be acted upon by the database object table process. Then, press the Build button. The columns assigned to the specific database object table process will appear in the lower right window.

8.5 Semantic Checking

Some operations can cross collections of data types and make semantic sense. Other such operations do not. For example, intervals can be added to time. But, Time multiplied by Distance makes no semantic sense. One time cannot be semantically divided by another time, and so on. While the Metabase System contains virtually all the necessary metadata to accomplish semantic level checking of data integrity rules, it does not contain some missing metadata or required semantic checking processes. This future extension was prototyped on a different Metabase System implementation with great success.

What is required to be added to the Metabase Systems meta model is a meta entity called Allowed Operations, and the set of data types allowed within those operations. The semantic checking process would "read" the code, look for variables and their data types and determine if these are in an allowed set of combinations. For example, a column, Start Time could be increased by an integer, such as add 1 hour to 9:15 AM to create 10:15 AM. Or, two times could be subtracted to create an interval.

Accomplishing this would require a significant extension to the notion of data types, well beyond the traditional set of Integer, Character, Decimal, and the like. These extensions are commonly called "strongly typed data." Some data type extensions have already been created by the ISO/ANSI standard SQL. These SQL extensions merely expand the existing notion of data type, e.g., Integer, to "type of data" that includes arrays, groups (called Row Types), and other complex structured data.

The extensions called for here are more about the object-oriented extensions that exist in SQL:1999, 2003, and 2008. These data type extensions allow for the definition of data types such as time, or distance, or volume, and so on. Each of these data type extensions permits the "branding" of a database column with the extension and the definition of an allowed set of operations on the extended data types.

Accomplishing the extensions contained in SQL:1999 and beyond requires an SQL:1999 and beyond DBMS. Nevertheless, if those do not exist, what can be done? That is where the Metabase System can help. Data type and allowed operations within the Metabase can identify the processes that can and cannot be performed against Metabase System defined facts. For example, what would be the meaning of adding Middle Name to Start Time? In contrast, adding 1 hour to 9:15 AM to create 10:15 AM has meaning and value.

These types of extensions to the Metabase System are clearly possible because of the existing infrastructure. While these facilities do not currently exist in the Metabase System employed within this book, extensions of this kind are contemplated for Metabase System Version 7. What has to be done, how it can be done, and all the infrastructure to accomplish it already exists. What has to be accomplished is first the implementation of the Data Integrity Rules meta models depicted in Figures 76 and 77. These meta models are practical to implement only because of all the already implemented meta models for data elements, Specified Data Models, Implemented Data Models, Operational Data Models, and View Data Models.

8.6 Error Classes

The fact specification chapter enabled the specification of facts in forms ranging from single valued facts to complex and combination facts. If facts exist in combinations, then, given that DBMSs are employed as the "protectors" of data, processes can be created and embedded in the DBMS schema language that enforce many different classes of fact based rules.

This chapter also showed how all the fact cases are defined within the Data Element, Specified, Implemented, Operational, and View Data Models. This chapter also provided the strategies for semantic allocation to the fact's common business name, and automatic name and definition construction.

Because of these Metabase System automatic features, there can be a significant level of name and meaning standardization across projects, organizations, functions, and the enterprise. There can also be these same characteristics across all the data model levels from the Data Element Model down through the View Data Model.

SQL:1986 through SQL:1992 were simple. Each column represented a single atomic value. SQL:1999 and more recent SQL versions started to allow more sophisticated data structures. Returned were the "good old days." Some, however say those days were not all good. For example, in SQL:1999 an array has persistent cell value ordering. Thus, there might be semantics inferred by being the "first" or the "last" value in an array. If so, that semantic has to be defined, stored, and retrievable somewhere. Each of the fact cases, if properly specified and enforced can reduce error quantities. For example, by never allowing a U.S. Address to be entered without a valid combination of City-State-Zip.

Fact specification cases beyond the simple, single value facts are best specified in a business rules component of the Metabase System. Currently (as of this book's publication), these rules are set out within the database object classes model that is the subject of Chapter 12. Properly engineered and embedded in sophisticated DBMSs, both classes of errors can be comprehensively addressed.

8.7 Fact Specification Cases Summary

This chapter focused on understanding all the different cases associated with fact specification. Some cases are simple, for example, single values for fields like Birth Date. Other cases require groups of fact collections such as a Full Name (First Name, Middle Name, and Last Name). Still others required facts like Life Span that would be from Birth to Death. Many of these fact cases result in virtual values such as a person's age ((Today - Birthdate)/365.25).

Others require sophisticated processes. Examples of all these data integrity rules that govern the fact cases are set out.

This chapter not only identified the different business fact cases but also showed how these facts are defined atomically within data elements, attribute, columns, and DBMS columns. Also shown was how to display traceability across all these contextual definitions. This chapter showed how fact cases can be bound into SQL Views where a given SQL column maps to multiple DBMS columns, and vice versa. Detailed also is how view columns can be mapped to both compound data elements, and to derived data elements.

This chapter set out the different classes of data-based data integrity rules, sometimes also known as business rules. Since the rules in this book are data-centric, and since business rules are usually contained in business information systems, the name used in this book remains Data Integrity Rules.

Shown in this chapter also are two meta-models, one for specifying business rules, and another for binding business rules in a Metabase System environment. The reader should be note, however, that as of publication of this book, neither of these meta models exists in the Metabase System. Both models can be implemented in just a few staff-months and are scheduled to be done during 2008.

This chapter concluded with a look into the near future where data integrity rules can be subject to semantic compiling such that rules that multiply distance by time would be automatically prevented.

The next chapter, Chapter 9, Data Element Model, is the next "can do" chapter. Data elements are not facts that are "valued." Instead, data elements are the specifications of fact-based semantics that are, in turn, employed to "brand" attributes of entities, and columns of database tables. Data elements are also not simple. Data elements can be set within derived and compound data elements. They are set within data element concepts value domains. Finally, data element concepts are set within concepts and conceptual value domains. This is quite an extensive context for each data element.

8.8 Questions and Exercises

1. What are the six different fact cases? Are there any missing? Why isn't derived one of the fact cases?

2. What is the purpose of having fact cases? Why are they important to recognize when accomplishing data semantics management?

3 Examine each fact case and identify these within your own project, function, or organization. Have these been recognized and managed?

4. Section 8.2.1 describes simple fact cases. Do these simple fact cases meet all your needs? Find examples other fact cases that are not satisfied by simple facts. How does your organization address simple fact cases. What problems occur when all facts are considered as simple facts?

5. Section 8.2.2 deals with compound facts. The example given concerns case-code. It shows that there are multiple simple facts involved in this compound fact. What other compound facts in your project, function, or organization are like this? How have you identified, created, and managed these fact cases? How do you ensure that these fact cases are implemented in a standardized way throughout your project, function and organization? What is the consequence of not managing these in this way? What is the resource cost?

6. Section 8.2.3 deals with group facts. The example given concerns Address. It shows that there needs to be a network structure that then governs the instances of actual addresses. What other group facts in your project, function, or organization are like this? How have you identified, created, and managed these group fact cases? How do you ensure that these group fact cases are implemented in a standardized way throughout your project, function and organization? What is the consequence of not managing these in this way? What is the resource cost?

7. Section 8.2.4 deals with pair facts. The example given concerns Marital status. It shows that there needs to be two facts that in combination provide a complete semantic fact. This is not the same, for example, as Department Code and Department Name. These are separate facts. What other pair facts in your project, function, or organization are like this? How have you identified, created, and managed these pair fact cases? How do you ensure that these pair fact cases are implemented in a standardized way throughout your project, function and

organization? What is the consequence of not managing these in this way? What is the resource cost?

8. Section 8.2.5 deals with related facts. The example given concerns Contract Dates. It shows that a contract has two dates that brackets its duration. What other related facts in your project, function, or organization are like this? How have you identified, created, and managed these related fact cases? How do you ensure that these related fact cases are implemented in a standardized way throughout your project, function and organization? What is the consequence of not managing these in this way? What is the resource cost?

9. Section 8.2.6 deals with complex facts. The example given concerns a Product Number field. In this example, some of the contained fields are codes that, in turn, require other fields to present its full meaning. Other fields can be a compound field within a complex fact. What other complex facts in your project, function, or organization are like this? How have you identified, created, and managed these complex fact cases? How do you ensure that these complex fact cases are implemented in a standardized way throughout your project, function and organization? What is the consequence of not managing these in this way? What is the resource cost?

10. Section 8.3 addresses data integrity rules. How does this book's specification of data integrity rule differ from a general definition of business rules? Does the business rule metadata model in Figure 77 and Table 32 provide the necessary basic information for data integrity rule specification and binding within, for example, a column? What's missing?

11. Note that in Figure 77, the data integrity rules are diagramed as a network. Why is that necessary? Why not just flat or a hierarchy?

12. Figure 78 presents the biding metadata model for data integrity rules. Does this meta model represent an adequate solution? If not, why not. What's missing?

13. Given that Figures 77 and 78 are sufficient for data integrity rule specification and binding, how does your organization deal with this whole issue? How are your data integrity rules identified, created, and managed within projects, functions, and across the enterprise? Can you prevent redundancies and conflicts?

14. Section 8.3 identifies and defines seven different classes of data integrity rules. Are there any missing? Have you, within your project, function or organization systematically identified, specified, and managed data integrity rules? What has been the cost and benefit from doing so? What has been the risks and costs from not managing data integrity rules?

15. Section 8.4 contains screen shots and supporting text for the definition of facts within the various places in the Metabase System. Examine each one and how it is defined for the different types of fact cases. Review the applicability notes. Are they correct? If so, why? If not, why not?

16. The subsections of Section 8.4 have meta models for each type of defined fact. Are the meta-entities what define the semantic context of these fact valuable? Why? Given that the definitions of these meta-entity fact contexts contribute to the fact's auto-generated definition, what else would be missing from a good definition? What methods and strategies does your organization employ to ensure good definitions? How well are these strategies practices within projects, functions and across the enterprise?

17. Section 8.5 sets out a strategy for semantic checking. How valuable would such a strategy be? Would you implement it differently? How?

18. The Error Classes section outlines the types of errors addressed by this chapter's materials. Do you agree that the error classes are addressed? How? What could be different? Has this chapter's contents and strategy provided an overall justification for its implementation? If yes, why? If now, why not? How would you quantity the benefits from having these error classes resolved? How would you present the approach, costs, and benefits to management in order to justify funding?

Appendix 1
Database Architecture Classes

An enterprise-wide data architecture contains five different classes of databases. These are:

- Reference databases.
- Original data capture databases.
- Transaction data staging area databases.
- Operational data store databases.
- Warehouse databases.

Figure 120 presents all five of these database classes within the context of an overall enterprise data architecture. At the bottom of each of the second to the fifth data class are its most significant characteristics. The remaining part of this appendix contains a six column table that describes each data class in terms of:

- Persistent data collection name.
- Persistent data characteristics.
- Process characteristics.
- User considerations.
- Technical considerations.
- Examples.

A description of each of these five data classifications follows.

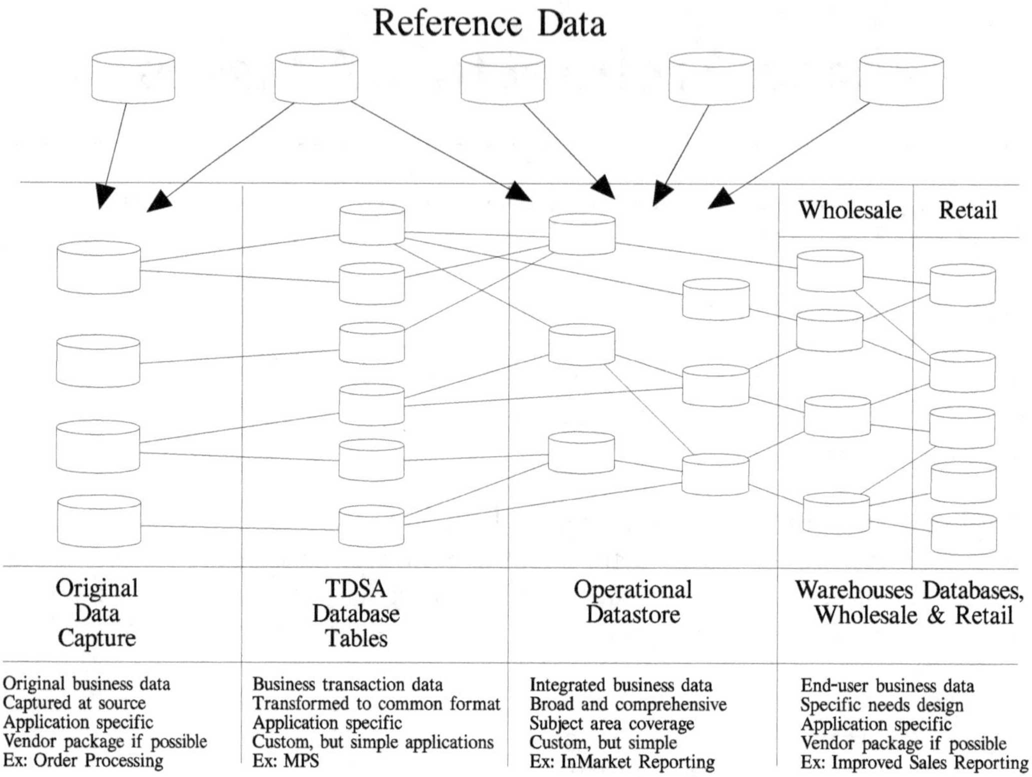

Figure 120. Data architecture classes topology.

Data Architecture Persistent Data Classifications and Characteristics				
Persistent Data Classification	**Persistent Data Characteristics**	**Process Characteristics**	**User Considerations**	**Technical Considerations**
Original Data Capture	Detailed atomic data Accurate as of the last update Well-defined, long lasting database designs Normalized database designs Uses reference data No invalid data updates allowed	Tuned for transaction capture, storage and update Application oriented Transaction driven Processing supported by well known data integrity and business processing rules Understands, creates, and maintains TDSA databases through Original Data Capture system extract and TDSA loading software. Source data for TDSA	Original data source entry personnel High availability Supports day-to-day operations	Amount of data for processing is small Multiple vendor packages Package specific May or may not be controlled by SQL DBMS

Data Architecture Persistent Data Classifications and Characteristics				
Persistent Data Classification	**Persistent Data Characteristics**	**Process Characteristics**	**User Considerations**	**Technical Considerations**
TDSA: Transaction Data Staging Area	Transient data/short lived Foundation data source for all operational data systems Enterprise-wide standard semantics Package independent database designs Denormalized Full business transaction Does not use reference data	Accepts, stores, and pushes forward function Only refreshed with changes from previous version Operations application data system daily update event driven Translation and transformation	Users cannot access	Multiple platforms Interface monitoring Applications insulation SQL DBMS controlled

Data Architecture Persistent Data Classifications and Characteristics				
Persistent Data Classification	**Persistent Data Characteristics**	**Process Characteristics**	**User Considerations**	**Technical Considerations**
ODS: Operational Data Store "Subject Area Data Store"	Detail level data May be lightly summarized Current or nearly current Rolling histories Broad subject area database scope Normalized database designs Redundant data from across enterprise May contain derived data from "outside" Uses reference data May receive and/or send data to databases within class Data source for all warehouse databases	Updated daily via TDSA data transaction files Accepts and stores data from TDSA Supports comprehensive reporting and generalized ad hoc query	End-user detailed level analysis Used for up to the minute decisions Used for detailed decision making	Requires fast response time Large volume SQL DBMS controlled

Data Architecture Persistent Data Classifications and Characteristics				
Persistent Data Classification	**Persistent Data Characteristics**	**Process Characteristics**	**User Considerations**	**Technical Considerations**
Warehouse: Wholesale	Summarized and some detail Rolling Histories Load/replace, no end-user update Enterprise-wide standard semantics Narrow/subset of one or more subject areas Redundant data from across enterprise May contain internal derived data Reference data fully embedded May receive and/or send data to databases within class Data source for all retail data warehouses	No end-user updating Regular, periodic updates Supports standardized, on-demand reports Supports general complex business data analyses such as trends and forecasting Views data from multiple subject areas	Supports managerial community Used for broad direction and positioning Used to formulate and assess long term decisions	Availability not on business' critical path User workstation access Large data volumes per query High processing power required SQL DBMS controlled

Data Architecture Persistent Data Classifications and Characteristics				
Persistent Data Classification	**Persistent Data Characteristics**	**Process Characteristics**	**User Considerations**	**Technical Considerations**
Warehouse: Retail	Light to highly summarized and some detail Rolling histories Load/replace, no end-user update Enterprise-wide standard semantics Denormalized and highly designed to specifically favor one or more reporting formats Redundant data from across enterprise May contain internal derived data Reference data fully embedded May receive and/or send data to databases within class	Availability not on business' critical path Regular, periodic updates Highly designed, end-user on-demand reports Supports very specific simple to complex business data analyses Views data from multiple subject areas	Supports managerial community Cannot update Used for direction and positioning Used for long term decision making Specific reporting need	Relaxed availability User workstations Large volume High processing power SQL DBMS controlled

Data Architecture Persistent Data Classifications and Characteristics				
Persistent Data Classification	**Persistent Data Characteristics**	**Process Characteristics**	**User Considerations**	**Technical Considerations**
Reference Data	Durable codes and long value alternatives with policy definitions and full descriptions Enterprise-wide standard semantics Source of all valid and invalid values including alternatives for different countries and languages Multiple group data field constructors suitable for different countries and languages Definitive source for multi-use data in all other databases Changed data history supported by conversion mappings	Simple updates Update mappings required for reference data value migration	Needed by all levels in the organization Used by all systems Enables understanding and conversion of historical data	Supports the concept of single source Integration with all data store types

Reference Data

Reference data is that set of data that is commonly employed across the other different classes of data. Examples of reference data include sets of data like:

- Gender codes and names.
- State codes and names.
- Country codes and names.
- Units of measure and names.
- Currency codes, denominations, and names.
- Standard address parts such as street (ST), road (RD), etc.
- Manufacturing parts, names, and descriptions.

Reference data has distinct characteristics such as:

- Persistence.
- Employed in longitudinal policy formulation.
- Impacts multiple policy areas.
- Enables fact discrimination and/or selection.
- Break points for fact aggregation and statistical analysis.
- Requires higher level of management approval.
- Causes multiple negative impacts when changed.
- Is not easily melded with other strategic data (zip | telephone = ???).
- Gives meaning to quantities (physical dimension, time dimensions).

Reference data is established through formally created projects that cause the reference data to be created, stored, reported and maintained. Reference data projects are critical to the business in that the business' fact data that is stored in the Original Data Capture databases is categorized, summarized, validated, and reported by means of the reference data.

Supporting each type of reference data are full policy definitions and descriptions. Because of enforced reference data, a corporation can enjoy enterprise-wide semantics. The reference data must also support the full set of valid and invalid values including alternatives for different countries and languages. The key issue in representing multiple language reference data is whether all the short values, for example Female or Hembra is encoded as an F, or whether all encoded values are represented as numbers, for example, "1".

A critical characteristic of reference data is that it must support evolving data history. Reference data supported by dates tracks data value changes. Business data can thus be interpreted under either the current set of meanings or their original meanings.

Reference data is commonly seen as only business data that is used for references (gender codes and names). Reference data also embraces metadata, that is, the standard names and definitions for business data, including for example the definition of Country Code, its length, and the ISO standard from which its name, definition, and value set is derived. Metadata embraces not just data fields but also database tables, databases, subject areas, programs, systems, etc.

Original Data Capture Databases

Original Data Capture databases are those that contain all business facts. Prior to being stored in an Original Data Captures database, the business data is manual, that is, customer orders, employment applications, hours worked but not yet recorded, and the like. The characteristics of operational data are that its detailed atomic data rather than summarized. The data is accurate as of the last update. The databases that contain this type of data are stable and long lasting with normalized data structures. These databases make heavy use of reference data to ensure that it is understood across all users. Because of reference data and because of very extensive integrity checking no invalid data should ever be allowed.

Original Data Capture databases are generally tuned for the transaction capture, storage and update of data. Correctness and integrity is of paramount importance. Performance is clearly secondary. The databases are specifically oriented towards specific applications such as order capture, order processing, staff time recording and reporting, and payroll. Because these applications are very narrow in scope, they are commonly acquired as commercial off the shelf (COTS) packages.

Original Data Capture databases are must have 24 hours and 7 days availability as they are the source of the company's business data. Simply put, these databases support the company's day-to-day.

Transaction Data Staging Area Databases

Transaction data staging area (TDSA) databases are the beginning of integrated data semantics for the entire enterprise. While individual Original Data Capture database applications may have both specialized and localized semantics. These localized semantics may exist throughout the enterprise because of different languages, laws, and existing sets of application packages that existed prior to migrating to an enterprise-wide data architecture.

To handle all these situations, data translators are created that take localized data values and meanings and translate them to enterprise-wide values and meanings. The transaction data, now understood through enterprise-wide semantics is placed in a staging area for a specific period of time. At the end of the specified time period, the data is dropped from the staging area. Applications requiring data from the staging area only have to know where it is and the set of semantics that govern it.

The fundamental role of data within the TDSA is to relieve the individual Original Data Capture databases from having to contain enterprise-wide semantics and to relieve the operational data store databases from having to know the location of the individual Original Data Capture database applications and from having to know the values and semantics for each Original Data Capture database.

While the TDSA database applications represent another class of applications, the cost of its design, implementation and maintenance is more than offset by the reduced cost of having separately developed and maintained value and semantic translators in each of the operational data store databases. Additionally, this intermediate layer of value and semantic translators enable the organization to change the Original Data Capture database applications on an as needed basis to accommodate new DBMS, new operating system, and different classes of hardware that may be required to support organizational units of different sizes and locations.

Operational Data Store Databases

An operational data store (ODS) database embraces a wide area of a business such as all finance, all human resources, all sales, orders and marketing. The data for these integrated databases is taken from the various TDSA databases and is combined into one overall database that would then be the Sales and

Customer Management database. This database is an amalgamation of data from the databases:

- Sales organization management that defines and manages all sales regions, districts and territories.
- Human resources that provides the proper information about each salespersons who is assigned to a territory.
- Customer management who ensures that all the customers are properly identified, classified, and managed.
- Order processing which receives, tracks, and performs all order management through to actual shipment.
- Product inventory management which receives product into product warehouses, fills orders, and begins the process of product distribution.
- Manufacturing which transforms raw products into finished goods.
- Product distribution which receives filled orders from inventory and delivers them to the customers.
- Accounts payable and receivable which financially accounts for orders and invoicing.

All these Original Data Capture database systems provide transaction data to the enterprise TDSA. The data loading and update systems that feed the ODS databases take data from the TDSA and load and/or update the ODS databases.

The names of these eight systems might imply a corporate organization structure. That would be an unfortunate conclusion. The systems that could logically belong in one enterprise organization, Sales and Marketing would be:

- Sales organization management.
- Customer management.
- Order processing.

Another organization, manufacturing, inventory and distribution could undertake all three of those applications. Human resources is likely its own organization, and finally, accounts payable and receivable would be part of the finance department.

In addition to the data integration route, some data from one database may certainly be required in others. For example, sales organization management assigns customers and sales persons to specific sales organizations such as regions, districts and territories. The data from these "cousin" applications would be available in a read-only manner. Whether the references from one the sales organization management database that lists the current district manager as Bruce Baine does so through a "hot-link" or through a once a week integrity check routine is a matter of technology, not of fundamental design.

The key data characteristics of ODS database are that its data is at the detail level. The data may be lightly summarized by storing invoice totals, customer sales-to-date information and the like. The data, because of the TDSA is either current or nearly current. The database supports either endless histories or rolling histories. A rolling history might be the annual sales for prior years 5 and 4, quarterly sales for year 3, monthly sales for prior year 2, and daily order, shipment, and invoice line items for the current year.

The key uses of the ODS databases are detailed level analysis of current and historical trends, up to the minute decisions that require a complete picture. For example, a sales call might require:

- Customer orders.
- Sales organization assignments.
- Payables and receivables history.
- Recent history of sales.
- Deliveries and returns.

ODS databases may be interconnected. That is, they might be the source of data for other ODS databases. For example, if there was a ODS database for world-wide finances, summarized data from the various sales and customer management databases might be selected and forwarded onto the world-wide finance database.

ODS databases are typically very large but must provide a fast response time to ad hoc queries. Because of these characteristics, a common approach for implementing these databases is to use very large and fast servers or mainframes.

ODS databases are not however warehouse databases because of two major reasons:

- Updating other than summarization may occur. For example, a characterization of a customer's importance to a company may be stored, or a program might run through all a customer's records and assign an Key Customer Ranking code.

- All retained data is commonly at the transaction level such as order, shipment, invoice, payables, and receivables line items.

Warehouse Databases

A warehouse database is a special class of databases for two reasons:

- No updating other than summarization occurs.

- All data has its units and granularities synchronized.

Simply put, data, as a warehouse item represents a finished good. No more "manufacturing" or "finishing" is going on. In the case of data that means no more updating. The data (warehouse item) may be selected according to a specific order and may even be palletized with other items of the same or different types and sent to a more localized distribution point. Because of these similarities with the concept of product warehouse the concept of a data warehouse arose.

A very early proponent of data warehouses, Bill Inmon (Prism Solutions of Littleton, Colorado) defines it to be a collection of integrated, subject-oriented databases designed to support the decision support function, where each unit of data is relevant to some moment in time. The data warehouse thus contains atomic and lightly summarized data. Ron Ross, another data warehouse proponent defines a data warehouse to be a structured database that stores consolidated data for decision support from production-level files or databases. A more recent definition for warehouse comes from Vidette Poe who defines a data warehouse as the [data] foundation of a decision support system (Building a Data Warehouse, Prentice Hall, 1996).

Data warehouses come in two flavors: wholesale and retail. The critical differences between the two is in the area of size and scope. A retail data warehouse is clearly smaller because it contains less data. For example, there may be a customer profile retail data warehouse that contains all relevant customer information for the Yellow Knife sales territory. The wholesale data warehouse might be the one that contains all the customer sales information for all the Northwest Territories of Canada.

A retail data warehouse also differs in scope in that its focus may be narrower than that of its source wholesale warehouse. For example, all the customer profile information for hospital customers for a particular segment of a product line versus all the customers for all the product lines.

Because warehouse databases address well bounded subject areas as do ODS databases they are commonly know by the data they keep, e.g., Human Resources, Sales and Customer Management, Inventory Management, Finance, and Real Property Management.

A common technique employed in the development of retail data warehouses is to perform extensive summarizations that are related to less extensive (more detail) summaries that are, in turn, finally related to data that is almost transaction level data. Specialized query and reporting tools enable the end user to begin at the highest summary level and to "drill-down" to the ultimate level of detail. This technique is sometimes called "data mining."

The software systems that perform this type of analysis and reporting are called on-line analytic processors (OLAP). Traditional relational DBMS vendors such as Oracle have developed specialized report writers that also perform on-line analytic processing.

Warehouse databases often derive their data from multiple ODS databases. A given data item, for example, a travel expense voucher might have its data end up in multiple data warehouse databases dealing with finance, projects, airlines or lodging. Because of this redundancy, it is impossible to perform correct updates. If warehouse data is observed to be wrong, the original source data must be found and updates. Once the warehouse is refreshed or updated, it will be correct.

The principle reason for warehouse databases is to support business data analyses such as trends and forecasting. These analyses are possible because the data warehouse's database can include data from multiple ODS databases. As historical data grows it may be turned into high level summaries similar to the five years of history data described above. The main set of data warehouse users are the managerial community who both set broad direction and positioning, and formulate and assess long term decisions. Because of this type of use, data warehouse databases are generally not on a business' critical path, thus, while the need for data selection and processing power is extensive, response time is clearly not required to be "subsecond" as the decisions the data supports may take months to formulate, adjudicate, and promulgate.

Index

www.ingramcontent.com/pod-product-compliance
Lightning Source LLC
LaVergne TN
LVHW080013210926
848481LV00011B/3010